D1036058

Astronautics

**A Historical Perspective of Mankind's Efforts
to Conquer the Cosmos**

**Book 2 – To the Moon
and Towards the Future**

Ted Spitzmiller

Astronautics

**A Historical Perspective of Mankind's Efforts
to Conquer the Cosmos**

**Book 2 – To the Moon
and Towards the Future**

By
Ted Spitzmiller

An Apogee Books Publication

Dedication

This book is dedicated to the memory of my mother
Helen Elizabeth Eckert
who encouraged me to pursue my interest in aviation and space.

Acknowledgement

Many people have moved through my life and helped me along the way. Mr. O.F. Libby, of the Curtiss-Wright Corporation, took the time to answer queries from an eight-year-old boy. Mrs. Florence Rader and Mrs. Elizabeth Frazer were high school teachers who recognized that some kids do not perform to their potential and need a little straight talk. My wife, Donna, has been a constant source of strength through the years. Les From gave me the first opportunity to exercise my technical writing skills. Sid Gutierrez used his valuable time to review excerpts and provide important historical insights. Ramu Ramakesavan gave a critical assessment from an international perspective and Charles Gray provided a technical evaluation. Finally, my appreciation to Penny Rogers, who supplied the grammatical guidance to make up for my years of neglect of the English language.

Published by Apogee Books, Box 62034, Burlington, Ontario, Canada, L7R 4K2, http://www.apogeebooks.com Tel: 905 637 5737
Printed and bound in the USA

Astronautics (Book Two) by Ted Spitzmiller
ISBN 9781-894959-66-7

©2007 Apogee Books/Ted Spitzmiller Front cover by Robert Godwin
Back Cover pictures Wernher von Braun, Sergei Korolev, Valentin Glushko and Maxime Faget

Contents

Astronautics may be defined as "the study of the science and technology of space flight." *Astronautics* is also the title of this history of mankind's exploration of space. This account examines the epic events that shaped an era and provides appropriate insight into the wide-ranging impact that this endeavor has had on technology, politics, and society. It provides enough detail to satisfy the serious enthusiast, but without the minutia that so often blocks the casual reader. To accomplish this, *Astronautics* is presented as two books, each covering a particular phase of man's progress.

Book 1 — Dawn of the Space Age (Chapter 1-19) chronicles mankind's desire to know more about the cosmos and his dreams of reaching into its depths. It describes the initial discoveries, inventions, and engineering innovations that became the foundation of rocket technology. It follows the two preeminent countries in their quest for the 'ultimate weapon' that would provide the path to space. It describes the decisions that resulted in the first artificial satellite programs in the United States and the former Soviet Union. *Astronautics* chronicles the events that shaped the initial thrust into space as represented by the first Soviet Sputniks and the shocked response by the Americans. It details the belated and often failure prone launches that humbled a great nation. The book describes the first attempts to reach the Moon and the planets and explains the techniques and physics involved in easy to understand language. It illustrates the engineering requirements of the first manned spacecraft, Vostok and Mercury, and the selection and experiences of the first spacefarers.

Book 2 — To the Moon and Towards the Future (Chapter 20-39) examines the events leading to a commitment by the American President John F. Kennedy to land a man on the Moon within the decade of the 1960s and the affect of that venture on future space exploration. It details the Gemini, Voskhod, Soyuz and Apollo programs and the first exploration of the Moon. It reviews the development of the most complex machine devised by man—the Space Shuttle and follows the evolution of the space station. *Astronautics* highlights the effort to find extraterrestrial life and the exploration of the outer planets. It examines advanced propulsion technologies and speculates on what might lie ahead in space exploration.

Each chapter of each book analyzes a topic so that readers can achieve a relatively complete understanding of a special interest area without the need to ferret information from multiple chapters. However, each chapter and each book is linked to the whole by a careful interconnection of a set of themes.

Writing a concise history of space exploration is a daunting task. There are scores of books that detail various aspects of man's path to the stars, and there are two primary attributes that stand out when evaluating them: scope and technical accuracy. Scope is typically sacrificed, often because the subject matter offers a bewildering array of areas that can intimidate even the truly knowledgeable. In those few books where scope has successfully captured the essence, the level of technical detail is often overwhelming for all but the serious reader. *Astronautics* attempts to simplify and clarify technology, politics, and events to make them easier to comprehend.

The reader may note that the spelling of some names, particularly Russian, may be different from that seen elsewhere. The more common English spelling has been used here. Likewise, dates may vary by a day depending on the time and the location of the event.

In the 4-year period from 1957 to 1961, the war rocket became the vehicle that enabled the Space Race, a visible extension of the Cold War struggle between the two great powers and two diametrically opposed ideologies. This short period saw man move from his first small steps of instrumented Earth satellites to manned spacecraft The early phase highlighted the frustrations and failures that so often accompany hi-tech advances, especially those gated by political maneuvering such as Project Vanguard—what was to have been America's (and mankind's) first Earth satellite. It was a startling pace, as documented in Book One of *Astronautics—Dawn of the Space Age*, that amazed even the participants who made it happen.

Book Two of *Astronautics—To the Moon and Towards the Future* documents the period that followed the initial thrust into space. During this era, political urgencies caused limited technologies to be stretched to perform exciting feats as in the case of *Voskhod* by Soviet Premier Nikita Khrushchev and his Chief Designer, Sergey Korolev. While dramatic advances in technology were also achieved with such partnerships, a principle example being John F. Kennedy's 1961 challenge to *"put a man on the Moon and return him safely to the Earth."* President Dwight Eisenhower was hesitant to embark on an expanded manned space program in the latter part of his tenure before it was clear that there was a role for man in space. As his successor, Kennedy, was more politically inclined to initiate a massive program that would accelerate America past the Soviets in such a convincing manner that the "uncommitted" nations of the world, who were watching events in technology closely, would have no doubt as to the best form of government.

The assassination of President Kennedy in 1963, the growing U.S. involvement in Vietnam, and the surfacing racial tensions in the cities of America placed heavy burdens on Kennedy's successor, Lyndon B. Johnson. His vision of "The Great Society" was threatened by a budget squeezed by escalating costs of the unpopular war and continued funding of Kennedy's expensive commitment to send men to the Moon *"before this decade is out,"* a commitment that Johnson himself had helped bring about. Despite these problems, the Soviets and the Americans proceeded to engage in a spectacular race to the Moon during the 1960s. Many perceived it as a decisive win for America while others felt the Soviets were a "no show" and had never been in a race at all.

The "Moon Race," as it has often been categorized, caused two elements of the evolutionary process of mankind's move into space to be bypassed—the space station and the reusable rocket. As a result, the next advances in space that followed the Moon landing were to establish these capabilities. While the Space Shuttle, which promised inexpensive access to low Earth orbit, was a marvel of technology, its complexity escalated its costs dramatically. Its use in building a permanent space station in the first decade of the new millennium was crippled by its fragility as evidenced by the tragedy of the Shuttle *Columbia*.

With the fall of the Soviet Union in 1989, a new level of cooperation emerged between the United States and the largest surviving country of the former USSR—Russia. As a collaborative effort among many nations, the International Space Station (ISS) has become a costly and significant undertaking that will require more than a decade to complete. With the end of the Shuttle program, now slated for September 2010, and with the number of the Shuttle missions reduced because of its high cost and questioned reliability, the completion of the ISS as originally designed is uncertain.

The search for extraterrestrial life in the solar system, centered on Mars, has been the focus of unmanned planetary excursions over the past 30 years. What evidence of life has been found and what are the prospects for interstellar travel? Book Two of *Astronautics—To the Moon and Towards the Future* brings events in the quest for this "final frontier" to the present day and speculates on what the future of space travel may hold.

President Kennedy before congress, May 25th 1961

A Change in Leadership

The presidential race of 1960 had three key issues, two of which were interconnected: the Cold War and the Space Race. The Republican candidate, the incumbent Vice President Richard M. Nixon, had a strong record of anti-Communism but the indifferent endorsement given him by the outgoing President Dwight Eisenhower had hurt his standing. Being a part of the Eisenhower Administration that many believed had "allowed" the Russians to achieve the first spectacular successes in space continued to put Nixon on the defensive during the campaign. (The USSR was often referred to as 'Russia' in the popular press in those days). The Democratic challenger, Massachusetts Senator John F. Kennedy, hammered the elusive "missile gap" (the alleged disparity between the American rocket capability and that of the Soviet's) at every opportunity.

Kennedy's weakness, as perceived by the Republicans, was being soft on Communism. His apparent willingness not to defend two small rocky islands off the coast of the Chinese mainland, Quemoy and Matsui, from invasion by Communist China was held as an indication of this flaw. These rocky pieces of land were tentatively held by the Nationalist Chinese from their last stronghold on Taiwan and had been a point of contention between the two candidates during their Presidential Debates.

The election was one of the closest in history, and most analysts credit the "missile gap" issue as costing the Republicans the election (although voting fraud in Cook County, Illinois, was seen by some as the turning point). However, the most prominent aspect of the outcome was that a Catholic had been elected President in a predominantly Protestant nation—the third key issue. That the Space Race, the visible aspect of the missile gap, could have overshadowed the strong religious precedent was an indication of how grave the nation felt about the technological turn of events.

The new American President, John F. Kennedy, was sworn in on January 20, 1961. Amid what many considered the darkest period of the Cold War, he set the nation on notice during his inaugural address that its people should, *"ask not what your country can do for you, but ask what you can do for your country,"* (paraphrasing an 1884 speech by Supreme Court Justice Oliver Wendell Holmes Jr.). He clearly envisioned that the challenges that lay ahead for the American people would require some measure of sacrifice. He also issued an unyielding notice to the Soviets when he stated, *"Let every nation know, whether it wishes us well or ill, that we shall pay any price, bear any burden, meet any hardship, support any friend, oppose any foe, in order to assure the survival and the success of liberty."*

It was a powerful speech that set the tone for his brief but dynamic and event filled presidency. It was also a foreshadowing of a commitment he would make less than four months later to overcome the Soviet lead in the space race.

The outgoing President, Eisenhower, had issued his farewell address to the nation three days earlier. The current Cold War and the Space Race with the Soviets had played a dominant role in his presidency, and it was evident in his remarks. He cautioned America that it should not pursue solutions whose results might be detrimental to the nation's economic and social welfare. His cautious approach to space exploration was apparent.

> *"Crises there will continue to be. In meeting them... there is a recurring temptation to feel that some spectacular and costly action could become the miraculous solution to all current difficulties. A huge increase in newer elements of our defense; development of unrealistic programs to cure every ill..."*

He was clearly referring to the huge expenditures that were being proposed to overcome the perceived technological lead that the Soviets appeared to enjoy. He had labored diligently to thwart those in industry who would use the conflict with the Soviets as a means to profit themselves.

> *"Until the latest of our world conflicts* [a reference to World War II], *the United States had no armaments industry. American makers of plowshares could, with time and as required, make swords as well. But now we can no longer risk emergency improvisation of national defense; we have been compelled to create a permanent armaments industry of vast proportions.... We annually spend on military security more than the net income of all United States corporations*

> *"This conjunction of an immense military establishment and a large arms industry is new in the American experience...Yet we must not fail to comprehend its grave implications.*
>
> *"In the councils of government, we must guard against the acquisition of unwarranted influence, whether sought or unsought, by the military industrial complex. The potential for the disastrous rise of misplaced power exists and will persist...Only an alert and knowledgeable citizenry can compel the proper meshing of the huge industrial and military machinery of defense with our peaceful methods and goals, so that security and liberty may prosper together."*

Eisenhower also understood the impact that technology was having on our society.

> *"Akin to, and largely responsible for the sweeping changes in our industrial-military posture, has been the technological revolution during recent decades.*
>
> *In this revolution, research has become central; it also becomes more formalized, complex, and costly. A steadily increasing share is conducted for, by, or at the direction of, the Federal government."*

He also had a caution for the academic community:

> *"Today, the solitary inventor, tinkering in his shop, has been overshadowed by task forces of scientists in laboratories and testing fields. In the same fashion, the free university, historically the fountainhead of free ideas and scientific discovery, has experienced a revolution in the conduct of research. Partly because of the huge costs involved, a government contract becomes virtually a substitute for intellectual curiosity. For every old blackboard there are now hundreds of new electronic computers."*

It was very perceptive that Eisenhower even mentioned computers, as the computer revolution was more than a decade in the future and the personal computer would not make its mark for another quarter century.

> *"Yet, in holding scientific research and discovery in respect, as we should, we must also be alert to the equal and opposite danger that public policy could itself become the captive of a scientific technological elite."*

Here was a man who had a foreboding of the possible consequences that unbridled technology programs could unleash on an unsuspecting nation.

Searching for an Alternative

Shortly before the November 1960 elections, a Kennedy political speech had emphasized the Eisenhower Administration's *"failure to recognize the impact of being first in outer space"* and that the Soviets were *"on the march... [and] had definite goals..."* and that America was *"standing still."* He had no way of knowing that the Soviet space program was, for the most part, an Ad Hoc vision that existed only in the mind of Chief Designer Sergey Korolev. Far from "standing still," the newly created National Aeronautics and Space Administration (NASA), under the Eisenhower Administration, had been moving forward. It had begun the development of an advanced three man spacecraft called *Apollo* and had absorbed Wernher von Braun's Peenemünde team from the Army Ballistic Missile Agency (ABMA) in Huntsville, Alabama.

To overcome the Soviet lead in big boosters, the von Braun organization, for the first time, was now fully committed to work on space exploration and was well along with building the 1.5 million-pound-thrust *Saturn I* booster—a cluster of eight of America's largest rocket engines. NASA had begun development of a single chamber engine capable of 1.5 million pounds of thrust that would eventually serve as the basis for the conquest on the Moon. It sought to surmount the relatively small lifting capacity of the *Atlas* ICBM by developing a high-energy liquid-hydrogen-fuel (LH2) upper stage called *Centaur*. However, despite these advanced programs, Kennedy's comment about America *"standing still"* would be splitting hairs as campaign rhetoric goes.

The first NASA chief, Dr. Keith Glennan (President of Case Institute of Technology), had been the primary architect of the early space program that Kennedy inherited. Glennan's goal was the Moon, and under his direction, a preliminary plan for man in space had taken form. Without a firm understanding of the lunar environment (including its surface) and without a clear requirement of the technologies needed, a lunar exploration agenda had been mapped out by 1961. Glennan formed the Goett Committee (named for its chairman Dr. Harry Goett, the Director of NASA's Goddard Space flight Center) in May 1959. Its objective was to identify the key steps in a manned space flight schedule. Within two months it defined a manned lunar landing as a follow-on to America's first man-in-space effort, Project *Mercury.*

However, Eisenhower, in a move to emphasize his feelings that a space race was not the proper course of action, reduced NASA's budget request for fiscal year 1961. This left the fate and direction of America's space program up to the new administration with the admonition that *"further tests and experiments will be necessary to establish if there are any valid scientific reasons for extending manned space flight beyond the Mercury program."* It is interesting to note his use of the phrase *"valid scientific reasons."*

Even before taking office, Kennedy realized that the issue of the Soviet lead in space exploration would have to be high on his priority list. Although his campaign had promised a *new frontier*, Kennedy himself was not enamored with the issue of space flight and of the *Mercury* program in particular. Nevertheless, he recognized that in dealing with the Soviet Union only a position of unqualified strength would hold sway with the leaders in the Kremlin. He was keenly aware that opinions in many nations uncommitted to either Communism or Capitalism (thus prime targets for Soviet expansion)—believed that the Soviet Union had moved to a dominant position—and no one likes to be on the losing side.

President Kennedy's first few months in office were filled with challenges. Simply establishing a new administration is a daunting task, but to cope with the immediacy of international problems presented the young president with an event packed first 100 days—the measure by which a presidency has often been initially evaluated.

Now that he sat in the Oval Office, Kennedy had the responsibility to deal with these problems and to make the decisions that would affect the future of the United States, indeed the world, possibly for decades to come. Perhaps the most pressing concept that Kennedy had to wrestle with was just how important was the "conquest of space." The visionaries of the day were saying, without equivocation, that "the nation that controlled space controlled the world" in much the same way as aviation had proved a dominant military factor in World War II and the post war period. Moreover, they contended that it was imperative that this "control" be established within the next decade. There was little doubt in anyone's mind in the "free" world that control of "outer space" by the Soviets would represent a threat to democracy and personal liberty worldwide.

To help assess the situation, Kennedy had formed an Ad Hoc Committee on Space, chaired by his future Science Advisor and former Massachusetts Institute of Technology (MIT) president Dr. Jerome Weisner. This was but one of 29 transition committees that provided guidance to Kennedy on a variety of policy and program issues during the period between his election and inauguration. Interestingly, the committee was not impressed with the *Mercury* program and advised downplaying its significance. Weisner believed that the major benefit was simply political. To a large degree, these thoughts echoed what was in Kennedy's heart. Despite his campaign rhetoric about the missile gap, he was not a firm believer in the space program but understood the political aspect, which was to provide the major impetus for getting elected.

As it turned out, the committee had not really established a majority opinion as these views were Weisner's, and they were in conflict with other major participants in the technical arena including NASA and the military. Weisner was not in favor of a manned space program, which he felt was too risky and costly. He preferred an unmanned program that emphasized science.

The new Vice President, Lyndon Baines Johnson, had been a vocal critic of the Republicans since the first *Sputnik* and was eager to pursue a manned space program. His leadership as chairman of the Senate Space Committee was pivotal in providing the funding for many of the projects underway in the early years of the space race. He was eager to become a major player in the move into space, and as Vice President, he became the chairman of the National Aeronautics and Space Council, an organization formed under Eisenhower to assist in developing policy. As early as 1958, Johnson was quoted as saying, *"Control of space is control of the world."* He was of course parroting the visionaries such as the renown German rocket scientist Wernher von Braun.

The U.S. Air Force had initiated its own manned space program as an outgrowth of the X-20 DynaSoar Program and had pushed for the primary role. Under Eisenhower, they achieved only limited success, and with Kennedy they were lucky to hold their status quo. For Kennedy, the struggle for supremacy in space was to be a struggle not just with the Russians but also between NASA, the Air Force and the Vice President.

As the new president began to formulate the direction he would move the space program, two events unfolded in international politics that played a pivotal role in his decision. The first was in the small Southeast Asian country of Laos, where communist insurgents supported by the North Vietnamese were on the verge of toppling another American backed government. Here was Kennedy's first "indirect" confrontation with Soviet Premier Nikita Khrushchev. American military forces were put on alert, and the Joint Chiefs advised Kennedy that it might take a major military intervention that could escalate to the use of tactical nuclear weapons. Other than more verbal posturing and movement of the U.S. Seventh Fleet into the area, Kennedy elected not to provide any direct military support beyond the few "advisors" who were already in place.

The second and more significant international crises occurred on April 15, 1961, just three days after Soviet citizen Yuri Gagarin's historic first space flight—the Cuban Bay of Pigs debacle. America had been trying to destabilize Cuban dictator Fidel Castro's government since he had declared an alliance with the Soviet Union in 1959, following a successful revolution. Castro's bold swing to the Russians may have been strongly influenced by Soviet successes in space.

The Eisenhower Administration had accepted a Central Intelligence Agency (CIA) plan to equip and train a small force of Cuban volunteers to make an amphibious landing on the shores of Cuba. The CIA had assured the new Kennedy Administration that the Cuban population as well as some of its military leaders were prepared to turn against Castro at the first strong indication of a liberating invasion. Kennedy approved the plan which history has since showed was multi-flawed in concept as well as execution.

From the beginning, the invasion went poorly, and within hours, the leaders of the invasion force called upon Kennedy to provide American military forces to turn the tide. However, Khrushchev threatened to intervene if U.S. forces were employed. Kennedy backed down. Following a brief fight on the landing beaches of the Bay of Pigs, Castro's army captured the invading force of some 1,500 Cubans.

Occurring, as it did, almost 100 days into his administration, the event was humiliating for Kennedy personally. He was criticized from the Republican "right wing" for not providing crucial military support and from the Democratic "left wing" for even considering the action. Kennedy had suffered a major setback in his presidency. This event, coupled with Gagarin's flight, provided the key pressure elements on Kennedy to do something to regain the lost prestige of the United States.

The Thursday following Gagarin's flight, the president met with Weisner, the new NASA Director James Webb, and his deputy, Dr. Hugh Dryden. Webb and Dryden reviewed the status of America's space program, and the question of how to compete with the Soviets. They indicated that the Soviets would orbit the first multi-manned ship, establish a space station, and probably circumnavigate the Moon. Kennedy was insistent: *"Is there any place we can catch them? What can we do? Can we leapfrog?"* Dryden felt that a crash program, on the scale of the atomic bomb project of World War II, might land a man on the Moon in ten years—the only possible "space spectacular" that might be achieved before the Russians.

It was a technological gamble that could cost $20 billion or more. Kennedy was obviously expressing "sticker shock" when he replied, *"Can't you fellows invent some other race here on Earth that will do some good?"* His comment reflected his inner thought that man's venture into space was simply a grandstand act of little scientific importance. Nevertheless, the imagination of the world had been captured by the possibility of a flight to the Moon—and Kennedy knew it. The meeting ended with the president imploring that they provide a winning solution, and he added, *"There's nothing more important."* If Eisenhower had been blind-sided by *Sputnik*, Kennedy was trapped by the space race—a race in which the United States did not have the option of withdrawing. And, if that were so, then Kennedy had only one solution available to him—to win.

Kennedy sent a "Memorandum For The Vice President" dated April 20, 1961, that summarized the key questions and asked for *"an overall survey of where we stand in space."* He ended the memo with the request, *"I would appreciate a report on this at the earliest possible moment."*

Johnson wasted no time in preparing his five-page response on April 28, in which he succinctly and accurately summarized nine aspects of the problem:

Largely due to their concentrated efforts and their earlier emphasis upon the development of large rocket engines, the Soviets are ahead of the United States in world prestige attained through impressive technological accomplishments in space.

The U.S. has greater resources than the USSR for attaining space leadership but has failed to make the necessary hard decisions and to marshal those resources to achieve such leadership. This country should be realistic and recognize that other nations, regardless of their appreciation of our idealistic values, will tend to align themselves with the country which they believe will be the world leader—the winner in the long run. Dramatic accomplishments in space are being increasingly identified as a major indicator of world leadership.

The U.S. can, if it will, firm up its objectives and employ its resources with a reasonable chance of attaining world leadership in space during this decade. This will be difficult but can be made probable even recognizing the head start of the Soviets and the likelihood that they will continue to move forward with impressive successes. In certain areas, such as communications, navigation, weather, and mapping, the U.S. can and should exploit its existing advanced position.

If we do not make the strong effort now, the time will soon be reached when the margin of control over space and over men's minds through space accomplishments will have swung so far on the Russian side that we will not be able to catch up, let alone assume leadership.

Even in those areas in which the Soviets already have the capability to be first and are likely to improve upon such capability, the United States should make aggressive efforts as the technological gains as well as the international rewards are essential steps in eventually gaining leadership. The danger of long lags or outright omissions by this country is substantial in view of the possibility of great technological breakthroughs obtained from space exploration.

Manned exploration of the Moon, for example, is not only an achievement with great propaganda value, but it is essential as an objective whether or not we are first in its accomplishment—and we may be able to be first. We cannot leapfrog such accomplishments, as they are essential sources of knowledge and experience for even greater successes in space. We cannot expect the Russians to transfer the benefits of their experiences or the advantages of their capabilities to us. We must do these things ourselves.

The American public should be given the facts as to how we stand in the space race, told of our determination to lead in that race, and advised of the importance of such leadership to our future. More resources and more effort need to be put into our space program as soon as possible. We should move forward with a bold program, while at the same time taking every practical precaution for the safety of the persons actively participating in space flights.

The Questions that Kennedy posed were then answered point by point;

Q.1- Do we have a chance of beating the Soviets by putting a laboratory in space, or by a trip around the Moon, or by a rocket to land on the Moon, or by a rocket to go to the Moon and back with a man. Is there any other space program which promises dramatic results in which we could win?

A.1- ...As for a manned trip around the Moon or a safe landing and return by a man to the Moon, neither the U.S. nor the USSR has such capability at this time, so far as we know. The Russians have ... a time advantage in circumnavigation of the Moon and also in a manned trip to the Moon. However, with a strong effort, the United States could conceivably be first in those two accomplishments by 1966 or 1967...

Q.2- How much additional would it cost?

A.2- ...about $500 million would be needed for FY 1962 over and above the amount currently requested of the Congress. A program ... over a ten-year period, ...approximately $1 billion a year above the current estimates of the existing NASA program...

Q.3- Are we working 24 hours a day on existing programs? If not, why not? If not, will you make recommendations to me as to how work can be speeded up?

A.3- There is not a 24-hour-a-day work schedule ... except for selected areas in Project Mercury, the Saturn C-1 booster, the Centaur engines and the final launching phases of most flight missions... work can be speeded up through firm decisions to go ahead faster if accompanied by additional funds needed for the acceleration.

Q.4- In building large boosters should we put our emphasis on nuclear, chemical or liquid fuel, or a combination of these three?

A.4- It was the consensus that liquid, solid and nuclear boosters should all be accelerated...

Q.5- Are we making maximum effort? Are we achieving necessary results?

A.5- We are neither making maximum effort nor achieving results necessary if this country is to reach a position of leadership.

The two memos represented the sum total of the dilemma and provided Kennedy with all that he really needed to make his decision.

"Land a Man Safely on the Moon"

In assessing the impact of major space "firsts," the Soviets had now achieved the first three: the first satellite, the first unmanned Moon probe, and the first man in space. However, Kennedy pondered if the fourth major "first," landing a man on the Moon, could be achieved before the Soviets and what importance the world would place on that accomplishment. In addition, to be truly effective, this fourth accomplishment would have to demonstrate a commanding lead, or its impact might not be significant. It was understood that colonization of the Moon was expected as a possible immediate follow-on as well as expeditions to Mars.

While Kennedy pondered the lunar commitment, he did approve allocating money that Eisenhower had held back for *Saturn* and the new high thrust engine called the F-1, an increase of about 10 percent in the NASA budget. He was also presented with continued funding requirements for a more broad thrust into space for weather, communications, and navigation satellites that would serve both the military and civil sectors. Initial efforts in these areas had already been made, and the payback was promising.

Over all, a vigorous space program could be a significant shot in the arm for economic and technological growth in additional to its political implications. Despite future possible stimulants, the head of the Bureau of the Budget was opposed to an all-out program because of its drain on the federal resources.

The success of astronaut Alan Shepard's sub-orbital MR-3 flight on May 5, 1961, touched off a degree of euphoria in the country. Coupled with the political dilemma that Kennedy faced, his decision was not so much a surprise as it was a powerful visionary message delivered with a style that would make the pronouncement a landmark in American history.

On May 25, 1961, Kennedy delivered a special message to Congress on "urgent national needs." He stated: *"If we are to win the battle that is now going on around the world between freedom and tyranny, the dramatic achievements in space which occurred in recent weeks should have made clear to us all, as did the Sputnik in 1957, the impact of this adventure on the minds of men everywhere, who are attempting to make a determination from which road they should take."* There was no doubt Kennedy believed that, whatever his personal feelings about space, the duel between the United States and the Soviet Union compelled America to take a leadership role.

He then made the fateful commitment: *"First, I believe that this nation should commit itself to achieving the goal, before this decade is out, of landing a man on the Moon and returning him safely to the Earth. No single space project in this period will be more impressive to mankind, or more important for the long-range exploration of space; and none will be so difficult or expensive to accomplish."*

Of course, he also made it clear in the speech that it was up to Congress and the American people to validate his rational and the road he was mapping out. While some saw his careful wording as an indication of his lack of full commitment, others perceived it as a way of getting Congress and the people to "buy-in" to the commitment, to make them full partners in the decision. Whatever the case, it accomplished its goal. A careful reading of the challenge did not say that the United States would be first, only that an American would walk on the Moon within the next eight and one-half years—but the meaning was obvious. As it had at Pearl Harbor, the sleeping giant had been awakened and would ultimately emerge the victor.

If Kennedy still harbored a reluctance to proceed down the path he so eloquently defined, it was not evident to the electorate. With the goal clearly established, Congress approved the funding and progress began to happen—quickly. All this enthusiasm had been generated while an American had only 15 minutes in space.

America felt reasonably confident that it had the technological lead over the Soviets. Nevertheless, the ability to create the massive rockets, complex spacecraft, and computers to send men to the Moon would require new advances in the state-of-the-art in many areas. On top of this was the management of the projects and the people—perhaps the most difficult aspect of the endeavor.

However, the organization (NASA) that sought to plan and control this daunting project would have to grow from its role as an "advisory committee" of only a few years previous, to a complex structure capable of overseeing the activities of hundreds of contractors and tens of thousands of individuals— and billions of dollars. To manage the growing NASA bureaucracy, Kennedy sought a man with very special capabilities. It was felt that the first NASA administrator, Keith Glennan, appointed by Eisenhower, did not have the right attributes aside from being a Republican. What was needed was someone who could not only organize and budget a project that had few equals in history but who also had the political savvy to deal with the Congress. More than a dozen persons were approached who flatly rejected the job.

When James Webb was asked, he likewise declined. Here was a man in his mid-fifties who had performed successfully a variety of roles as a lawyer, Marine, and corporate executive. He had served his country in the Department of the Treasury, as Director of the Bureau of the Budget, and in the State Department. His North Carolina background added to his persona with a southern drawl that could soften his otherwise strong presence. He recognized that there would be an enormous amount of political pressure and he was his own man—he did not think he could run the program without interference. Nevertheless, with pressure applied by several influential people including Vice President Johnson, he finally accepted—but only after president Kennedy asked him personally and guaranteed his "independence."

Arguably, no one person shaped NASA nor created its image and influence more than James Webb. It has been said that it was only through his strong leadership that all of the elements needed to achieve the Kennedy goal could be welded into a cohesive and successful program.

In the Soviet Union there was little attention given to Kennedy's speech; after all, the Americans were good at speeches but poor on delivering their promises. Korolev in particular had no direct reaction, and there was no assessment of a plan to ensure that the first man on the Moon was Russian. In the years that immediately followed Kennedy's challenge, there would be several proposals from the Soviet Chief Designers to achieve a manned circumlunar flight. But none were embraced by the Soviet leadership until it would be too late to compete effectively.

Lingering Thoughts

President Kennedy continued to emphasize the importance of his commitment when he recognized John Glenn's accomplishment on becoming the first American to orbit the Earth the following February (1962). *"This is a new ocean, and I believe America must sail upon it."*

In a highly publicized speech at Rice University in September 1962 he said, *"The exploration of space will go ahead, whether we join in it or not. And, it is one of the great adventures of all time, and no nation which expects to be the leader of other nations can expect to stay behind in this space race. We mean to lead it, for the eyes of the world now look into space, to the Moon and to the planets beyond, and we have vowed that we shall not see it governed by a hostile flag of conquest, but by a banner of freedom and peace."* When he added, *"We choose to go to the Moon! We choose to go to the Moon in this decade and do the other things, not because they are easy, but because they are hard."* Few could have believe that he did not have a strong personal dedication.

A year after Kennedy made the commitment, NASA Chief Webb told Kennedy that there were still some in the scientific community who expressed doubt that it was possible to send men to the Moon. Kennedy responded that the Moon landing is NASA's top priority. *"This is, whether we like it or not, a race.... Everything we do* [in space] *ought to be tied into getting to the Moon ahead of the Russians... except for defense,* [it is] *the top priority of the United States government. ... Otherwise, we shouldn't be spending this kind of money,"* to which he surprisingly added, *"because I'm not that interested in space."*

While Kennedy might not have felt that the Moon race was truly the best alternative, its pursuit would provide a number of economic and technical advantages. The infusion of federal money into high technology projects would help an economy that had been in a recession for the past several years. Literally thousands of companies, small and large, would profit from the effort. Advanced research in power sources such as nuclear rocket engines would assure that America stayed abreast of possible break-through technologies that might otherwise go un-funded and result in yet another form of *Sputnik*-like surprise a few years down the road. There would be many other advances that would prove a vital stimulus to the American economy. These included the miniaturization of electronic systems that allowed America's small satellites to compete effectively with the giant *Sputnik*s on a scientific basis. This would lead to a semiconductor industry that, decades later, would put a computer in most homes in America. The planning and control of large complex projects would develop new management practices to help keep American industry efficient and competitive. However, these were but some of the nebulous unknowns of the future.

Less than three months after Kennedy's assassination, Congressman Thomas M. Pelly introduced a resolution asking that the Moon landing goal be extended five years—to 1975. He was aware of the affect that the time constraints were having on the engineering effort and the risks that were being accepted. There was also the question of extending the funding over a longer period to help ease the budget burden that was being squeezed even tighter by the escalating war in Vietnam.

It is interesting to conjecture what the future might have held if the United States had not pursued a vigorous manned space program in the 1960s, as the most significant advances for military and commercial use came from unmanned near-Earth satellite applications such as reconnaissance, communications, weather, and navigation.

Even before the first *Vostok* or *Mercury* spacecraft took men into orbit, the visionaries in both the Soviet Union and the United States were considering the next step. The initial planning had to be somewhat tentative because the effects of space flight on man's ability to function effectively in this unknown environment were yet to be discovered. The requirements for advancing mankind's reach into the cosmos were obvious; the technology and technique necessary to join two or more craft in orbit were needed to assemble a space station or to send men to the Moon and planets. This capability required levels of satellite tracking and computing that were essentially unknown in the early 1960s. Likewise, the ability to re-start reliably an upper stage in the vacuum of space was still being worked out. The terms that would move to the forefront in the ever-widening vocabulary of space were "rendezvous" and "docking." Rendezvous is the ability for two satellites, launched at different times, to be brought along side each other in space. Docking is the process of physically joining the two satellites.

Orbital Union

The basic physics involved in bringing two satellites into "union" to permit rendezvous and docking was defined by Walter Hohmann in a 1925 publication, and the technique is commonly referred to as the "Hohmann Transfer." It represents the most economical path (but not necessarily the shortest or fastest) for a spacecraft to transfer from one orbit to another. With respect to a rendezvous between two Earth satellites, the technique first requires that the two orbits have almost identical inclination and phase (the path of one satellite must be directly above or below the other). The ability to launch one satellite into the same plane and phasing as an established satellite is a significant accomplishment. It requires that the launch of the second satellite be accurately timed and guided to its pacing track. As the lower satellite has a shorter period, it will pass beneath the higher satellite numerous times over the course of several days. In practice, the new satellite is usually the "active" interceptor (in the lower orbit) while the existing satellite plays a passive role.

The transfer involves the creation of a new elliptical orbit for the active satellite, to make its apogee tangent to the orbit of the passive satellite (Figure 17). At the proper time, the active satellite is powered into this transfer path, and halfway through this new elliptical orbit, the active satellite intercepts the passive satellite and executes another "circularizing" burn to raise its perigee to achieve the same orbit as the passive target.

The initial burn used to create the "transfer" orbit also uses some energy to correct for slight differences in orbital inclination and phasing. The energy required for significant changes to these parameters is typically not available to the active satellite, thus the need for very accurate launch guidance into the initial orbit. To accomplish the Hohmann Transfer, guidance and tracking procedures had to be developed and the supporting radar systems and computer hardware and software built.

Soyuz Development

The primary mission of a second-generation spacecraft (as envisioned by Soviet Chief Designer Sergey Korolev) was for a manned circumlunar flight—although the Communist Central Committee had not yet approved a lunar program in 1962. However, Korolev also had to accommodate the military to assure funding, so military requirements were always paramount in the planning documents. For both missions the new ship had to be capable of rendezvous and docking to provide for orbital assembly of large space stations or refueling for lunar transfer missions. Several concepts were explored and heatedly debated among the various chief designers and the military before a final configuration was established.

The primary designers were again Mikhail Tikhonravov and Konstantin Feoktistov, of Korolev's team—the architects of the first Soviet manned space craft—*Vostok*. They proposed a craft that provided a ballistic reentry but had an offset center of gravity allowing some corrections in the descent to permit limited selection of the landing footprint and to provide lower-G reentry profiles. A wide range of possibilities was considered including an upgrade to the *Vostok*. Although winged vehicles had been explored, it was obvious that these were far too ambitious for the time schedule and state-of-the-art to

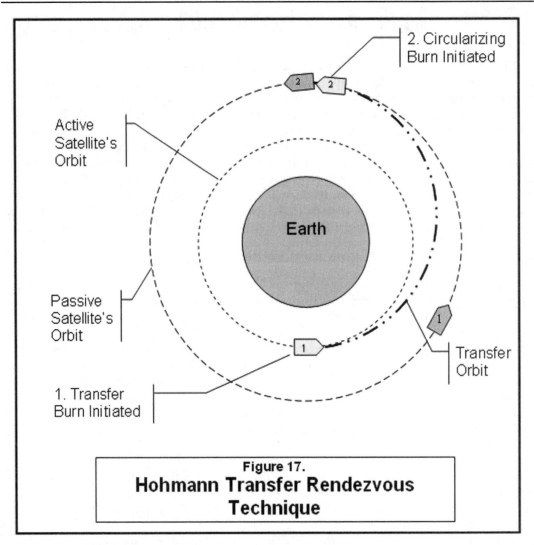

Figure 17.
Hohmann Transfer Rendezvous Technique

be considered. There were occasional differences of opinion between Korolev and his staff that made him periodically rethink and revise his plans. This would cause him to recycle some of his ideas and resulted in delays in the implementation.

Another factor that continually caused him to reevaluate the technical direction was the reports coming to him by way of the Western press, which frequently published optimistic schedules for the various projects being developed in the United States. While the "openness" of the American program was often an advantage to Korolev (allowing him to upstage the American effort), it also caused him to make premature decisions that would haunt the Soviet program in the long run.

The final second generation proposal, code named *Soyuz* (a Russian word for "union"), was submitted by Korolev in mid-1962. It consisted of three components: a three man "reentry module," a cylindrical "instrumentation and equipment module" (on one end of the reentry module), and a spherical "living quarters" and docking adapter on the other end (Figure 23). All modules were enclosed in an aerodynamic shroud that was jettisoned before achieving orbital velocity.

The reentry module now had a dish-shaped heat shield similar to the *Mercury* capsule to minimize the weight of the ablative material and, as a result, now required an attitude control system unlike the passive off-centered center of gravity that stabilized *Vostok* during its reentry. The rendezvous procedure between two satellites would be completely automated, although provisions were made for the crew to intervene. The fact that the docking adapter was not visible from the reentry module itself (from which the spacecraft was controlled) meant that electronic alignment of the two ships and television cameras would play a critical role in the final technique.

Ejection seats were no longer an option for either emergency escape or the final descent to land. Thus, a solid-fuel, tractor-type, emergency escape rocket (similar to the *Mercury* capsule) was configured to pull the spacecraft from the booster in the event of a malfunction. As the ship would again be recovered on dry land, some means of slowing its descent at touchdown was required to avoid a bone-crushing G-load on impact. This meant the inclusion of another solid-fuel retrorocket suspended from the parachute assembly and triggered by a feeler probe that sensed when the craft was several feet above the surface to soften the impact with the ground. Like *Vostok*, *Soyuz* would be totally enclosed in an aerodynamic shroud during the launch phase.

While *Vostok* had weighed in at 10,400 pounds, *Soyuz* would tip the scales at 15,000 pounds. This would require elongating the upper stage of the *R-7* to provide for more fuel for Earth orbital flights and a new bigger booster (still several years from being tested) for the manned circumlunar flights. The initial schedule called for the first flight of the *Soyuz* in early 1965 (Figure 18).

Mercury II

The Goett Committee convened in 1959-60 to outline America's plan to regain the technical lead in space exploration. One of its recommendations was the development of a second-generation spacecraft following *Mercury*. It would carry a three-man crew and would serve as the basis for a possible (but at the time unapproved) manned lunar program. In July of 1960, the project was formally named *Apollo*. Although work proceeded on the *Apollo* spacecraft, its role and schedule was still in doubt as the Kennedy Moon challenge had yet to be issued. It was also recognized that *Apollo* (like *Soyuz*) represented a quantum leap in many areas of systems technologies that had yet to be defined. Apollo also required a million-pound-thrust booster that was still years from being available.

As Project *Mercury* moved towards its first manned flight in early1961, the McDonnell Aircraft Company, in concert with NASA's Space Technology Group (in the person of spacecraft designer Maxime Faget), performed a study of an enlarged version called *Mercury* II. The McDonnell Company was hopeful of continuing its role in providing spacecraft to NASA.

As Project *Mercury* had been established on a crash basis and within the limits of the *Atlas* booster, this new round of planning was predicated on the availability of the *Atlas Centaur* with its liquid hydrogen second stage. It would allow double the weight of the spacecraft and provide many more capabilities. Several approaches were examined. The first was simply to expand some of the consumables so that the spacecraft could perform a one-day (18-orbit) mission (this would eventually be accomplished by upgrading the basic *Mercury* spacecraft). A second involved some structural changes to move many of the systems outside of the pressure hull so that they could be more easily accessed for problem isolation and engineering change modification. These latter operations had been consuming a significant amount of time and had put the basic *Mercury* program well behind its initial schedule. The most ambitious of the possible configurations was a two-man capsule that would be more than twice as heavy as *Mercury*.

With the advent of the Kennedy challenge, the design of the *Apollo* was mired in the complexities of its lunar mission, with many of its subsystems representing unknowns. What was needed was an interim spacecraft that bridged the gap both technologically and chronologically between the anticipated end of the original *Mercury* program in 1963 and the projected start of *Apollo* in 1965. As the various possibilities were explored, a version that could be the basis for a manned circumlunar flight was also suggested, but that would put the whole existence of *Apollo* in question and introduce more complexities. The *Mercury II* proposal that was finally approved was to build a scaled up version of *Mercury*. It would have a set of attached segments on the aft end to house a hypergolic propulsion unit to provide orbital maneuvering and rendezvous. It would also have provisions for long duration flights of up to two weeks and features that would allow the crew to accomplish extra-vehicular activities—a space walk.

The *Atlas Centaur*, on which the increased weight would be accommodated, was another obstacle as its availability continued to slip. Its high-tech nature also made its ability to launch with precise timing for the needs of rendezvous, problematical. If *Mercury* II was to become a NASA program, another launch vehicle would have to be found.

Titan II: A More Capable Launch Vehicle

During the early years of the *Atlas* project, a recommendation was made from several sources that a second Intercontinental Ballistic Missile (ICBM) be developed, not only as a back up, but to provide a heavier lift and longer range capability. This proposal was partially prompted by the radical pressurized "balloon" structure of the *Atlas*. Its thin pressurized tank walls were to provide the required stiffening. The analogy is a common soda-pop can, which, before it is opened, is pressurized and cannot be easily distorted. However, the strength of this type of structure had been questioned by many knowledgeable engineers.

The Glenn L. Martin Company, had just begun work on the *Vanguard*, was chosen in September 1955 to produce WS 107A-2—the *Titan* missile. At the time, it was envisioned that only one ICBM (either *Atlas* or *Titan*) would actually become operational—the first one to complete its development program and become available for deployment. With a first stage powered by a set of Aerojet General LR-87-AJ-1 engines of 150,000 pounds thrust each, and a second stage hosting a single LR-91-AJ-1 of 80,000 lbs thrust, the *Titan* was a true two-stage missile. Unlike the *Atlas, Titan* had conventional aircraft-like monocoque construction where the metal skin is tightly coupled (welded/riveted) to underlying structural members to provide the support.

Titan flight-testing began in February 1959, almost two years after *Atlas*. The first four flights were successful, but then a string of failures set the program back. With the cold war and Khrushchev's threats (among them the "we will bury you" speech) still placing a burden on military preparedness, it was decided to deploy both missiles, with 54 *Titan*s being in place by 1962. However, *Titan* was powered by liquid oxygen (LOX) and kerosene, a combination that was not conducive to quick response nor one that could be held in a high state of readiness for a prolonged period of time.

The impetus for a larger and more powerful version of the *Titan* was born of the concept of multiple reentry vehicles and the need for more flexible reaction time in the launch sequence—enter Titan II. With lighter weight warheads and extreme accuracy (arriving within one mile of the target after a flight of more than 6,000 miles) being achieved, it became more economical for each missile to carry several warheads and have them target objectives within a few hundred miles of each other with a single launch. This would also thwart possible anti-missile defense systems by presenting multiple threats. To accomplish the improved performance, the eight-foot diameter of the second stage was widened to match the ten-foot first stage, and the total length went from 98 to 104 feet. A second major change was to use storable (but less energetic) propellants so that the missile could remain fueled for an extensive period and achieve a quick launch.

The combination of A-50 hydrazine and N204 nitrogen tetroxide was teamed with a new set of engines that produced more thrust than the Titan I to make up for their lack efficiency as compared to the use of the favored cryogenic oxidizer—liquid oxygen. The efficiency of a rocket engine is typically measured by its Specific Impulse (Isp)—the thrust produced relative to the fuel consumed per second. The first stage now had 430,000 pounds thrust, and the second, 100,000. This quicker launch capability and greater weight lifting of the new *Titan II* also proved to be the selling point for the *Mercury* II application. However, combustion instability with the second stage engine resulted in a high frequency vibration (25,000 hertz) that literally shook the engine to destruction. A second problem related to the high altitude start for the second stage propellant pumps. Both problems were ultimately addressed as the *Titan* II moved through its development phase that was completed with its 32nd flight in April 1964.

The proposed allocation of a dozen *Titan* IIs to NASA was also fraught with political bickering between NASA and the military, who were moving forward with yet another version—*Titan* III. NASA wanted Martin to enlarge the propellant tanks of the *Titan*, but this would impact the military program. After much debate, NASA finally settled for the standard *Titan* II with some special upgrades for the process of "man-rating" the booster. These began with structural modifications to the second stage to permit the mating of the heavy 7,000 pound *Gemini* spacecraft (Figure 18). The inertial guidance was exchanged for one more appropriate for a satellite trajectory along with the installation of redundant control actuators to improve reliability. The single use batteries were replaced with rechargeable units, and a Malfunction Detection System (MDS) was installed that monitored critical aspects of the launch vehicle to command an abort sequence if that became necessary. These aspects included fuel tank and combustion chamber pressures, attitude rate changes, and range safety officer commands.

Evolution of Gemini

As the planning for *Mercury* II became more defined in scope and schedule, it was separated from the *Mercury* program in the first days of 1962 and given a new name: *Gemini* (a constellation of stars represented as "The Twins"). Independent funding (apart from Project Mercury) and a planned 10 manned flights would develop the technologies that would take America to the Moon and ultimately cost almost one billion dollars (more than ten times the original *Mercury* budget which itself ended after spending $350 million). The *Gemini* program had five primary objectives: (1) develop spacecraft systems for long duration Earth-orbital flights of up to 14 days, (2) determine man's ability to function during these extended missions, (3) demonstrate rendezvous and docking with a target vehicle, (4) develop improved launch procedures and techniques for rendezvous, and (5) move the primary recovery mode to land based techniques. The original (and typically optimistic) schedule called for completion of all objectives by October 1965.

The McDonnell Aircraft Company (again the prime contractor as it had been for *Mercury*) moved swiftly not only to establish a modular approach to building the spacecraft but also to refine their newly developed project management methods. These techniques would allow them to move more quickly to isolate problem areas and ensure resources were effectively brought to bear. They also bypassed the subcontracting process by using existing suppliers with whom they had established good working relationships. Nevertheless, they were to discover again that the development schedule and budget could not be easily controlled.

The new spacecraft consisted of three modules, the first being the 12 foot long scaled-up crew module that would provide for two astronauts (hence the appropriate reference to the twins of *Gemini*). Virtually identical in basic shape to *Mercury*, it had a maximum diameter of 7.5 feet at the base of the heat shield. The second module, forming a part of the widening flange to interface to the 10-foot diameter *Titan* upper stage, contained four 1,200 pound thrust solid-fuel retrorockets. The third segment contained the on-orbit electrical power, and the Orbital Attitude and Maneuvering System (OAMS— pronounced ohms) in addition to other equipment (Figure 22).

The OAMS had four functions: 1) to provide thrust to enable the spacecraft to rendezvous with the target vehicle, 2) to control the attitude of the spacecraft, 3) to separate the spacecraft from the second stage of the launch vehicle, and 4) to provide abort capability at altitudes above 300,000 feet. OAMS comprised 16 thrust chambers: eight 25-pound attitude thrusters for pitch, yaw, and roll, and eight 100-pound thrusters to maneuver the spacecraft fore and aft, up and down, and left and right with respect to its longitudinal axis.

Because the crew module required attitude control (pitch, yaw, and roll) on its return to Earth, it contained two redundant sets of 25 pound thrusters and fuel for an independent Reaction Control System (RCS). It also contained the computerized guidance system that would allow the spacecraft to use its offset center of gravity to change its landing point by taking advantage of its positive lift-to-drag ratio.

Electrical power for the spacecraft presented a new set of problems. *Mercury* and *Vostok* had used sets of batteries. Batteries would be a part of the crew module as they would provide the power during reentry. However, the mission duration for *Gemini* would extend up to two weeks, and there was no way that much battery power could be allocated. Given the high levels of electrical power for the many new systems, including the radar needed for the rendezvous operation, and the weight restrictions, a new solution had to be found. The obvious second choice was solar power. However, with the expected maneuvering that *Gemini* would have to perform, the structure of the solar panels would have to be quite robust or capable of retracting during the powered phases. This was deemed unacceptable. What was needed was a revolutionary capability to generate electricity.

Some research had been done on an electrical generation device known as the "fuel cell." Working on the reverse principle of electrolysis (in which electrical power is used to separate water into its two primary elements of oxygen and hydrogen), the fuel cell combines hydrogen and oxygen to produce electrical energy with water as a by-product. The process involves a simple chemical reaction of hydrogen with a catalyst to free electrons. Fuel cells had never been used in practical applications in the early 1960s, and they required the spacecraft to carry liquid hydrogen in addition to liquid oxygen. The latter was already a component needed to supply the breathing oxygen for the crew. Extensive work was required to make this power source (which would be located in the third adapter) a viable option. General Electric, whose fuel cell design used ion-exchange membranes rather than gas-diffusion electrodes, was chosen to develop two units for each spacecraft

The pure oxygen cabin environment was retained for *Gemini* along with the lithium hydroxide for removal of carbon dioxide. An omen of future consequences of this volatile atmosphere occurred in September of 1962 with a fire in a simulated space cabin at the Air Force School of Aerospace Medicine, Brooks Air Force Base, Texas. The experiment was designed to provide information on possible effects of the crew breathing pure oxygen in an environment of five pounds-per-square-inch for a duration of 14 days. One of the two participants (non-astronauts) was severely burned. Had the fire occurred in a sea-level environment of 14 pounds-per-square-inch, the event would have been catastrophic.

One of the main problems with *Mercury* was the inability of the Environmental Control System (ECS) to manage the heat generated by the astronaut, the electronic equipment, and the friction during the initial passage of the spacecraft through the atmosphere on its ascent into orbit. *Gemini* would relocate many of the electrical systems outside the pressure hull, but it would still create more than three times the heat of *Mercury*. This problem was addressed by a heat exchange system located on the skin of the adapter section which provided ample area for cooling.

Many of the systems were segmented into two independent (or semi-independent) parts, including the battery power, environment, and attitude control as these capabilities had to be provided to the spacecraft, following its separation from the adapters, for reentry. A new set of mission rules dictated that should the spacecraft have to switch to any of these reentry systems during a mission, then the mission had to be terminated. Those *Mercury* flights that ended with depletion of the attitude control fuel during reentry would not be tolerated with *Gemini*.

With the move to the *Titan* II and its hypergolic propellants, the emergency escape system during launch was revisited. Originally, it was envisioned that *Gemini* would use an escape tower powered by a solid-fuel rocket, as had been the case with *Mercury*. However, a catastrophic explosion of the *Titan* would not present as lethal a fireball in size or temperature as the *Atlas Centaur*. This would allow for the use of ejection seats for low altitude abort profiles up to 60,000 feet (including pad aborts). The retrograde rockets in the forward adapter (for reentry) would be used to separate the spacecraft from the booster from 60,000 feet to normal second stage burn completion. This approach also allowed the seats to be used after the capsule returned to the lower levels of the atmosphere following high altitude aborts. Likewise, following reentry, if there were any malfunction with the landing system, the crew could elect to eject.

To achieve the required ejection timing, the hatches had to open in less than one-third of a second. The system allowed the astronauts to make the abort decision but provided a ground activated backup. It was also determined that if either pilot elected to eject, the system would eject both, using a rocket propelled seat which was set at a slight angle to each other to aid in assuring there would be no trajectory conflict between the seats during the process. While the ejection method had been carefully thought out and engineered, the astronauts held the system with suspicion.

The landing system was the subject of much activity. Originally, two *Mercury* type parachutes would lower the craft, following reentry, to a water landing as had been done with *Mercury*. However, a NASA engineer named Francis M. Rogallo of the Langley Research Center, proposed a large inflatable delta wing and tricycle landing skids that would allow the spacecraft to glide to a landing on hard surfaced terrain. Although its glide ratio was limited to about 5-to-1, it would provide limited atmospheric maneuvering capability and a more conventional landing. North American Aviation received a contract to pursue the "paraglider" concept and to expand its scope to other applications to include returning the first stage of large boosters back to Earth to be reused—although *Gemini* remained the prime focus for the paraglider development team.

Repeated failures in a wide range of hardware and schedule slippages, coupled with opposition from several quarters within the Manned Spacecraft Center's management, doomed the paraglider effort. Even if the technical difficulties could have been overcome, there was still the weight—the paraglider system required an added 790 pounds (10 percent of the weight of the entire spacecraft). The inability to provide adequate adaptation of the concept for a contingency water landing, along with the weight, complexity and need for back-up parachutes, finally dictated a cancellation of the endeavor. A second effort was made to replace the Rogallo wing with a parasail—a steerable parachute similar to what sport parachutists now use. Time was growing short as 1964 came to a close. With the first flights nearing, the tried and true method would be used, and *Gemini* would descend under a single 80-foot parachute for a water landing.

One aspect of the paraglider concept was retained, however; unlike *Mercury*, which landed on it back, *Gemini* would transition during its final descent to a more horizontal position of 35 degrees relative pitch up to the horizon. This would allow the capsule to impact with the crew in an upright position and the spacecraft to float like a boat. This was also more desirable during crew egress in that the large crew access doors opened out and allowed some protection from wave-action during the process by "splash curtains" that could be raised when the hatches were opened.

Rogallo and his revolutionary wing did not disappear, however. Amateur hang gliding enthusiasts adopted the lightweight concept, and the design spawned a renewed interest in inexpensive sport gliding.

Two horizon sensors (a primary and back up) were part of the spacecraft's Guidance and Control System (GCS). They identified the Earth's infrared horizon to provide reference for aligning the inertial guidance platform and to provide commands to the control system for the spacecraft's attitude in pitch and roll axes.

In developing the rendezvous capability, several approaches were employed. An optical (and manual) method used a flashing light on the target vehicle and visual alignment of the spacecraft to perform the final rendezvous and docking. A completely automated system used radar, a computer, and a gyroscopically stabilized platform to allow the spacecraft and the target vehicle to electronically lock on to each other at distances up to 500 miles and maneuver automatically from that point on. The automated method was the most efficient in terms of both time and fuel used; it was also the most complex. Because of the possibility of failure of one or more components in the fully automated system, a series of integrated procedures were produced that would allow the completion of the rendezvous using a combination of the two methods. *Gemini* would allow experimenting with several rendezvous techniques.

With respect to the docking requirement, NASA planners decided to use a target vehicle that was more than just a satellite to which *Gemini* could link itself. A fully stabilized unit including a restartable engine would be employed to provide experience in coordinating the activities of two semi-autonomous vehicles, the mechanical linkage, the electrical interfaces and control transfer techniques. Next to the rendezvous capability itself, the linking with another "live" rocket that contained propellant was the most daring of the *Gemini* objectives. The *Agena* (launched by the *Atlas*), was selected as the target satellite because it possessed all of the required capabilities and was in the process of achieving a high level of reliability. Eleven *Atlas-Agena* vehicles were procured as rendezvous targets with modifications that included radar and visual navigation and tracking aids as well as an external rendezvous-docking unit.

The rendezvous radar, situated in the forward section of *Gemini*, would locate and track the target vehicle during rendezvous maneuvers. A transponder beacon (a receiver/transmitter) located in the *Agena* would reply to an "interrogation" signal from the *Gemini* radar. A digital computer in *Gemini* would integrate all of the data (position, distance, velocity, etc.) and compute the required burn of the OAMS system. It provided an incremental velocity indicator (which visually displayed changes in spacecraft velocity), a manual data insertion unit (a keypad), and a display to readout the computer solutions.

The total weight of all three segments of *Gemini* would eventually grow to 8,355 pounds in orbit with a total length of 18 feet. Following separation of the various adapters during the reentry process, the spacecraft would weigh 4,840 pounds at splashdown. It was a very capable spacecraft in all respects.

As an extension of the goal of determining man's ability to function during long duration missions, the possibility of one of the astronauts leaving the spacecraft while it was in orbit was discussed in March 1961. The event was termed an "extra-vehicular activity" or EVA. While this originally did not take a high profile, it was seen as an opportunity to develop a spacesuit capable of providing more mobility than simply an emergency response to cabin depressurization. The ability of the spacecraft hatch to be repeatedly opened and to reseal effectively in weightless and vacuum conditions would provide techniques needed on the lunar surface with *Apollo*, and by February 1963, the EVA was a full operational objective of *Gemini*.

With the manned space flight program now expanded to *Gemini*, the need for more astronauts led to a new recruiting effort in which nine candidates were selected: Neil A. Armstrong, Frank Borman, Charles Conrad, James A. Lovell Jr., James A. McDivitt, Elliot M. See Jr., Thomas P. Stafford, Edward H. White II, and John W. Young.

As the program progressed into 1963, it was evident that both the initial schedule and the price had been overly optimistic. Budget overruns caused the program to exceed the one billion dollar level, and there was an almost frantic effort to stem the rising costs. The number of ground and flight tests for a variety of systems, including the *Titan* II booster and *Agena* target vehicles, was reduced. There was serious talk of eliminating the *Agena* altogether as the *Apollo* could accomplish these tasks. However, the *Apollo* and its *Saturn* booster would cost significantly more and would extend that schedule out to accommodate the required orbital experiments. Therefore, the original objectives of *Gemini* remained, as the effort to minimize costs became almost an obsession.

The first orbital, unmanned *Gemini-Titan* flight slipped a full year from 1963 to the morning of April 8, 1964. It was not an operational spacecraft but was simply to test its structural integrity during the launch phase and to verify that the modified *Titan* could loft the weight into orbit. As such, GT-1, as it was called, had the normal weight, center of gravity, and moment of inertia. It was not intended to be recovered, and it did not have an ablative heat shield. In fact, four large holes in the aft end of the spacecraft ensured its destruction when it reentered the atmosphere. GT-1 achieved an orbit of 100 by 200 miles. At the end of three orbits, four hours and 50 minutes after launch, the first *Gemini* mission was completed, although it actually stayed up for nearly four days allowing the world-wide tracking stations, controlled from Goddard Space Flight Center in Maryland, to follow the vehicle by radar. It reentered over the South Atlantic following its 64th orbit.

The next *Gemini-Titan* (GT-2) was an unmanned, sub-orbital flight to qualify the heat shield. Originally scheduled for August 1964, a series of storms, including three hurricanes, overtook the Cape and forced continual postponements and subsequent re-tests of systems to assure that lightening had not damaged any of the electrical circuits. McDonnell was having significant problems with spacecraft serial number 2 and probably would not have met any of the earlier schedules. The second flight was set for mid-November 1964. The third, and perhaps the first manned flight with GT-3, was scheduled for the end of January 1965.

While NASA and its contractors labored to get the second unmanned *Gemini* aloft, a new Soviet announcement once again upstaged the American effort. On October 12, 1964, the Soviet Union orbited *Voskhod* I. By all accounts, it was not just the Soviet's second-generation spacecraft; it was in a class with *Apollo* with a crew of three cosmonauts. The crew flew in a "shirtsleeve" environment (flight coveralls rather than space suits), and all remained in the spacecraft to a landing on terra firma. The implication was that the Soviet's had made the spacecraft environment so safe that there was no need for the "backup" protection of the space suit.

Could it be that the Soviets had been able to move directly to a lunar capable vehicle so quickly? All indicators seemed to point in that direction. The Chief Designer had once again been able to "pull one more rabbit from his hat" (as one of his colleagues recalled) and out-engineer the Americans, who were still months from flying its interim two-man *Gemini* and several years from its three-man *Apollo*. However, there was much more to the story than the terse TASS announcements of another Soviet "first." While the world press was heralding the "advanced second generation" *Voskhod*, the Soviets were breathing a sigh of relief, reveling in their slight of hand, but concerned over their indeterminate future goals.

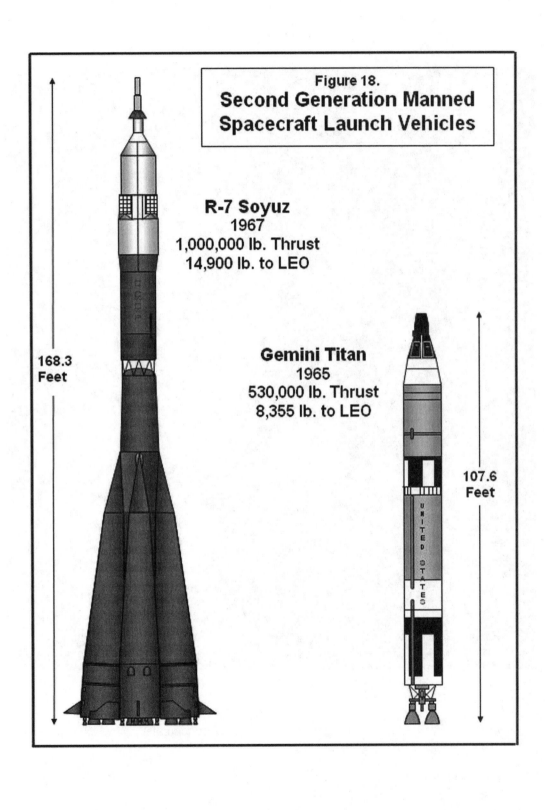

Figure 18.
Second Generation Manned Spacecraft Launch Vehicles

R-7 Soyuz
1967
1,000,000 lb. Thrust
14,900 lb. to LEO

Gemini Titan
1965
530,000 lb. Thrust
8,355 lb. to LEO

168.3
Feet

107.6
Feet

The United States' Ranger 7 spacecraft

America's first lunar exploration attempts in 1958 and 1959 were quite modest compared to what the Soviets had accomplished with *Luna* 3 in October of 1959, and there was little scientific or propaganda return for the effort. The notable exception to America's less than successful nine attempts was the discovery by Pioneer I of the outer Van Allen radiation belt that surrounded the Earth, with confirmation and mapping by Pioneers III and IV. (Some spacecraft were referred to with Arabic numbering while others were given Roman numerals.)

With the commitment by President Kennedy in May of 1961 to land a man on the Moon before the end of 1969, there was much to be learned about the Moon before man could take his first step on its surface. There were many ideas as to what that surface might consist of. These ranged from fine powdery sand that might be hundreds of feet deep into which a lunar lander might sink, to a rock strewn facade that offered little open area for a spacecraft to set down. Thus, the next set of "firsts" with respect to lunar exploration was to get close-up pictures of the surface and to soft-land an unmanned probe. NASA established three programs to accomplish these goals.

The first American program was a series of spacecraft called *Ranger*, which would send respectable sized payloads to the Moon where cameras would transmit pictures of the surface back to Earth before crashing onto the lunar surface. Later missions would "rough land" some basic instruments to try to assess the Moon's possible seismic activity and the basic conditions on the surface.

The second and much more ambitious program was *Surveyor*, a large sophisticated lander that would "soft land" on the surface and transmit back pictures and make detailed analysis of the surface conditions. The third program was *Orbiter*, which would do just that—orbit the Moon with high-resolution cameras and totally map its surface.

Although the Soviets had succeeded in only three of nine lunar attempts, they had hidden their failures and achieved spectacular results with their successes: passing close to the Moon with *Luna* 1, impacting the Moon with *Luna* 2 and photographing the far side with *Luna* 3. Their next step would involve a lander (Ye-6) and an orbiter (Ye-7). Permission to proceed with these projects was granted to Chief Designer Sergey Korolev's design team in 1960. However, their efforts would face a series of set-backs over the next five years as no less than eleven launches between January 1963 and December 1965 ended in failure for a variety of reasons that included booster unreliability, navigation inaccuracy, and a breakdown in the spacecraft communications.

Ranger: Off to a Bad Start

The spy satellite program that used the *Discoverer* project as its cover had provided the foundation for the development of the second stage *Agena B* that could be restarted in space. When paired with the *Atlas*, it provided a workhorse that would rival the initial payload capability of the Soviets. The R-7 with the RD-105 upper stage had sent 800 pounds to the Moon. The first *Atlas Agena-B* would send 675 pounds to the Moon, and this would grow to over 800 pounds as the program progressed. However, the new *Ranger* program would experience significant teething problems, as had all high-tech programs of both the United States and the Soviet Union.

Ranger 1 was launched on August 23, 1961, with the objective of testing the basic systems of a standard structural "bus". This "bus" would provide the essential functions for cis-lunar (the space between the Earth and the Moon) and interplanetary missions although it did carry some scientific instrumentation. The first version was referred to as the Block 1 configuration and consisted of a hexagonal base four feet across. Extending upward from the center of the base was a cone-shaped six-foot high aluminum structure. A high-gain directional dish antenna on the bottom of the base and an omni-directional medium gain antenna were provided for two 3-Watt transmitters. Two solar panel wings measuring 16 feet across extended from each side of the base. Electrical power was supplied by 8,680 solar cells on the two panels recharging a 130 pound silver-zinc battery.

A 100 pound thrust monopropellant hydrazine engine provided mid-course trajectory correction with four jet-vane vector-control thrusters. Attitude control about all three axis used 12 cold gas (nitrogen) pitch and roll jets under the direction of three gyros and a solid-state timing controller, six sun sensors and an Earth sensor. Temperature within the craft was regulated passively by gold plating, white paint, and polished aluminum surfaces.

An optimum period exists each month when a spacecraft can be sent to the Moon with minimum energy. With the restartable engine of the *Agena*, this period could be further optimized by first placing the payload in a 100-mile high Earth orbit called a "parking orbit" (also called a transfer orbit). Then, at the proper point, a second burn of the *Agena* propelled the spacecraft into a highly elliptical orbit that intercepted the path of the Moon about 60 hours later. Although an initial velocity of about 25,000 mph was imparted to the spacecraft, the Earth's gravity quickly slowed its speed so that by the time it entered the Moon's gravitational influence at a distance of 200,000 miles from the Earth it was traveling at about 3,000 mph.

While the first launch of *Ranger* 1 was primarily to test the various attitude control and mid-course correction motor systems, it was hoped that a lunar impact would occur. However, the *Agena* B failed to restart, and it re-entered and burned up in the Earth's atmosphere a week later.

Ranger 2, launched in November 1961, was almost identical to the first but weighed 726 pounds. The spacecraft was again put into a low Earth parking orbit, but an inoperative roll gyro prevented the *Agena* restart, and the spacecraft reentered Earth's atmosphere two days later.

Ranger 3, the first Block II version with complete instrumentation, launched two months later in January 1962. The vidicon television camera aboard *Ranger* had a scan mechanism that would yield one complete frame every 10 seconds. The objective was to transmit pictures of the lunar surface back to Earth during the final 10 minutes of flight prior to impact on the Moon. An instrument capsule containing a seismometer embedded in a two-foot diameter impact-limiting balsawood sphere would be slowed by a 5,000 pound thrust solid-fuel retrorocket. At the appropriate altitude, as determined by a radar altimeter, the capsule would separate from the "bus," and the retrorocket would ignite to reduce the impact speed from 5,000 mph to about 80 mph. A 50-milliwatt transmitter using silver-cadmium batteries provided for 30 days of operation on the lunar surface

It was an ambitious mission, but a series of malfunctions doomed the effort. A fault in the booster guidance system resulted in excessive spacecraft speed. Then, reversed command signals caused the spacecraft to pitch in the wrong direction, and the telemetry antenna lost its lock on the Earth. As a result, mid-course correction was not possible, and *Ranger* 3 missed the Moon by approximately 22,000 miles on January 28 and went into orbit around the sun. Finally, a spurious signal during the terminal maneuver prevented transmission of useful TV pictures. Some engineering data was obtained from the flight, but America was growing impatient with the steady string of failures.

Ranger 4 was launched in April 1962, but an onboard computer failure inhibited the solar panels from extending. The instrumentation ceased operation after the batteries depleted their charge about 10 hours into the flight. The spacecraft was tracked by the smaller battery-powered 50-milliwatt transmitter in the lunar landing capsule. *Ranger* 4 impacted the far side of the Moon after 64 hours of flight—the first American spacecraft to hit the Moon—but no pictures were transmitted.

Ranger 5, was launched in October 1962 and experienced an electrical malfunction, and the spacecraft ceased operation after about eight hours and passed within 500 miles of the Moon before going into a solar orbit.

Ranger 6, the first of the Block 3 versions, was launched in January 1964. The 838-pound spacecraft carried six television cameras, two wide-angle and four narrow-angle, arranged as two independent units with separate power supplies, timers, and transmitters to afford the greatest reliability and probability of obtaining high-quality video pictures. Higher video bandwidth was provided to allow for rapid framing sequences. The planned midcourse trajectory correction was accomplished early in the flight by ground control.

As *Ranger* 6 approached the Moon 65 hours after launch, for a planned impact at the eastern edge of the Sea of Tranquility, the orientation of the spacecraft to the surface during descent was correct. However, as the multi-million dollar package hurled towards the surface, the cameras inexplicitly failed to turn on, and no images were returned before its destruction on impact. A review board determined the most likely cause of failure was due to electrical arcing in the TV power system when it inadvertently turned on for 67 seconds approximately two minutes after launch—probably because of vibration caused by the booster-engine separation.

NASA instituted a thorough investigation into the Jet Propulsion Laboratory (JPL), *Ranger* program, in an attempt to determine the cause of the failures. Several procedural and quality control steps were established to improve the reliability of the spacecraft. One positive element of these failed Ranger missions was the high reliability of the Atlas booster.

Soviet Frustration

Luna 4, a configuration called the Ye-6 within Korolev's group, was designed to land an instrument package on the Moon. Launched April 2, 1963, it was the USSR's first successful spacecraft of their second-generation lunar program (two previous attempts the preceding January and February had failed and were thus unnamed). The spacecraft was first placed in an Earth "parking orbit" and then the upper stage performed a re-start to send the craft towards the Moon. The total mass of the vehicle was 3,100 pounds (including the spent upper stage), but it missed the Moon by 18,000 miles on April 5, 1963, and ceased transmitting the following day.

Four more failures through 1964 and into 1965, which were again unannounced and thus unnamed, continued to plague the Soviet program. An evaluation of the entire program was undertaken and changes were made to increase the reliability of several components.

Luna 5, yet another attempt at a lunar "rough" landing, was launched on May 9, 1965, and all appeared to go well until it was time for the retro-rocket to slow the descent to the Moon's surface. The sequence failed, and the spacecraft impacted the lunar surface in the Sea of Clouds. *Luna 6*, launched in June 1965, missed the Moon because of a failed midcourse correction. *Luna 7* in October 1965 experienced premature retrofire, and *Luna 8* in December 1965 had a late retro-fire—both spacecraft were destroyed on the lunar surface in the Sea of Storms. If there was any bright spot in the Soviet program, it was that the star-orientation system and flight trajectories were flawless.

Both America and the Soviets were building increasingly sophisticated spacecraft. With so many imponderables in the alien environment of space, every minor flaw is magnified and success becomes probabilistic. Hence, success depended on providing tight quality control, careful fabrication, and above all, redundancy.

Ranger 7: A Lunar Close-up

Atlas 250D and its *Agena* B, inserted its *Ranger* 7 payload into a 120-mile high parking orbit on July 28, 1964. Half an hour later, a second burn of the *Agena* engine injected the spacecraft into a lunar intercept trajectory, and the spacecraft separated from the *Agena*. At 68 hours into the flight, the cameras began their one-minute warm up 18 minutes before impact. The first image returned was from an altitude of 5,000 miles above the surface of the Moon. A total of 4,308 photographs of excellent quality were transmitted over the final 17 minutes of flight. The last image had a resolution of 18 inches as *Ranger* 7 impacted in an area between Mare Nubium and Oceanus Procellarum (subsequently named Mare Cognitum).

After 13 attempts over a period of six years, America had finally completed a fully successful lunar mission. The pictures showed a surface scarred by an infinite number of impact craters of widely varying sizes. Of great interest was the revelation of a series of cracks or "rills" across portions of its surface that raised many questions and prompted the Director of JPL, Dr. William Pickering, to comment, *"…that's pretty fascinating!"*

Ranger 8, launched in February 1965, was a duplicate of the previous mission but targeted a different spot on the Moon. At a distance of 100,000 miles from Earth, the planned mid-course maneuver took place, involving reorientation of the spacecraft and a 59-second burn of the 100 pound thrust monopropellant motor. A planned terminal sequence to point the cameras more in the direction of flight just before reaching the Moon was cancelled to allow the cameras to cover a greater area of the lunar surface. The first of 7,137 images was taken at an altitude of 6,500 miles and continued over the final 23 minutes of flight—a second success.

Ranger 9, the final mission of the program, was launched on March 21, 1965. A terminal maneuver oriented the spacecraft so the cameras were more in line with the direction of flight to improve the image resolution. The first of 5,814 photographs appeared 19 minutes before impact. What made this flight so astonishing was that network television broadcast the pictures as they were received. It was an extraordinary event for several million early morning viewers to sit in their living rooms and watch the Moon grow larger and larger until the spacecraft finally made impact. The last image had a resolution that could distinguish surface features as small as one foot.

Total cost for the *Ranger* series of spacecraft was approximately $170 million. Through a size-frequency distribution analysis of the craters, *Ranger's* photos allowed statistical extrapolation to provide assurance that the surface was clear enough of rocks and craters to permit a manned spacecraft to find sufficient room to land. *Ranger* increased man's visual understanding of the Moon's surface by a factor of 1,000 over Earth-based telescopes! However, the actual composition of the surface would have to wait for the *Surveyor* series. The *Atlas* itself performed almost flawlessly for the nine *Ranger* shots.

Luna 9: Pictures from the Lunar Surface

Eleven consecutive failures marred the Soviet lunar exploration agenda and cast a shadow on the once indomitable Russian space program. *Luna 9* launched on January 31, 1966, with a translunar injection weight of 3,476 pounds consisting of three major segments. The retrorocket portion that would retard the descent contained a 10,000 pound thrust engine using an amine-based fuel and nitric acid oxidizer. The central part of the ship contained the enroute communications and control, and the landing sequencer included four thrusters mounted on arms for attitude control. Attached to the sides of the central section were two modules which contained the radar altimeter that would initiate the retro-sequence. These 600 pound units were jettisoned (to reduce the weight to be decelerated) after they served their purpose. The final segment was the 220 pound lunar instrumentation container.

At an altitude of 20,000 miles from the lunar surface, the attitude control thrusters oriented the assembly so the retrorocket faced the direction of travel—toward the Moon that was growing steadily larger. The retrorocket began its burn while the two modules on either side, their job now done, were released to reduce the weight. At fifteen feet above the lunar surface and at a speed of less than 50 mph, a mechanical probe extending down from the spacecraft encountered the lunar surface and triggered the cutoff of the retrorocket. The instrumented payload was simultaneously ejected upward and slightly outward from the top of the delivery vehicle, which then proceeded to fall the remaining distance as it had completed its task.

The weak gravitational field of the Moon (one sixth that of the Earth), coupled with internal shock absorbers, provided a high level of surety that the "rough" landing would not damage the 23-inch spheroid instrument container. Just over four minutes after landing, a timer extended four spring-loaded legs, which acted like petals of a flower. These appendages would turn the module upright if it landed upside down. The radio antennas were then extended, and the first transmission of data from the lunar surface began.

While the probe was enroute, the British radio telescope at Jodrell Bank had closely monitored its progress. This high sensitivity receiver had played an important role since the first *Sputnik*. It allowed recording the signals of both American and Russian spacecraft, and provided improved tracking. As the scientists at Jodrell Bank observed the signals being transmitted by *Luna 9* (now sitting on the Moon's surface) on an oscilloscope, one of them noted that the electrical patterns appear similar to the standard television broadcast signal. The incoming data was literally plugged into a television, and there before them was a picture of the lunar landscape being transmitted by *Luna 9* from the Ocean of Storms—it was February 3, 1966. Scooping the Soviets with their own data, the British released the pictures to the world (the frame size was somewhat distorted because of differences in video standards). Seven radio sessions, totaling just over 8 hours, transmitted three series of pictures which provided a panoramic view of the nearby rocks and of the horizon about one mile from the spacecraft—an astounding success!

Surveyor: A Soft Landing

Surveyor 1 (like *Ranger* 1) was primarily an engineering test vehicle designed and built by the Hughes Aircraft Company under the technical direction of JPL. Its objectives included the first lunar soft landing and television pictures of the surface. Launched four months after *Luna 9* on May 30, 1966, (and three years after it was first expected to fly) the three legged triangular structured 642-pound aluminum craft was the first mission assigned to the "hopefully debugged" *Atlas Centaur* of which this was its eighth flight. The spacecraft was accelerated to 34,630 feet-per-second (23,600 mph)—only 165 fps greater than nominal. The guidance system performed so accurately that, had a mid-course correction not been performed, the spacecraft would have missed its desired touchdown point by only 250 miles. The 20-second mid-course maneuver took place about four hours after launch and slowed the craft by about 40 mph.

Unlike the Soviet *Luna 9*, *Surveyor* was a true soft-lander, powering its payload intact to the surface. The landing sequence began at 46 miles above the Moon (as determined by a radar altimeter) with the spacecraft traveling at 6,000 miles per hour. It was initially slowed by a solid-fuel rocket, which then was jettisoned, and the deceleration continued by a set of three liquid-fuel vernier thrusters to an altitude of ten feet and four miles per hour. The verniers then shut down and the craft dropped to the surface, landing at about ten miles per hour. It was believed that by cutting off the engines at this altitude their exhaust would not disturb the surface dust. Spacecraft control was so precise that the three footpads touched down within 19 milliseconds of each other in the Oceanus Procellarum. It was the first attempt to soft land on the Moon—and it was a success.

The 600-line television system began transmitting the first of 11,150 pictures of lunar terrain and surface material to JPL's Deep Space Facility in Goldstone, California. Although no scientific instrumentation was aboard this first flight, considerable engineering information was obtained that could be extrapolated to a scientific survey that included data on the radar reflectivity and bearing strength of the lunar surface. Spacecraft temperatures were used to determine the lunar surface temperatures.

Photos of the immediate terrain and of the impression of the footpad, which had made a depression of only a few inches into the surface, confirmed that the surface could support the *Apollo* Lunar Module (LM). (By this time the design of the Lunar Module had been completed based on estimated values from *Ranger* data.) Resolution of the TV allowed objects on the lunar terrain as small as .02 inches to be discerned, an order of magnitude greater than *Luna* 9. A series of lens filters on the camera allowed for color images to be produced, and more than 100,000 commands were exchanged with the spacecraft during its first few days on the surface. The quantity and quality of the photos were impressive and, along with other events of the preceding year (*Gemini* and *Ranger*), gave the world its first indications that the Russians now had serious competition.

The rock-strewn surface was a surprise to the scientists and presented an *"interesting mystery"* as noted by JPL scientist Dr. Leonard Jaffe. Although the rocks in the immediate vicinity appeared no larger than a foot, their presence (which had been observed by *Luna* 9) was not expected. Larger boulders that were over three feet in size could be seen in the distance. The source of this debris and the fact that it set literally on the very top what was thought to be a fairly deep layer of lunar dust was the enigma. Dr. Eugene Shoemaker (who was to achieve added notoriety two decades later with a comet) speculated that the rocks meant that the manned lunar lander would have to be able to hover and move laterally to assure a safe landing. This was a very prophetic statement.

Using solar cells to recharge its batteries, the *Surveyor* spacecraft transmitted data until July 14, 1966, and then shut down for the lunar night (June 14 to July 7, 1966). Data continued until January 7, 1967, although the original design required that the spacecraft only operate for the two week lunar day following its landing.

Surveyor 2 launched in September 1966 but failed when one of the vernier engines of the spacecraft did not fire during a mid-course maneuver, resulting in unbalanced thrust that caused it to tumble. *Surveyor 3* included a soil surface sampler, and strain gauges on the spacecraft landing legs. The spacecraft was launched in April 1967 to the southeastern part of Oceanus Procellarum where it arrived 66 hours later. Touchdown of the 664-pound spacecraft actually occurred three times because the vernier engines failed to shut down at the 10 foot height causing the spacecraft to lift from the surface after each successive touchdown until the fuel was exhausted.

Significant data on the strength, texture, and structure of lunar material was received from the spacecraft in addition to the lunar photography. (Two and a half years later, in November 1969, the *Apollo* 12 Lunar Module landed about 500 feet from the craft, and Astronauts Pete Conrad and Alan Bean removed about 20 pounds of parts, including the TV camera, for analysis back on Earth. The camera is now on display in the Smithsonian National Air and Space Museum in Washington, D.C.)

Surveyor 4 functioned normally after its July 1967 launch, but radio signals from the craft ceased during the terminal-descent phase, about two minutes before touchdown, and the mission was unsuccessful.

Surveyor 5, launched in September 1967, achieved a successful lunar soft landing in Mare Tranquillitatis on the edge of a small crater with a slope of about 20 degrees. During its lifetime, it used its vernier engines in an attempt to stir-up the lunar soil to allow for a more detailed examination of its properties. Because of the adhesion properties of the lunar soil, the engines failed to create a dust plume. A chemical analysis of the lunar soil was performed, and data on the thermal and radar reflectivity of the surface were obtained. Among the experiments was a small bar magnet attached to one footpad to detect the presence of magnetic material in the lunar soil. Except for the lunar nights, transmissions were received for three months until December 17, 1967.

Surveyor 6 landed in November1967 in Sinus Medii, in the center of the Moon's visible hemisphere. Following an initial probing of the lunar surface, this spacecraft performed a successful "hop," rising approximately 12 feet and moving laterally about eight feet to a new location on the lunar surface using the descent thrusters.

Surveyor 7 was the fifth and final spacecraft of the program and achieved a soft landing following its launch on January 7, 1968. The objectives of the 673-pound craft included exploring an area other than the smoother maria regions to provide a significantly different and higher risk landing environment from the other missions. Stereoscopic views of the surface area were made on the outer rim of the crater Tycho.

Lunar Orbiters

The Soviet *Luna 10* spacecraft launched March 31, 1966, out of a parking orbit and was the first spacecraft to achieve a lunar orbit on April 3, 1966, with a period of 180 minutes. A wide array of scientific instruments aboard included a gamma-ray spectrometer, a magnetometer, a meteorite detector, instruments for solar-plasma studies, and devices for measuring infrared emissions from the Moon and radiation conditions of the lunar environment. Gravitational studies were also conducted. The spacecraft played back to Earth the communist anthem "Internationale" during the Twenty-third Congress of the Communist Party of the Soviet Union.

Luna 10 was battery powered and operated about two months before radio signals ceased. There was no equipment for photographic data. The next launch on May 1, 1966, failed to leave its parking orbit and was identified by the Soviets only as Kosmos 111 with no other information being provided.

The American *Lunar Orbiter* spacecraft was designed primarily to photograph the lunar surface for selection and verification of safe landing sites for the *Apollo* missions. The craft had the general shape of a truncated cone, five feet tall and four feet in diameter at the base. A Canopus star tracker was employed with five sun sensors for orientation as well as internal gyros. Four solar panels that unfolded to a span of eleven feet provided 375 Watts of power from 10,856 solar cells. Batteries were used during periods when the Moon's shadow shielded the spacecraft from the sun. Propulsion was provided by a gimbaled hypergolic 100-pound thrust rocket motor. Three-axis stabilization and attitude control were provided by four one-pound nitrogen-gas jets. Communications used a ten-Watt transmitter, and thermal control was maintained by a multi-layer aluminized Mylar and Dacron thermal blanket, special paint, insulation, and small heaters.

The *Lunar Orbiter 1* was placed in a parking orbit by an *Atlas Agena* on August 10, 1966, with TLI (translunar injection) occurring 30 minutes later. The 625 pound spacecraft experienced a temporary failure of the Canopus star tracker and overheated during its cruise to the Moon. The star tracker problem was overcome by navigating using the Moon as a reference, and re-orienting the spacecraft relative to the sun solved the overheating. The spacecraft achieved a lunar orbit 92 hours after launch; its perilune was 118 miles (closest point to the lunar surface) and apolune 1,200 miles (farthest point from the lunar surface) with a period of 227 minutes and an inclination of 12 degrees to the lunar equator. The spacecraft used a photographic system to record a series of frames on film that were developed on board and then transmitted by a video scanner back to the Earth. A total of 42 high resolution and 187 medium resolution frames were transmitted that covered over 2 million square miles of the Moon's surface, although a number of the first high-resolution photos were unusable due to movement of the spacecraft. *Lunar Orbiter 1* also took the first two pictures of the Earth from the Moon.

The orbital track indicated high mass concentrations in isolated areas of the Moon. These would become known as *mascons* and caused some concern as to the ability of the manned lunar landings to calculate accurately lunar rendezvous. On August 21, the perilune was dropped to 36 miles and a week later to 26 miles. With the absence of an atmosphere, the only problem with a low orbit was the possibility of hitting the mountainous terrain, compounded by the inconsistencies in the orbit due to the mascons, and the blurring of the pictures by the speed of the craft—approximately 2,000 mph. It was decided to end the intended year-long mission early when the attitude control gas unexpectedly showed a low level. The spacecraft orbit was finally lowered to impact the lunar surface on the Moon's far side on October 29, 1966, on its 577th orbit to avoid possible interference with the next Lunar Orbiter.

The next Soviet attempt, *Luna 11,* launched during the same launch window as America's *Lunar Orbiter 1* on August 24, 1966, with a variety of scientific instruments but no photographic results were ever published. A total of 277 orbits of the Moon were completed before transmission ceased on October 1, 1966.

Luna 12 was launched in October 1966 with a television system that obtained and transmitted photographs of the lunar surface. The photographs contained 1,100 scan lines with a maximum resolution of 60 feet. Pictures of the lunar surface were returned on October 27, 1966, but, because they were of poor quality compared to those taken by the American *Lunar Orbiter,* few were released.

Lunar Orbiter 2 was launched in November 1966 on a virtually identical mission to *Lunar Orbiter 1.* A total of 609 high resolution and 208 medium resolution frames were returned, most of excellent quality with resolutions down to three feet. One of the most spectacular photos returned was a low angle picture of the crater Copernicus. The photo was hailed by the international press as one of the great pictures of the century. On December 8, 1966, the spacecraft's inclination was changed to 17.5 degrees to provide new data on lunar mascons. *Lunar Orbiter 3* acquired photographic data on February 15, 1967, that provided a frame of the *Surveyor 1* landing site that permitted locating the spacecraft on the surface.

By the time *Lunar Orbiter 4* launched, the three previous missions had satisfied the needs for *Apollo* mapping and site selection. It was given a more scientific objective: to perform a broad systematic photographic survey of almost all lunar surface features. To accomplish this, it was injected into an elliptical polar orbit that was 200 miles by 3,800 miles with an inclination of 85.5 degrees and a period of 12 hours.

Lunar Orbiter 5, the last of the series, was tasked to take additional *Apollo* and *Surveyor* landing site photography and to complete a detailed photographic survey of the Moon's far side. The spacecraft, launched in August 1967, was placed in an elliptical polar lunar orbit. A total of 633 high resolution (three feet) and 211 medium resolution (six feet) frames were acquired, bringing the total photographic coverage of the Moon's surface to 99 percent by the 5 Lunar Orbiters.

The micrometeoroid instruments aboard the Lunar Orbiters recorded 22 impacts indicating that the average micrometeoroid flux near the Moon was about two orders of magnitude greater than in interplanetary space but slightly less than the near Earth environment. The radiation experiments confirmed that the design of *Apollo* spacecraft would protect the astronauts from short-term exposure to solar radiation. The *Lunar Orbiters* were also used to evaluate the Manned Space Flight Network tracking stations, and the program was completed at a total cost of $200 million.

The *Ranger* and *Surveyor* projects had been focused by NASA to provide engineering data on the lunar surface. However, the time that had elapsed between their initial schedule and final success was such that the manned *Apollo* Program was unable to benefit directly from their revelations except to confirm that surface composition estimates used for building the Lunar Module were within acceptable tolerances. The Soviets and the Americans now had enough data about the lunar environment, and a complete map of the Moon's surface from their lunar orbiters, to permit their manned exploration programs to proceed with confidence.

Dr Wernher von Braun, architect of the Saturn booster

Chapter 23 — Building the Super Booster

Of all the surprises that the Russians had sprung upon the United States in the fall of 1957, the weight of their *Sputnik*s was the most worrisome. It not only gave clear evidence that they had developed an ICBM (the *R-7*) but that its development schedule was ahead of America's *Atlas* program. Therefore, it was natural that one of the first objectives for the United States was to accelerate the *Atlas* program and develop new boosters of greater power. The magic number, in terms of thrust, was one million pounds. Estimates at the time placed the Russian booster at about one-half million pounds when in fact it was closer to one million. It was their conservative engineering and resulting greater structural weight that prevented even larger payloads in the early *Sputnik*s. However, by 1960 development of upper stages for the *R-7* revealed its true launch capability with satellites of 10,000 pounds.

If America had been caught napping by *Sputnik*, it was not from lack of vision by men in key technology positions. Even before the *Atlas* started taking shape in the Convair manufacturing facility, Air Force research and development began studies for larger rocket engines. A 1955 contract with the Rocketdyne division of North American Aviation examined the feasibility of a single-chamber engine with a thrust of up to 400,000 pounds which was designated the E-1, and an even larger engine of one million pounds was considered possible. Although the Air Force had no specific role for such large engines, it was prudent to continue to see where the state-of-the-art could be directed.

A November 1956 meeting of the USAF Scientific Advisory Board recommended engines of 5 million pounds of thrust. However, it was recognized that, should the need occur for such power, a better interim solution would probably be the clustering (grouping) of smaller engines. A design competition for a single-chamber engine of one million pounds of thrust resulted in a contract to Rocketdyne in June 1958 for such an engine designated F-1.

The von Braun Team

During the 1950s, the Army Ballistic Missile Agency (ABMA) at the Redstone Arsenal in Huntsville, Alabama, was under the command of General John B. Medaris. The Army had managed to retain the core of the German rocket team that had developed the V-2 during Word War II. Under the technical direction of Wernher von Braun, the team continued to operate organizationally in much the same manner as they had in Germany. Each of the ten laboratories that represented the various aspects of missile development (engines, structures, fuels systems, etc.) was headed by one of the Paperclip (the code name of the program that brought them to the United States in 1945) technologists. By 1958, ABMA staff numbered about 2,600 with mostly American technologists in subordinate positions including 350 American military officers.

The management style employed by von Braun was "participatory" in that he relied on his laboratory heads to discuss and debate critical issues while he observed and patiently listened and prodded to guarantee that he was getting the best input from each. He worked towards an amiable consensus, assuring that there were no critical exceptions that might later divide the team. Because of the leadership he had provided in the stimulating years at Peenemünde, and the uncertain times that followed the war, von Braun had unquestioned loyalty as well as exceptional competence. The team had established itself as being innovative yet conservative in its engineering practices.

In part, because of his assimilation into the U.S. Army missile command after World War II, von Braun's ability to move his vision of space travel forward had been impeded by the lack of a viable role for the Army. The stereotypical image of the infantryman, as well as the prejudice that continued to stalk these former members of Nazi Germany's elite technical rocket team, hindered their contribution. While von Braun could provide a vision of the coming space age with clarity and persuasion, his ability to loosen the purse strings of the government was still lacking.

Shut out of contention for orbiting the first satellite (Project Orbiter) and lofting the first man in space (Project Adam), the team was desperate to find an application for its expertise. One area that held promise was a Department of Defense (DoD) proposal circulated in April of 1957 for a large communications satellite placed in geosynchronous orbit. As envisioned in a time before the miniaturization of electronics, this 10,000-pound giant would require a million pound thrust class rocket. However, this pre-*Sputnik* booster, referred to as the "Super *Jupiter*" was still just a paper study.

Creating Saturn I

With the launching of *Sputnik* I in October of 1957, the von Braun team intensified its possible options for building a large booster, looking to the Rocketdyne E-1 that promised almost half a million pounds of thrust and the F-1 with almost three times that power. But, these were still paper projects. The need to move quickly to catch-up with the Russians required that existing technology be employed. Thus, while the E-1 and F-1 offered promising futures, the "future" for the von Braun team was already upon them.

The Advanced Research Projects Agency (ARPA), formed in early 1958 to coordinate DoD space activities in response to the *Sputnik* challenge, began discussions with ABMA to examine their heavy-lift concepts. ARPA encouraged von Braun to consider clustering existing engines to shorten the development cycle. The obvious choice of an engine was the Rocketdyne LR-89 (LOX/RP-1 propellants) that powered the *Jupiter* IRBM (as well as the booster stage for the *Atlas*). It was the largest U.S. engine that had any track record.

However, to achieve the required thrust, at least eight of these engines would have to be used, and this number of engines presented many engineering issues. The anticipated savings of time and money loomed as an enticing motivation. Given the initial designation of *Juno V*, funding for the project was approved in August 1958 with the goal of achieving the first static test within 18 months, and, in keeping with the optimistic outlook of the times, the first flight test was anticipated by September 1960.

For the first time the von Braun team had been turned loose on a project of greater potential than any since the V-2 almost 25 years earlier. To provide a strong differentiation from its origin, the name of the vehicle was changed from *Juno V* to *Saturn*. Two objectives were immediately established: to simplify the LR-89 to improve reliability and performance, and to minimize the time necessary to assemble the propellant tank structure.

The thrust chamber of the LR-89 that powered the *Jupiter* IRBM used a revolutionary tubular construction instead of the double walled configuration that had been the hallmark of the V-2 and its *Redstone* descendant. Even the contour of the nozzle now reflected the need to extract higher exhaust velocities by changing from a linear conical design to a bell shaped curve. The expansion ratio increased from 5:1 to 8:1 to achieve greater efficiencies at the higher operating altitudes.

However, to design and construct a totally new first stage structure would require engineering a 20-foot diameter assembly and creating new tools and fixtures in addition to novel manufacturing processes and techniques. Moreover, transporting such a huge stage would be another obstacle. To avoid this problem, the existing eight-foot diameter tank assembly of the *Jupiter* IRBM was lengthened six feet to 52 feet and was used as the central structure around which eight *Redstone* tanks of equal length were assembled. A radial I-beam structure at either end provided the interface to the engines and upper stages. Although an "overbuild" with respect to its engineering, the assembly would provide the required first stage propellant tanks using existing tooling and could be readily disassembled and transported by air for reassembly at the test and flight facilities.

With the formation of NASA in October of 1958, it was obvious that it was just a matter of time before von Braun's ABMA team would be transferred to that new organization. The team continued to work with both ARPA and the fledgling NASA on an equal basis until that time arrived. A "Special Committee on Space Technology" was formed by NASA that included a "Working Group on Vehicular Programs" chaired by von Braun. In a report issued by that group, five classes of boosters where identified. These included the *Vanguard* and *Jupiter-C* on the low end (Class I), followed by the *Jupiter* and *Thor* as Class II, and the *Atlas* and *Titan* as Class III. The eight-engine *Saturn* was considered Class IV, while the truly gigantic Class V, which would have perhaps a much a six to ten million pounds of thrust, was called *Nova*.

Upper stages for the *Saturn* were also a part of the planning process, and the *Jupiter, Thor, Atlas,* and *Titan* were all potential candidates to take advantage of existing hardware. However, adapting a rocket that was designed as a first stage to a second stage function presented many problems. Moreover, even if these problems could be resolved, the lack of performance when compared to the use of liquid-hydrogen (LH2) engines was limited. However, the LH2 Centaur program was still in its teething stages in 1959 and success was not assured.

Although the military roles and missions for the big booster were still somewhat nebulous, there was a general agreement that the *Saturn* booster was a key ingredient in catching the Russians. Optimistic estimates placed a four-man space station in orbit by 1961 and a manned lunar flight by 1966, with a permanent Moon base by 1974. The key to these programs would be the Class IV and Class V boosters. And, while virtually everyone believed these estimates, there was yet no funded programs for the *Saturn* rocket to support.

Before ABMA's *Saturn* had progressed very far, Dr. Herbert York, the Director of DoD Defense Research and Engineering, proposed canceling it. As he viewed the various programs that were vying for the defense dollar, the DoD did not have a requirement for *Saturn*. In a September 1959 meeting, the paper super boosters, including *Saturn*, *Nova* and an advanced version of the *Titan*, were reviewed, and there was no mission for *Saturn* or *Nova* in military planning. It was this pronouncement that prompted NASA finally to incorporate the ABMA team since *Saturn* and *Nova* played a prominent role in their space planning at that time.

Some viewed York's proclamation as simply political arm-twisting to get von Braun to move to NASA—as he had held back, seeking the best situation for insertion of the ABMA team into the NASA structure. If he continued to balk, von Braun realized that his team might be left without a mission to support his rockets. It was a power play reminiscent of von Braun's dealings with Operation Paperclip and Colonel Holgar N. Toftoy after the war to assure the best possible arrangement for his team. However, there were voices in the upper echelons of government that recognized the unique talents of the von Braun team and cautioned NASA that it was vital to keep the team together and continue their work on big boosters. The transfer would take the entire team and its facilities.

There was however, some validity to York's reasoning. In the late 1950s the Soviet threat, as evidenced by *Sputnik*, was evoking a military response that envisioned space as an immediate and critical battlefield environment. For the DoD the storable propellant of the *Titan* II, coupled with large solid-fuel boosters affixed to its sides, offered a heavy-lift capability that was more launch responsive than the liquid-oxygen-based *Saturn*.

Although a manned lunar mission had yet to be authorized, President Eisenhower understood the need for a capable heavy-lift booster program and authorized additional funding for *Saturn* and the coveted DX priority—the highest priority that a government project could receive, in January of 1960.

By March of 1960, the move of the von Braun team from the Army to NASA was completed with the bulk of the ABMA facilities becoming the George C. Marshall Space Flight Center (MSFC). Some believe the choice of the name was President Eisenhower's way of mitigating the role of the German team by overlaying the name of the only professional soldier to win the Nobel Peace prize—and the man who was a primary architect in the defeat of Nazi Germany and its post-war reconstruction.

As the need to establish the configuration of the upper stages of *Saturn* became more critical, a decision had to be made regarding the role of LH2 (liquid hydrogen) in any lunar program. The von Braun team was less than enthusiastic about its use. Their conservative approach, coupled with the problems being encountered by the pioneering efforts of Convair's *Centaur*, seemed to support their view. However, even the ten-foot diameter of either the *Atlas* or *Titan* was too restrictive for the large payloads being envisioned. This issue, together with NASA's Abe Silverstein's continued push for LH2 upper stages, finally turned the tide. Although the risks of the highly volatile fuel had yet to be mastered, the decision was made in the spring of 1960 to go with LH2-powered upper stages for *Saturn*.

With the Pratt & Whitney RL-10 being the only LH2 engine then under development, it was readily determined that at least six of these engines would need to be configured (90,000 lbs. of thrust) as a first step in providing a potentially useful upper stage for *Saturn*. This new LH2 stage was given the designation of S-IV. The designation reflected it as the fourth stage of the proposed *Nova*. The von Braun team, now fully a part of MSFC, understood that engineering that number of questionable engines was a challenge of even greater magnitude than their eight-engine booster.

As the various future configurations of *Saturn* began to proliferate, and the massive *Nova* rocket began to take shape on the drawing boards, it was equally obvious that a new LH2 engine of 200,000 lbs. would be needed, and that it, too, would eventually be clustered. The dependence on LH2 to provide a heavy-lift program now became vital. While science and the military continued to be used as the dominant reason for America's move into space, the President's Scientific Advisory Committee stated that *"at present the most impelling reason for our effort has been the international political situation which demands that we demonstrate our technological capabilities if we are to maintain our position of leadership."* The report emphasized the importance of LH2 development.

Of all the possible configurations put forth for *Saturn*, two moved forward to the flight test stage. The first was designated *Saturn* I with the six RL-10 engines as a second stage; the second, called *Saturn* IB, used a new LH2 engine (designated the J-2) as its second stage.

The engine chosen for the *Saturn* first stage was based on the LR-89 and was re-designated as the H-1. Its 150,000 pound thrust would be upgraded over several years to 165,000, 188,000 and finally to 205,000 pounds with a Specific Impulse of 261 Seconds. As each of the thrust upgrades was engineered, the ever present combustion instability had to be addressed. The phenomena can be equated to the pinging that occurs in an automobile engine. Most typically, this problem relates to the area within the combustion chamber where the "flame front" occurs. The consistency of the propellant injector to achieve the desired atomization determines where the "flame front" physically occurs—and its relative movement fore or aft of that point. Changes in the position of the "flame front" caused pressure variations that affected the delivery of the propellants which could (and often did) aggravate the combustion zone. If the instability became divergent in its movement, it could (and occasionally did) destroy the engine.

While engineering design was always a key in the performance and reliability of the engine, quality control was also an important factor. In one example, a vendor had manufactured turbine pump blades from the wrong material, and a static test of the engine resulted in catastrophic failure. A follow-up analysis discovered the cause, and procedures resulted in locating all of the other turbo-pumps that contained the flawed blades.

The arrangement of the eight H-1 engines required that special attention be given to how the heat flowed up into the area above the engines where fuel, oxidizer, and hydraulic lines had proved vulnerable in all three earlier applications (*Atlas, Thor,* and *Jupiter*). The engines were thus arranged as an inner group of four and a second outer group of four offset by 90 degrees (Figure 19). The aerodynamic shroud around the lower part of the engines was contoured to route the airflow to dissipate the heat.

Another interesting aspect of the H-1 modifications was the elimination of a separate lubricant (and its added weight) for the propellant turbo-pumps. A small amount of the rocket fuel (a refined form of kerosene called RP-1) was mixed with an additive and routed to the pump bearings and then back to the fuel inlet for the combustion chamber where it was consumed.

The *Saturn* I had the ability to orbit 22,000 pounds, while the enhanced *Saturn* IB with its LH2 second stage would deliver 35,000 pounds to low Earth orbit. In July of 1962, the *Saturn* IB configuration (called the C-1B at that time) was chosen by NASA to provide the Earth orbital testing of the *Apollo* hardware.

The von Braun team made extraordinary progress in assembling the big *Saturn* booster, and by early 1960, the first of these hung in the static test stand at the MSFC in Huntsville. Initially only two engines of the cluster of eight were fired in March as the test schedule worked its way towards an eight-second firing of the full cluster on April 29. There were still those who felt that the basic concept of clustering eight engines was flawed, but of course, no one outside of the USSR had yet viewed the 32-engine cluster that was producing the space spectaculars for the Russians. By June a full two-minute run had been achieved, and it was only after the full cluster had been tested that the true affect of vibration and acoustic impact could be assessed and potential problem areas addressed—there were few!

One of the first advances made because of clustering was that of automated checkout of the various systems. In the past, manual readings and verification of instrumentation and switch settings preceded the various steps in the countdown. However, with so many items comprising this new rocket, techniques for sequencing through the various checklists had to be automated. Large banks of electromechanical stepping switches provided the early concept of programmed instructions that would soon give way to digital computerization.

The Saturn I Flies

Static testing continued throughout 1960, and it was not until October 27, 1961, that the first *Saturn* I (SA-1 denoted the first Saturn/Apollo test) took flight with inert upper stages. It towered above any previous rocket with a length of over 165 feet and carried the structure of an inert *Jupiter* IRBM as its second stage to complete the physical configuration (and to save money). The primary objective was to test the S-1 first stage structure and propulsion system. All objectives were met, although some problems were encountered such as the sloshing of the propellants, a situation which required the installation of baffles in subsequent vehicles. For the first time in over four years, America had launched a rocket with more power than the Soviets—who were not sitting on their laurels.

Vehicle SA-4, in March of 1963, demonstrated that the configuration could continue to fly should one engine shut down during the flight. Depending on when that shutdown occurred, the vehicle could still complete its mission by extending the firing time of the operating engines using the unburned fuel that had been allocated for the dead engine. A deliberate engine cut was made 100 seconds into the flight, and the rocket continued straight and true with the control movement of the seven good engines providing a slight thrust vector bias. SA-4 also marked the completion of the Block I configuration that had no fins. None of these first four flights involved stage separation and were only tests of the first stage.

Over the next two years that followed the first flight, the *Saturn* I program proceeded at what some considered a snail's pace compared to its first 18 months. As von Braun had predicted, the LH2 upper stage development had consumed much more time than its optimistic supporters had projected. The first of the Block II vehicles, SA-5, with the LH2 S-IV upper stage and eight aerodynamic fins to aid flight stabilization, flew in January 1964. The Block II also used the up-rated 188,000 pound H-1 engines that now gave the *Saturn* a total of 1.5 million lbs. of thrust. The propellant tanks were lengthened and had a simplified propellant interchange system to ensure that little was left unburned in any tank. Over the test period of the ten *Saturn* I vehicles, the propellant utilization increased from 96.1 to 99.3 percent. The hold-down time during engine start and stabilization was reduced from 3.6 seconds to 3.1 seconds providing yet another performance improvement.

With a launch every six months, it was not until SA-6 on May 28, 1964, that an unmanned, boilerplate *Apollo* spacecraft payload was finally orbited using a live 6-engine S-IV LH2 upper stage. During this flight one of the first stage engines shut down prematurely due to a problem with engine No. 8 turbo pump at T+117 seconds, but the flight continued to a successful conclusion. Video and recoverable film cameras were used to view what was happening within the fuel tanks as well as to record the external events of staging.

The last three *Saturn* I flights (SA-8, 9, and 10) were used to orbit large wing-like satellites called Pegasus (in reference to its resemblance to the winged horse of Greek mythology). Using the inert segment of the *Apollo* Service Module (SM) for its enclosure, Pegasus remained attached to the expended S-IV stage after it achieved orbit. The 4,000-pound satellite unfolded in space to a 96-foot span and a 14-foot width to present a variety of materials to the hazards of micrometeorite impact. A television camera provided pictures of the satellite deploying its wings which exposed more than 2,300 square feet of instrumented surface, with thickness varying up to 16/1000 of an inch.

Numerous satellites had been instrumented to detect the impact of these super-high speed meteorite bullets of space. However, the large surface area of Pegasus provided more complete statistical data on the dispersion and affect of these particles in low Earth orbit. The 208 panels electronically reported punctures by micrometeoroids since spacecraft designers needed to know how likely this occurrence would happen. The sensors successfully measured the frequency, size, direction, and penetration of hundreds of impacts.

The predicted useful lifetime of Pegasus, powered by solar cells, was 6 months. However, the satellites reported data for three years before finally being turned off in August 1968. The data confirmed that the structure of the *Apollo* spacecraft was adequate for all but the largest meteoroids, whose probability of impact was negligible.

A 10,000-pound boilerplate *Apollo* spacecraft was a part of the launch configuration of SA-8 through SA-10, along with the inert Service Module and the spacecraft escape system, which was jettisoned as a test of that subsystem during launch. The Command Module (CM) was separated after orbital insertion.

The total mass placed in orbit with each Pegasus launch was 33,895 pounds. The satellites had a typical perigee of 307 miles, apogee of 461 miles and orbital inclination of 31 degrees resulting in an orbital period of 97 minutes. It was by far the largest object placed in orbit by either the Soviets or Americans. However, the heavy lift of *Saturn* I was not used during this period other than the initial *Apollo* boilerplate tests and the Pegasus experiments, and the program was concluded after the tenth test flight as the next version of *Saturn* provided significantly more capability.

What was extraordinary about the entire *Saturn* I test flight regime was that not one catastrophic failure had been experienced. Virtually all other large rockets had undergone the ignominious fate of a fiery explosion, often accounting for 50 percent of the first dozen or more. Thorough ground testing and the ability to simulate and understand the flight environment had paid enormous dividends.

Selecting a Path to the Moon

As the *Saturn* I began to take shape and its upper stages were defined, the commitment by President Kennedy in May of 1961 to land a man on the Moon within the decade required that one of several possible paths to the Moon be chosen. The specific configuration and size of the rockets developed after *Saturn* I would be determined by this mode. By the spring of 1962, work had progressed as far as it could on the lunar program without this decision being made. There were seven variations of achieving a manned lunar landing, which could be categorized into three primary modes. Depending on which mode was chosen, one or more of a series of increasingly more powerful boosters that ranged in designation from C-1 through C-8 would be developed. (All of them were paper proposals at this point except the C-1 which was the *Saturn* I.) Each increment in the numerical designator represented an increase of 1.5 million pounds of thrust.

The first and most obvious path (or mode) was direct ascent. Its basis is the popular notion that a rocket would send its occupants from the Earth directly to the Moon, as was generally depicted in the science fiction movies of the day. The entire spacecraft would land on the lunar surface and return, perhaps leaving a portion of its structure there, such as the descent engine and landing gear, and would weigh an estimated 140,000 pounds. This would require a Class V vehicle such as the proposed C-8 *Nova*—a five hundred foot tall beast with at least eight F-1 engines to provide 12 million pounds of thrust. This method would not require any form of rendezvous and docking (which had yet to be demonstrated) and was considered to present the least technological risk as it would involve a single multi-stage rocket. Development of *Nova* was initially backed by MSFC (the von Braun team). However, the energy required would demand the use of LH2 upper stages which had yet to prove themselves. The sheer size was also of great concern since fabrication and movement of such a large and heavy structure would present great problems. The projected (and typically optimistic) schedule for *Nova* showed its first launch in the autumn of 1967.

The cost of *Nova* (perhaps as much as 200 million dollars per launch) sent shock waves through NASA management. Using the existing concepts of flight testing it was estimated that perhaps as many as 100 launches of the smaller *Saturn* and the proposed *Nova* would be required to prove and man-rate the spacecraft and the rocket.

A second mode to reach the Moon was to use a somewhat smaller launch vehicle of the *Saturn* C-4 type (5-million pound thrust) and use a low Earth orbit to assemble or refuel a translunar stage that would then launch to the Moon from its parking orbit. This method (termed Earth Orbit Rendezvous—EOR) had significant technological risk as it required several rockets be launched, rendezvoused, and docked. Von Braun had originally envisioned this method in his Collier's magazine articles some ten years earlier when a large space station was seen as the jumping off point for lunar and planetary missions. The price and the time schedule again made heads shake.

There were also variations of both the direct ascent and EOR that would involve the placing of equipment on the Moon in advance of the astronauts. This too required multiple launches and the ability to accurately pre-select and land next to these supply depots.

A third path to the Moon resulted from an evaluation of the roles of each piece of hardware and the energy needed to transport that hardware to only those points in space where their function was needed. For example, the primary role of the "mother ship," as it was sometimes called, was to transport the crew through the atmosphere (up and back) as well as provide crew radiation shielding during the 2-3 day trip to the Moon. However, there was no need to take all the weight of the thick heat shield and all three crewmembers to the surface of the Moon. The mother ship (which the *Apollo* spacecraft was evolving into) could be left in lunar orbit while one or two crewmembers descended to the surface in a much lighter lunar lander. Likewise, the ascent back up from the lunar surface could be done with only a portion of the lander, leaving the descent engine, its empty propellant tanks, and the landing legs (which were no longer needed) on the surface. Chance-Vought Aircraft had done a significant amount of in-house research on this in hopes of finding a place in the lunar landing program. They estimated that a lander of this type might weigh only 10,000 pounds—far less than a 140,000 pound direct ascent lander.

This latter method, termed Lunar Orbit Rendezvous (LOR) had many attractive benefits: one rocket (smaller than *Nova*), a shorter development period, and less cost. LOR would reduce the total weight accelerated to escape velocity from 100 tons to 45 tons. However, as conceived in 1961, it represented the highest risk with respect to the need to develop rendezvous and docking. It also meant that many critical activities during the mission would be performed a quarter million miles from Earth where the round trip time to send and receive electronic signals took three seconds—an eternity if a real-time life-threatening condition had to be evaluated during landing, take-off, or rendezvous.

The ability to estimate time, money, and product performance (the triad of project management) was facilitated during these evaluations by a method known as the Program Evaluation Review Technique (PERT). Developed for the Polaris submarine-launched- ballistic-missile project by the Navy, PERT allowed the many variables of a project to be quickly assimilated by a computer and the output presented in a manner that facilitated "what if" scheduling scenarios. This capability was to prove pivotal in establishing the advantages and disadvantages of the various modes in terms of the three primary attributes of time, cost, and performance.

The problem of two or more NASA organizations handling different aspects of the hardware came to a head in 1960— the responsibilities of the booster development team and the spacecraft team were in conflict. The new MSFC (headed by von Braun) was charged with developing the booster, while the Space Task Group was given the spacecraft itself. To assure that both groups were operating as equals, the Space Task Group was renamed the Manned Spacecraft Center (MSC) with Robert Gilruth as its director. This essentially elevated the spacecraft development to the same organizational level as the rocket itself.

NASA management continued to block EOR and the most promising method, LOR. Gilruth felt that the need to rendezvous (LOR in particular) compromised mission reliability and flight safety. He believed that EOR and LOR were being advocated primarily because they appeared to promise the least cost of achieving Kennedy's goal, not because they represented the safest path to the Moon. John Houbolt, an engineer at NASA's Langley Research Center, had thoroughly evaluated LOR and was firmly convinced that it was the only path that offered a lunar landing within the decade, and that the other methods of direct ascent and EOR would require excessively complex hardware. In a rare and bold move borne of frustration, Houbolt wrote directly to NASA Associate Administrator Robert Seamans in November 1960 detailing the pros and cons of all the possible paths with LOR being the preferred choice.

Because of the Houbolt letter, Milt Rosen, Director of Launch Vehicles and Propulsion (and a Direct Ascent advocate), was tasked by Seamans to prepare a report on the various boosters and paths to lunar landing. Working with a group of representatives from the major NASA centers, the report ambiguously concluded that only by using rendezvous techniques would there be a reasonable chance of meeting the Kennedy goal. However, the feasibility of rendezvous would not be demonstrated for several more years (until at least 1964)—too late to redirect efforts to another mode (such as Direct Ascent) if rendezvous proved difficult to master. There were also proposals for a costly parallel development as a hedge against failure of one method. The report, submitted in late 1960, stated that LOR presented the highest technical risk and the group essentially recommended the Direct Ascent and development of the C-8 *Nova*.

The logic within the report, however, made more converts (Gilruth in particular) to LOR, and now it was Milt Rosen's turn to see the light. The decision had become a critical issue, and the two often debated the merits. On one occasion, Gilruth made a concluding point that centered on the massive size of a Direct Ascent lander and the power and complexity that it required. He contrasted it with an LOR lander that would be less than 10 percent of the weight and would provide much more maneuverability and abort options. The basic argument was sound, and Rosen now stood in agreement on LOR.

However, von Braun was not yet in the LOR camp. He had vacillated between Direct Ascent and EOR, but following an intense meeting, he decided that the issue needed to be resolved and that LOR was as good a choice as any. If the other centers and NASA management felt LOR was the way to go, he would concur and commented, *"We are already losing time in our overall program as a result of lacking a mode decision."* He also noted that a rocket the size of C-8 *Nova* would present some very difficult fabrication and transportation issues in addition to the technology itself. It was also at about this time (1962) that the von Braun team, in its continuing effort to configure "paper" boosters, discovered that a five F-1 engine configuration known as the C-5 might just provide the required power for LOR with a single launch.

The decision was made; America would place its prestige and the lunar landing commitment on Lunar Orbit Rendezvous—a mode that would require building a rocket five times as powerful as the *Saturn* I. Called the *Saturn* C-5, its name was later revised to simply *Saturn* V (Figure 20).

Origins of the Soviet Lunar Program

The desire to send men to the Moon and to Mars had always been a part of Korolev's personal plan that had to be subordinate to the interests of the military and the realities and expediencies of the economic and political climate. In fact, Korolev felt that a manned Mars mission was a higher priority than a lunar landing. The Heavy Interplanetary Ship project (designated TMK), which he proposed in the early 1960s, provided preliminary planning for a three man crew to assemble an interplanetary mission in Earth orbit (EOR) using multiple launches of a large booster. It was envisioned that by 1967 a Mars landing could be achieved. These goals, however, seemed to ignore the many problems of long duration manned flight that included the requirement for nuclear reactors and electric engines. Moving his vision into the "official" arena had to wait for proper timing.

Early in 1960, while at the peak of his prestige with the Soviet power structure (although some felt his influence was on the downward trend), Korolev and his key engineering staff created what was termed the "Big Space Plan." Among the projects was a 3.5 million-pound-thrust booster with nuclear upper stages which would provide the capability to send men to the Moon and Mars using EOR.

By mid-1960, the Communist Central Committee had essentially approved the essence of Korolev's plan with a decree entitled "On the Creation of Powerful Carrier-Rockets, Satellites, Space Ships and Mastery of the Cosmic Space 1960-67." Among the specifications was the requirement to launch payloads of up to 200,000 pounds to low Earth orbit (LEO). The decree allowed Korolev's organization to begin designing a rocket that would be known as the *N1* with the ability to send 100,000 pounds to LEO and the N2 with a 160,000-pound capability. The initial schedules for these projects were just as overly optimistic (1962 and 1967 respectively) as their American counterparts. The euphoria of the post-*Sputnik* period generated a plethora of visions from virtually all of the Chief Designers, most of which would never become a reality. With few exceptions, most would be canceled by political maneuvering and/or lack of funding before achieving their goals.

As reports of America's progress began to show concerted and successful results (specifically with the *Saturn* I), a meeting was held with many of the top Chief Designers and scientific leadership in April 1963 to assess the possible implications of the Kennedy challenge that had been issued almost two years earlier. Korolev reported that the *N1* would be capable of accomplishing the lunar landing. It was at about this point that Korolev's emphasis shifted from a Mars landing to a lunar landing as he recognized that the political leadership was concerned by the American effort.

Toward the end of July 1963, a letter from the British astronomer Sir Bernard Lovell to NASA's Deputy Administrator Hugh Dryden set off an interesting debate in the media, most particularly in America. The letter stated that Lovell had met personally with several top-level Soviet academicians (including Academy President Mystislav Keldysh) who informed him that there were no current plans for a Soviet manned lunar program. The concept of a manned space program had been initially criticized by some in the American scientific community when the *Mercury* program was begun in 1958. Each major new program (*Gemini* and *Apollo*) again received that same denunciation as a waste of taxpayer money. The Lovell letter brought that debate to the forefront once again. The position of those opposing the lunar landing plan in particular was that unmanned probes could do as well for a lot less money. It was recognized by many in the government, as well as NASA, that one of the prime reasons for the expensive manned space program was the competition with the Soviets—not science. If indeed there was no Soviet program to send men to the Moon, then the United States was spending billions racing against itself.

Khrushchev added to the debate by publicly proclaiming in October 1963, *"At the present time we do not plan flights of cosmonauts to the Moon by 1970."* The implications of his statement reinforced the notion that there was no space race. Although the *Apollo* lunar program continued, the events at the close of the decade of the 1960s would lead many to speculate that indeed there had been no intent on the part of the Soviets to reach the Moon before the Americans. Of course, Keldysh and Khrushchev knew that the Soviet objective was to beat the Americans to the Moon. By refusing to acknowledge their goal, the Soviets could not be "beaten" politically.

Kennedy seemed to reassess his own commitment in a September 1963 speech when he called for a joint effort with the Soviets to conquer the Moon together. Khrushchev seemed to open the Iron Curtain a crack a few weeks later when, in another statement to the press, he said (with regard to Kennedy's offer), *"It would be useful if the USSR and the United States pooled their efforts in exploring outer space... specifically for arranging a joint flight to the Moon."*

Here was a golden moment for both Kennedy (who had never liked the Moon race option) and Khrushchev (who, while wanting to beat the Americans, could not really afford the price tag) to reach a new accord. With the assassination of Kennedy three weeks later and the Soviet military's recalcitrance to reveal any aspect of their technology, the opportunity passed.

However, the reality of a Soviet lunar program was assured during this same period with Korolev's latest proposal for a comprehensive five point program that included a manned circumlunar flight, an unmanned lunar rover, and a manned landing by 1968. The mode for each was to be Earth Orbit Rendezvous (EOR). The plan called for the *Soyuz* (called product 7K), along with the 9K translunar injection stage, and a series of 11K tankers to rendezvous in Earth orbit to refuel the 9K for its flight to the Moon with 7K as its payload. The advantage of EOR was the ability to assemble a large (200-ton) structure that would provide a high degree of reliability and redundancy. The success that the Americans were achieving with their big boosters (the first flights of the *Saturn* I and the live static firings of the F-1 engine) had finally moved the almost insolvent Soviet government to approve a manned lunar landing plan—but the final decree was slow in coming.

Early in 1964 General Nikolai Petrovich Kamanin, the cosmonaut chief, summarized the effort in his diary: *"The Central Committee is approving a plan for sending an expedition to the Moon in 1968-1970. The N1 rocket... will be used for this purpose. The mass of all the systems... will comprise about 200 tons. The plan is still only on paper, while the Americans already have done much for carrying out flights to the Moon."* Kamanin headed the cosmonaut office from 1960 until 1970. He was a vocal proponent of the manned space program and kept a series of diaries that have revealed much about the Soviet space program.

A meeting with Khrushchev in March 1964, in which Korolev apparently received his Lunar program approval, suggests that Khrushchev was somewhat ambiguous in his own desires. He wanted the Soviets to remain in a commanding lead in the space program, but at the same time, he lamented the costs and the sacrifices that he was forcing on the peasants who made up the vast majority of the populace. Even with Khrushchev's reluctant approval, it was still several months before a complete plan and funding was forthcoming.

Finally, more than three years after Kennedy's challenge, the Soviets had an approved and funded program for sending men to the Moon. However, the competition was not simply between the USSR and the USA. Korolev also had to compete with Mikhail Yangel, another Chief Designer who had taken over Korolev's ICBM role and who had a manned lunar proposal, the R-56—which would ultimately wither for lack of official approval. And, yet another program was also receiving funding— Chief Designer Vladimir Chelomey's manned circumlunar effort with the *UR-500*.

Chelomey's UR-500 Proton

With Chief Designer Sergey Korolev's *R-7* ICBM falling from favor as a weapon by 1960, the door had been opened for other Chief Designers to pursue their projects. Emerging to prominent visibility was Vladimir Chelomey who had Soviet Premier Khrushchev's son, Sergei as a member of his design group. In addition to the *UR-200* ICBM that used storable propellants and was thus more flexible and responsive to the military needs, Chelomey also had plans for a space booster of much greater capability than Korolev's *R-7*. As expected, the design was also touted to the military as a heavy-lift ICBM called the GR-2. Chelomey's proposal was approved in April 1962, with the first flight scheduled in 1965. With funding always restricted in the hard economic times that surrounded this era, Chelomey's venture would divert precious rubles from Korolev's projects.

Using the designation *UR-500*, the heavy-lift ICBM (and space booster), as it began to emerge in its 1963 design, had many of the structural attributes similar to *Saturn*. A 13.5-foot-diameter, 65-feet-tall, cylindrical oxidizer tank provided a central core around which six 5.25-foot-diameter fuel tanks were clustered. Using a design rational similar to the German V-2, the diameter of the core tank size was the maximum limit of the Soviet railway transportation capability. This allowed the booster to be disassembled for transportation from the factory to the test site.

Six RD-253 engines of 330,000 pounds of thrust each provided for a total of two million pounds for a duration of 127 seconds. The second stage used a single 540,000 pound thrust engine for 212 seconds. Each of the main engines was set on a gimbal to provide control. As a space booster, a variety of upper stages would be configured to provide for both orbital and planetary missions (Figure 19).

Although the configuration was visually similar to the *R-7*, the surrounding tanks did not "stage" but remained a part of the booster throughout the first stage flight as with the *Saturn*. The engines used the familiar storable propellants of nitrogen tetroxide and unsymmetrical dimethyl hydrazine (UDMH). Using an advanced form of "staged combustion cycle" engines, that allowed for very high combustion pressures and efficient use of the fuel, the design avoided much of the combustion oscillations that led to destructive instabilities.

Chelomey used the *UR-500* as a basis for a manned circumlunar project, designated "Lunar Ship 1" (LK-1). It included a manned spacecraft that Sergei Khrushchev noted was "similar in outward appearance to the American *Gemini*." The program received approval in an official decree in May 1964. Carrying only one cosmonaut, Chelomey's LK-1 was scheduled to make its historic flight in time for the fiftieth anniversary of the Russian revolution—November 1967.

Critical analysis of Korolev's *Soyuz* and Chelomey's LK-1 presents an interesting schism in view of the American lunar landing program. Under what rational did the Soviet's believe that a manned circumlunar program would compete with *Apollo*? Did the Soviets discount America's abilities at this point? The Soviet leadership, in particular Premier Khrushchev, seemed to regard America's efforts with disdain. While the Soviet space program gave the impression of a carefully planned and vigorous effort, its ability to vie with the Soviet military needs always left it on the short end of funding save for the creative manipulations of Korolev and Chelomey.

Although Chelomey envisioned the *UR-500* as a means to a manned circumlunar flight, he received the requisite military support from the Soviet Air Force, which had been beaten out of one project after another by the Soviet Strategic Missile Forces. They looked to the *UR-500* as the basis for competing with the American DynaSoar project. This was another indicator that the openness of American missile developments played a strong role in directing the Soviet efforts. Likewise, the cancellation of DynaSoar in late 1963 eventually redirected Soviet military interest away from the *UR-500*.

Despite all the political intrigue that surrounded the various projects being developed by the Chief Designers, the *UR-500* found its way to the launch pad in July 1965. The initial two-stage version placed an 18,000-pound scientific satellite in orbit on its very first test. Of course, the Soviet press played up the weight, which, when the final stage was included, amounted to 26,000 pounds in orbit. The American *Saturn* S-IB (with its LH2 S-IV second stage) had exceeded that more than a year earlier with the SA-5 flight that placed a total of 37,000 pounds in orbit.

The Soviet satellite was named Proton I, and the launcher was identified with that same nomenclature by the Soviet press, although the Chelomey team had been calling the big rocket "Gerkules" (Hercules). While the booster performed its first flight flawlessly, a malfunction in the payload resulted in no radio contact with the onboard experiments.

Three more tests of the *UR-500* over the next 18 months resulted in two successes. Here was a rocket that would play an important role as a large space booster which would endure for over 50 years. It would also be a threat to beat the Americans to the first manned circumlunar flight.

N1: Korolev's Giant

The key to any manned lunar or planetary program is the development of a very large booster. The primary components of that booster are the engines. By late 1960, the Central Committee's decree that addressed "Creation of Powerful Carrier-Rockets" had resulted in a formal specification to four design bureaus for a series of closed cycle, high-pressure combustion chambers with high Specific Impulse. Engines for each of the four stages of the "Powerful Carrier-Rocket" were to be developed, with high-energy fuels (LH2) being the focus of the top two stages. Such a rocket would rival the size of the *Nova* being considered in the United States.

Valentin Glushko, who had been the premier engine provider, had moved away from large LOX based chambers because of combustion instabilities and the desire of the military for storable fuels. It was Glushko's engines which propelled Chelomey's two million pound thrust *UR-500*. A confrontation between Glushko and Korolev effectively divided the Soviet program into two opposing camps (storable vs. cryogenic propellants). Korolev prevailed, but only marginally, with his selection of LOX for the first two stages of the new generation super booster designated the *N1*. Formal approval came in September 1962 from the Council of Ministers and the Central Committee with the objective to *"ensure the leading position of the Soviet Union in the exploration of space."* Specifically, the new booster should initially be capable of placing 150,000 pounds in low Earth orbit and 200,000 pounds with the availability of LH2 in the upper stages. The first launch was scheduled for 1965.

Khrushchev recognized the divisiveness between Korolev and Glushko, and, in an effort to try to pull together perhaps the two strongest space technologists in the Soviet Union, invited them and their wives to his dacha for a weekend of relaxation and technical discussions. Korolev put together a very impressive presentation for a proposed lunar landing that had Khrushchev spellbound. As the discussions moved to the price tag, it was obvious that the 12 billion ruble cost was beyond what Khrushchev was willing to consider. The poor showing of the current agricultural program, coupled with an increasingly unstable economy, put many programs (including the space initiative) in jeopardy and required a careful analysis of where the Soviets would put their rubles in the coming years. It was obvious at this point that the Soviet's ability to keep up with the American's ever expanding space program was going to exact a high price on an already troubled economy. The weekend did not produce the desired thaw between Korolev and Glushko.

Whereas the *Saturn* V was being configured specifically for the lunar mission (and the LOR mode in particular), the *N1* initially was not optimized for any specific role. This may have been another political maneuver to ensure that the military supported the project. The development of these super boosters again pointed up the fundamental differences between the American and Soviet programs. With Eisenhower's insistence that the basic exploration of space be carried out as a civilian activity under the auspices of NASA, the battle for funding occurred in the House and Senate, allowing the military and civilian programs to be more clearly delineated.

For the Soviet military, the *N1* would provide the ability to orbit large space stations that could be used for reconnaissance or for command and control, directing orbital weapons back towards targets on the Earth. From a scientific perspective, the *N1* offered the opportunity to explore Mars, which was still seen by Korolev as a more desirable goal than the Moon.

Because of his falling out with Glushko, Korolev had to turn to Nikolai Kuznetsov, (a leading designer of aircraft engines) for the *N1* rocket motors. Kuznetsov fortunes had begun to wane with the emphasis placed on missiles in the late 1950s, and he was forced to move his expertise into the design of rocket engines. With Kuznetsov's help, by the end of 1962 the basic configuration of the *N1* was established. The first stage would consist of 24 engines of 360,000 pounds thrust each, in a circular arrangement, providing a total of over 8 million pounds of thrust. The second stage would employ eight 390,000 pound thrust engines with over three 3 million pounds of thrust. The second stage engines were similar to the first stage except they were optimized for the vacuum conditions of space and thus achieved higher thrust levels. The engines that Kuznetsov had designed were among the most advanced in the world.

Technology aside, a major obstacle in building the *N1* (or any rocket of that size) was the ability to move such a large structure from its manufacturing facility to the static test stands and to the launch site itself. After much thought it was decided that the rocket had to be designed in such a way that it could be transported in parts. No other transport capability was feasible. The United States also faced that problem but built all the *Saturn* V stages as complete units and transported the first and second stages by barge, using a manufacturing facility (Michoud, Mississippi) that was co-located with a navigable waterway. The Kennedy Space Center (KSC) provided the receiving terminus on the Atlantic.

Continued lack of funding, coupled with the still lagging industrial infrastructure of the Soviet Union, resulted in progress on the *N1* falling behind continually revised projections. Whether it was the financial situation, technical considerations, or just the strange politics of the Soviet space program, EOR was abandoned for LOR by mid-1964, and the configuration of the *N1* had to change to accommodate the heavier lifting requirement.

Although committed in spirit to LH2 upper stages, the resources of the design bureaus were such that progress was slow and would never provide an LH2 engine for decades. While there was much speculation and tentative planning for nuclear and electric propulsion, that technology, too, would have to wait many years for progress. By late 1965 it was realized that the third and fourth stages would have to be LOX-kerosene, at least for the initial versions of the *N1*. To make up for the lack of high energy upper stages, six more engines were added in a smaller concentric ring within the first stage complement of 24 bringing the total to 30—and providing almost 30 percent more thrust than the *Saturn* V.

In an effort to sidetrack Korolev's *N1*, Chelomey produced yet another design even larger than the *N1*. Designated the *UR-700*, this rocket would have had twelve million pounds of thrust. But, with the investment that had already been made in the *N1*, it was not possible for Chelomey to interest the Soviet hierarchy in yet another project—regardless of what the Americans were doing.

The *N1* booster that finally took shape had a take-off thrust of 10 million pounds. The diameter at the base of the rocket was 55 feet as compared to the 33 feet of the *Saturn* V. Pitch and yaw of the first and second stages were controlled by reducing power on one side of the thrust circle as none of the engines were gimbaled. Roll control was achieved by four 15,000 pound thrust engines (Figure 20).

However, the performance calculations for the *N1* initially showed only 175,000 pounds could be lifted to low Earth orbit compared to 250,000 for the *Saturn* V. The impact of the LH2 development program in the United States was the significant difference. The huge *N1* rocket was approximately 20 feet shorter than the 363-foot *Saturn* V. Because of the massive size of the propellant tanks, the *N1* first and second stage tanks required separate structures that were not load bearing—adding a weight penalty. The inability of manufacturing processes, coupled with metallurgical limitations of Soviet industry, continued to plague the effort. An attempt to increase the performance of the engines, while shaving weight from the rocket itself, became an obsession. Likewise, the payload underwent intense scrutiny. Even the initial parking orbit was lowered 50 miles and the launch azimuth was reduced to 51 degrees to get the maximum velocity from the Earth's rotation. The final payload moved to slightly more than 200,000 pounds—perhaps enough if the crew size was reduced from three cosmonauts to two.

F-1: The Largest Single Chamber Engine

Although Rocketdyne's F-1 began its development cycle as an attempt to scale-up the H-1, it was quickly learned that the temperatures and pressures involved did not extrapolate to the same solutions. Attempting to increase thrust by an order-of-magnitude (ten fold) resulted in problems that could not be addressed in the same manner as in the H-1.

The emphasis in development of the F-1 focused on reliability and the desire to remain within the state-of-the-art whenever possible. This did not imply that new materials, methods, and techniques could not be innovative. In fact, it was only through unique solutions that many of the problems were resolved. One example of innovation was the requirement of the hydraulic system to gimbal (move) the engines in precise directions and angles. As with the use of the RP-1 fuel as a lubricant for the turbo-pumps, the fuel was also used as hydraulic fluid, eliminating yet another special use fluid. (However, RP-1 is not a good hydraulic fluid as it is less viscous, more corrosive, and is prone to contamination.)

The construction of the F-1 thrust chamber, like the H-1, was a longitudinally contoured set of tubes brazed together to form the chamber, throat, and a portion of the flared bell-shaped exhaust nozzle. The RP-1 fuel flowed through these tubes to cool the engine before being injected into the combustion chamber to be burned. The latter eight feet of the exhaust nozzle, which was not constructed of the tubular process, was cooled by routing the relatively cool exhaust gases from the turbo-pump into the exhaust. This created an insulating "gaseous fluid" between the nozzle structure and the ultra hot primary exhaust of the thrust chamber.

The injector design was a critical part of the F-1 if the nemesis of combustion instability (also called combustion oscillation) was to be resolved. Attempts to simply scale-up the H-1's injector did not work because of all the variables involved. At one point early in 1961, there were serious concerns that an engine the size of the F-1 could not be made to work. NASA Deputy Administrator Hugh Dryden was convinced that all of the problems had an effective engineering solution. The injector problem reached critical proportions in July 1962 with an explosion during a static test, destroying the F-1 and some of the stand.

At this point it was suggested that an alternative to the F-1 should be pursued in light of the apparent progress the Soviets seem to be making. Nevertheless, von Braun was as equally convinced as Dryden that more engineering research would yield a solution. Rocketdyne assigned almost 20 percent of its engineering talent (175 persons) to address the problem.

Because combustion instabilities occurred at what appeared to be random times (eight of 310 tests), observing and recording the phenomena was difficult. In an effort to force an instability to occur so that it could be more effectively studied, small explosive charges were set off within the chamber during actual hot firings of the engine.

Eventually a series of small and relatively simple changes provided the solution set. The orifice size of the injector was increased, and the angle at which the fuel and oxidizer impinged on each other was optimized. The ultimate design of the injector was so effective that the engine would recover from intentional divergent instabilities (produced by the small explosive charges) and return to normal operation within a fraction of a second. Design changes to the injector required more than 18 months of effort, and the engine was not "flight rated" until January 1965.

In the final design, the head of the combustion chamber consisted of a series of concentric rings alternating between fuel and oxidizer. A total of 3,700 holes delivered fuel while 2,600 provided for the oxidizer. A series of copper baffles directed and contained the flow to ensure uniform distribution. Copper, a low-melting point metal, was acceptable at this point in the combustion chamber because the flame front was further aft in the chamber. The F-1 burned 2,200 pounds of fuel and 4,400 pounds of oxidizer each second under a pressure of 1,200 pounds to achieve the rated thrust of 1,500,000 pounds.

Developing the J-2

The development of a significantly larger liquid-hydrogen (LH2) engine for use as an upper stage began even before the smaller 15,000 pound thrust Pratt & Whitney RL-10 had been test fired. Studies by NASA in 1959 showed that not only was LH2 a key ingredient but that its full importance would be revealed in engines of 150,000 to 200,000 pounds of thrust—and clustered for even greater power levels. Pratt & Whitney's pioneering expertise in LH2 not withstanding, the Rocketdyne Division of North American Aviation, an early participant in LOX/RP-1 development, was awarded a contract in June of 1960 to proceed with the development of such an engine—to be designated the J-2. It was felt that Pratt & Whitney had its hands full with the RL-10, which at the time was a critical ingredient in the *Centaur* program. Within 18 months, Rocketdyne was test firing an LH2 thrust chamber, and by October 1962, full thrust tests of a complete engine were accomplished for periods of up to 250 seconds.

With the anticipated availability of the J-2, planning moved ahead with its use as the second stage of the *Saturn* IB and as the third stage of the *Saturn* V Moon Rocket. It was designated as the S-IVB. A challenge beyond the ability to fashion such an advanced power system was the requirement of the S-IVB to perform a restart, since it would not only provide the final thrust to the parking orbit for a lunar mission but it would also have to restart to supply the power for translunar escape velocity.

Pratt & Whitney provided Rocketdyne with data on the innovative engineering concepts it had developed for the RL-10—an important legacy of information. Of course, the injector and the ever-present combustion instabilities were carefully addressed. One engineer noted that, when observing the normal (almost clear) exhaust gasses during a static test, its sudden change of color to green indicated that the flame front had moved forward and was now devouring the copper baffles in the injector area. The resolution to the problem had been worked out by Pratt & Whitney with a porous injector face, but Rocketdyne had been reluctant to incorporate it. NASA finally had to assert its influence to move Rocketdyne to accept the Pratt & Whitney innovation.

The turbo-pump employed a start-up technique called "bootstrapping," which was similar to the method used in the RL-10 engine of the *Centaur*. When the LH2 valves were opened for engine start, the liquid flowed under gravitational pressure (assisted by its own tank pressurization) through the engine's regenerative cooling structure. The ambient temperatures of the engine structure caused the LH2 to rapidly gasify, and some of this was then routed through the turbine stage of the propellant pump generating enough force to cause it to begin to spin up. This in-turn forced initial quantities of fuel and oxidizer into the combustion chamber where ignition would then occur. As the combustion chamber walls began to absorb the heat, the LH2 within the cooling tubes would gasify more rapidly causing the pumps to come up to operating speed after a few seconds.

Because of the difference in fluid dynamics of the cryogenics, two different turbo pump designs were used—a radial pump at 6,000 RPM for the LOX and an axial-flow pump at 25,000 RPM for the LH2. To provide for the restart capability, the gas generator that powered the turbine employed a unique method of using hydrogen gas tapped from the propellant supply and storing it in a small pressure vessel to begin its spin-up. The gas reservoir was then restored from the fuel tank so that it would have the ability to perform the second start function.

The first production engines were delivered in April 1964. By the end of that year, full-duration static tests of up to 410 seconds were achieved—although each engine was required to have a usable life of 3,750 seconds to provide for a high level of reliability. Computers performed the pre-firing checkout, propellant loading, and ignition sequence.

The super-cold LH2 posed many special problems ranging from seals to insulation. One problem involved the LH2 gasifying too quickly during engine start. The resolution was to "chill-down" (pre-cool) critical portions of the engine. Each problem was methodically addressed, and the program moved forward more rapidly than many had predicted. In the end, the J-2 went through a series of upgrades such that the latter *Saturn* Vs had improved performance with the J-2 providing 230,000 pounds of thrust and a Specific Impulse of 421 lb/sec. The development of the J-2 was one of the most significant engineering feats of the Apollo program.

The first flight of the *Saturn* IB occurred in February 1966 with the new Rocketdyne 200,000 pound LOX/LH2 J-2 engine serving as the S-IVB second stage. The launch vehicle was designated AS-201 (*Apollo Saturn*—the word Apollo now preceding the word Saturn). It carried a more complete but yet unmanned *Apollo* Command and Service Module (CSM) through the entire sub-orbital launch and reentry sequence which lasted a total of 34 minutes as it arched over 1,000 miles into space to impact 5,000 miles down range.

AS-203 followed (numerically out of sequence) to flight test the second stage restart process, although a restart was not actually attempted. The ullage motors and the seating of the propellants (the LH2 in particular) were viewed with video cameras (in flight) to validate the technique. Following a simulated restart sequence, the S-IVB stage was then intentionally over-pressurized to destruction to record the points of failure and confirm design limits. A full four orbits were performed before it was destroyed.

AS-202, launched in August 1966, performed another sub orbital test of the heat shield using a "skipping" profile to expose it to high levels of thermal stress. The spacecraft was recovered near Wake Island in the Pacific after traveling three-quarters of the way around the world. The *Saturn* IB, now 224-foot-tall, and the *Apollo* spacecraft were ready for their first manned space flight. However, tragic events waited to intervene in the progress of the space program.

Saturn V Takes Shape

To achieve the LOR path to the Moon, the *Saturn* V required the ability to orbit 250,000 pounds and send almost 100,000 pounds to the Moon. Fifty pounds of booster weight was required to accelerate each pound of payload to escape velocity. To accomplish this task, the first stage would generate 7.5 million pounds of thrust—five clustered F-1 engines. Designated the S-1C, the construction of the first stage (33 feet in diameter and 75 feet in length) was contracted to Boeing. The S-II second stage, with its five J-2 engines, was built by Douglas Aircraft. The contractors participated in the design but the essence was provided by von Braun's MSFC team. It is interesting to note that, although NASA (in the person of Administrator James Webb) wanted to assure that the most qualified contractor was selected for the various stages and systems, it also desired to spread technology development among a vast number of manufacturers.

The first stage of the *Saturn* V emerged from the design period with four large fins at its base. The aerodynamic stability of the fins eased the work of the control unit in gimbaling the huge F-1 engines during most of the flight and provided a more stable platform during staging. The S-1C stage was static fired for the first time in 1965.

Because the propellant tanks were pressurized, they provided some of the structural integrity (although they were not the "balloon" structure of the *Atlas*). The interstage structure, not having the pressurized stiffening, required the use of a corrugated skin and internal stringers to provide the necessary strength.

If LH2 was superior to kerosene as a fuel, why wasn't it used in the first stage? As LH2 has only half the density of kerosene, the size of the first stage fuel tank would have been quite large, resulting in significant structural weight penalties and possible aerodynamic problems. It would have required 37 J-2 engines or the design of yet another larger generation LH2 engine.

A cluster of five J-2s provided one million pounds of thrust for the Saturn V second stage designated as the S-II. This second stage had the same 33-foot diameter of the first stage (Figure 21). The tank structure accounted for only three percent of the overall weight of the fully fueled assembly—a significant engineering feat.

Several factors required that the LH2 tanks be insulated. The most significant of these was the boil-off and venting of the fuel. This could result in the loss of up to 300 gallons per minute for the second stage tank. To reduce the boil-off factor, the tanks had to be "chilled down" over time during which the process itself consumed an equal amount of LH2 as the tank held. Another consideration was the extreme cold on the outside of the tank. This would actually cause liquefaction of the ambient air components of nitrogen and oxygen which would then run down the sides of the rocket. The problem of insulating the LH2 tanks resulted in considering either external or internal designs.

Internal insulation minimized possible damage to the material during routine handling and transportation of the stage. For a variety of reasons, Douglas chose to use external insulation. As for the material itself, balsa wood was considered almost ideal for its excellent insulating properties, light weight, and ease of shaping to the required contours. However, its lack of high volume availability gave way to a special process that used a polyurethane foam spray (although panels were originally tried). The problem of retaining a uniform and secure bond between the foam and the metallic tank was an ever-present problem that would come back to haunt NASA in the next generation of rockets.

Yet another problem was the thermal stress on the thin steel tank material which had a very high coefficient of expansion. The adhesives used to bind the insulation to the tank wall were also more amenable to less exaggerated variations of the binding surface. Finally, the bulkhead between the LH2 and LOX had to be insulated to keep the LH2 from freezing the adjacent LOX solid.

Even the filling of the cryogenic tanks required a purge, chill-down, and fill cycle using four separate flow rates: slow fill, fast fill, slow fill to capacity, and replenish. So critical was the desire to have topped-off tanks, the propellant fill lines remained attached to the rocket right up to the point of ignition and were automatically disconnected at lift-off.

Hydrogen flames are difficult to detect because they are virtually invisible. Because of the possibility that undetected leaks might cause a combustible mixture in the skirt area of the booster and the interstage areas, gaseous nitrogen was used to flood these areas.

To assure that all propellants were consumed, a propellant utilization system (PU) was devised that allowed for a change in engine mixture ratio (from the optimum) towards the end of the burn cycle.

Assembling the giant rocket posed yet another set of engineering problems. Traditionally, American rockets had been assembled "on the pad:" hoisted from a trailer-transporter to the vertical position, one stage at a time by a crane atop a gantry that would then enclose the rocket. The various electrical, hydraulic, and cryogenic connections would then be made to the Ground Support Equipment (GSE), and the gantry would be rolled back before launch to leave the rocket standing free.

However, this method posed many problems for a rocket as large as the *Saturn* V. There were two launch pads, and that would have required building two massive gantries—literally buildings that would have to be movable. In addition, the proximity of the launch pad to the gantry meant that it would be exposed to damage should the rocket fail during the initial phase of launch.

The answer to the problem came from the 1928 science fiction movie *Girl in the Moon* that was created by the famous German director Fritz Lang. In the movie, the massive rocket is assembled in a large building and then transported to the launch pad on a set of rails. The movie's technical advisor (rocket pioneer Hermann Oberth) had devised this concept. As Oberth had been von Braun's mentor in the early years, the arrangement of having a Vehicle Assembly Building (VAB) and a transporter to move the rocket to the actual firing site was adopted for the *Saturn* V.

A single 525 foot tall Vehicle Assembly Building was constructed that, at the time, was the largest enclosed building in the world. It had more volume than three Empire State Buildings—130 million cubic feet of space covering eight acres. With four "High Bays," each with a 456 foot vertically sliding door, four completely assembled *Saturn* Vs could be accommodated.

The rocket was assembled on one of three Mobile Launch Pads (MLP) that were 25 feet high, 160 feet long, and 135 feet wide. Each had their own 400-foot umbilical tower that provided connections for fuel and oxidizer, ancillary gases (such as nitrogen), ground electrical power and communications links—on big swing arms. One large opening in the base of the platform accommodated the flames of the five F-1 engine exhausts. Each Mobile Launch Pad weighed ten million pounds.

The mobile pad was set on six 22-foot-high steel pedestals when in the Vehicle Assembly Building or at the launch site. When on the firing site, four additional extensible columns were used to stiffen the Mobile Launch Pad against rebound loads, should engine cutoff occur right after launch commit.

To move the Mobile Launch Pad and the *Saturn* V to the firing site, yet another gigantic machine was created—the crawler-transporter. Here was a vehicle that could transport over 12 million pounds and travel at one mile per hour for the five-mile trip from the Vehicle Assembly Building to the launch site. Yet it was gentle enough to place the Mobile Launch Pad to within a fraction of an inch and keep its surface level so its towering cargo stayed perfectly vertical throughout the six-hour ride.

The crawler transporter was 131 feet long and 113 feet wide, with a flat-upper-deck carrying surface of 900 feet square. It moved on four double-tracked tread belts where each of the 57 "shoes" that comprised the track weighed more than a ton apiece.

When assembly and checkout of the rocket was complete in the Vehicle Assembly Building, the crawler-transporter was driven into the High Bay to a position under the Mobile Launch Pad. The crawler lifted the support pedestal and the *Saturn* V and carried them to the launch pad. Powered by two 2,750-horsepower diesel engines, the Mobile Launch Pad has a hydraulic system for jacking, equalizing, and leveling the crawler at all times, even as it climbed the short hill to the launch pad. This leveling system kept the rocket stable, although the very top could sway by as much as two feet during the journey.

Automated testing of a wide variety of the rocket's systems, such as the monitoring of temperatures and pressures during flight, became another critical aspect of being able to move the project along and to achieve a successful flight test program. More than 2,500 test points were monitored on the *Apollo* spacecraft, CSM and LM, while the *Saturn* V had more than 5,000 additional points. The emphasis on automated testing was predicated on the complexity and of the rocket itself, as well as the lives of the three astronauts. Human error and the relatively slow speed at which a technician could read and interpret a test point (coupled with the volume of information) would have dramatically increased the time required to prepare and launch the rocket. Finally, by using an automated system, the critical data could be electronically (and immediately) shared among several test conductors for evaluation.

Guidance and control of the *Saturn* V was managed by the Instrument Unit (IU), which occupied the upper three feet of the S-IVB stage. It was essentially an inertial unit that used a set of gyroscopes, accelerometers, a digital computer, and a variety of other components to monitor and control the giant on its journey. A direct lineage of the system could be found in the *Jupiter, Redstone*, and even the crude stepping switches of the V-2. The IBM computer provided triple redundancy with critical functions presenting their data for comparison. If one set of inputs disagreed with the other two, the majority ruled, and the third set of results was disregarded.

While progress with all of the stages was generally good, there were several significant technical as well as managerial problems that caused periods of considerable concern. Flight test dates continued to slip from 1966 into 1967.

All Up Testing

When the new Director of the Office of Manned Space Flight, George Mueller (pronounced Miller), reviewed his domain in September 1963, two project management factors stood out: time and money (quality was not considered a variable). The schedule for flight-testing of the *Saturn* V reflected the step-by-step approach that had been the traditional manner for proving new rockets in this high-risk environment. The plan called for several tests of the S-1C first stage before a "live" (fueled) S-II second stage would be incorporated for a series of tests on it. Only after proving the second stage would a "live" third stage be added. In all it was estimated that at least ten flights would be needed to man-rate the *Saturn* I and another ten for *Saturn* V. But if the *Saturn* continued to employ this conservative method, not only would it be virtually impossible to meet the lunar landing goal by the end of the decade but the cost of the number of huge rockets required (which was approaching 150 million dollars per launch) would be prohibitive.

Mueller issued a proposal early in November 1963 in which he stated his belief that an accelerated schedule could be achieved that would result in flying the first manned *Saturn* IB after only its third flight. More surprisingly, he proposed the same three-flight goal for the mammoth *Saturn* V. The concept was known as "all up testing." The pressures of time and money could be mitigated, he felt, by the use of new and innovative testing techniques. These procedures would use automated sequences enhanced by computers to achieve a high degree of assurance in comprehensive ground testing—before the rocket lifted into the air for the first time.

After recovering from the initial shock of the proposal, von Braun countered that the second and third stages used the yet untested LH2 J2 engine. To launch the first *Saturn* V with all live stages was an invitation to an explosive disaster. However, as he evaluated the traditional schedule, he recognized that Mueller's approach was the only one that would allow any possibility of achieving Kennedy's goal. Perhaps more importantly for von Braun, it might be the only way of beating the Russians to what might be the final objective in the space race between the two nations. For a man who had already been "robbed" of the accolades of the first artificial Earth satellite and the first man in space, the loss of the Moon would leave him as an "also ran" in the annals of history.

Mueller received von Braun's assent to the new plan only after an alternative schedule was drawn-up as a fall back in case of catastrophic failure. Von Braun also faced a hard sell with his team, who were adamantly opposed to the concept. As it would turn out, all-up-testing would be the final critical process that would allow the United States to achieve the Kennedy goal.

Meanwhile, as money and time began to tighten around the Soviet's own *N1* Moon rocket project, Korolev proposed to move directly to a flight version of the first stage—without static testing. In addition, to add even more risk, he proposed to send aloft the upper stages on the first flight as well—all-up-testing. While components of the third and fourth stages would be existing rockets that had undergone preliminary tests, the first and second stages would present a high degree of unknowns. To address this risk factor was a system designed to detect a malfunction in a particular engine and shut it down along with the corresponding "good" engine on the opposite side of the outer thrust circle that composed the 24 engines (to avoid asymmetrical thrust). The remaining engines would then fire for a longer duration—consuming the propellant allocated to the dead engines. The system, whose acronym of abbreviation was KORD, consisted of a complex sensing algorithm that was beyond what could be expected of computer analysis of the time. KORD would also function on the clustered second and third stages

To add some rational to the testing scheme was the fact that the second stage would undergo complete static testing. As it used essentially the same engines as the first stage, its development would provide a reasonable understanding of the basic problems that would arise and their solutions. Yet another decision that would weaken reliability was that not every first stage engine would undergo static firing before being assembled into its cluster. Using statistical analysis, a group of six engines would be randomly selected from a manufacturing batch of which two would be chosen to undergo static firing. If they successfully completed the test, the other four would be accepted for flight.

All-up-testing would figure prominently for both nations efforts; one would achieve world acclaim while the other would hide its failure for years to come.

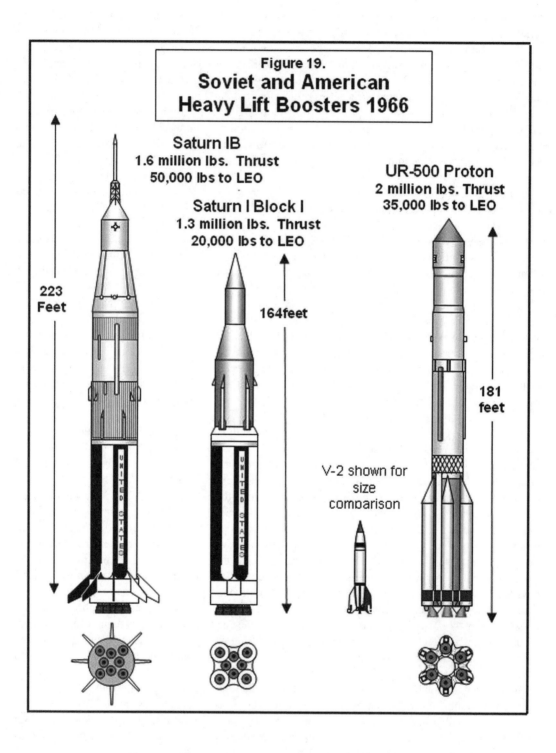

Figure 19.
Soviet and American
Heavy Lift Boosters 1966

Saturn IB
1.6 million lbs. Thrust
50,000 lbs to LEO

UR-500 Proton
2 million lbs. Thrust
35,000 lbs to LEO

Saturn I Block I
1.3 million lbs. Thrust
20,000 lbs to LEO

223 Feet

164feet

181 feet

V-2 shown for size comparison

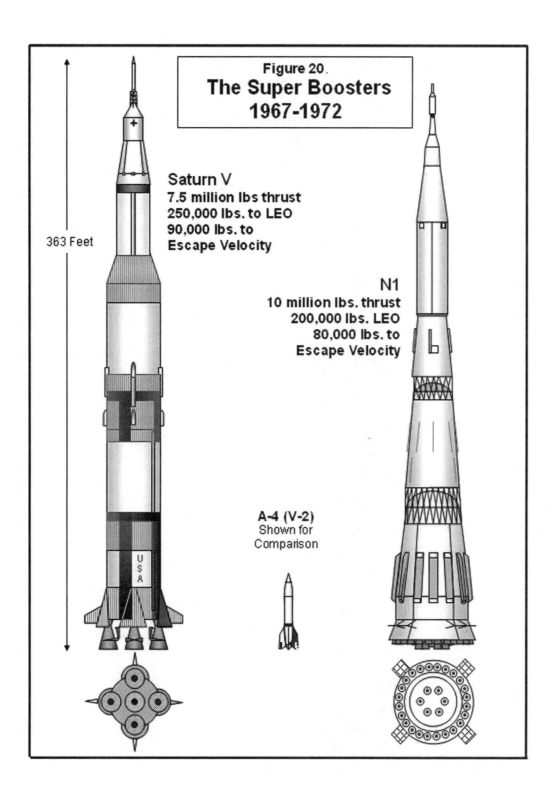

Figure 20.
The Super Boosters
1967-1972

363 Feet

Saturn V
**7.5 million lbs thrust
250,000 lbs. to LEO
90,000 lbs. to
Escape Velocity**

N1
**10 million lbs. thrust
200,000 lbs. LEO
80,000 lbs. to
Escape Velocity**

A-4 (V-2)
Shown for
Comparison

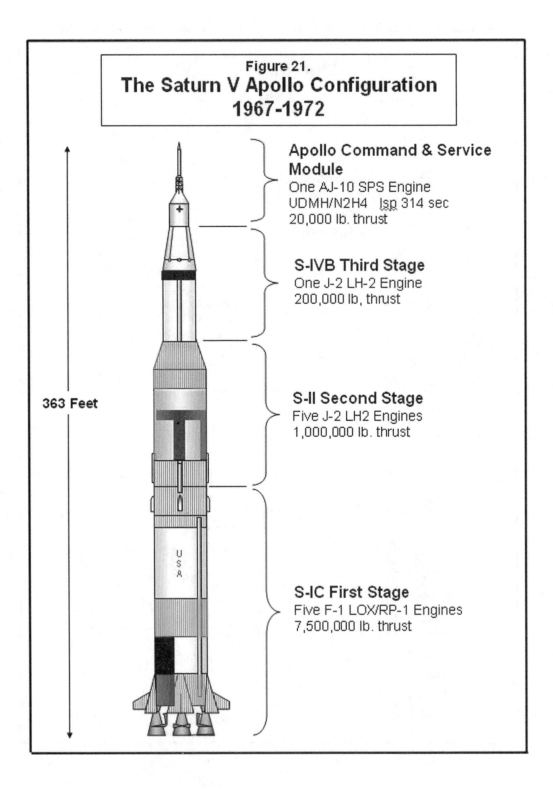

Figure 21.

The Saturn V Apollo Configuration
1967-1972

363 Feet

Apollo Command & Service Module
One AJ-10 SPS Engine
UDMH/N2H4   Isp 314 sec
20,000 lb. thrust

S-IVB Third Stage
One J-2 LH-2 Engine
200,000 lb, thrust

S-II Second Stage
Five J-2 LH2 Engines
1,000,000 lb. thrust

S-IC First Stage
Five F-1 LOX/RP-1 Engines
7,500,000 lb. thrust

Voskhod I: Korolev's Last Rabbit

If America had established an optimistic schedule for its *Gemini* and *Apollo* spacecraft, Korolev likewise had visions of the near impossible for his three-man *Soyuz*—the follow-on to *Vostok*. He had originally established a timetable that called for its first flight in May 1963. However, the level of effort demanded of his design bureau was far beyond what his limited staff and budget could sustain. Work on the Molniya and Zenit satellites, along with the unmanned probes to Mars and Venus, not to mention his commitment to advanced weapons to replace the *R-7*, stretched his staff's engineering ability to the limit.

A program to launch 12 more *Vostok*s, which had been approved in December 1962, had been cancelled when it was realized that the *Vostok* represented a dead-end technology. Yet the new *Soyuz* had been delayed by many technical problems and the inability of the Soviets to manage effectively the multitude of programs. Moreover, like the Americans, it was realized that some interim solution had to be defined that would not only keep Soviet men in space but would ensure that the perceived lead in technology was maintained over the Americans. Tentative goals of higher altitudes and longer durations, while important, did not represent advancing technology in the eyes of the world.

Korolev's manned circumlunar proposal, accepted by the government in December 1963 and pushed by the optimistic schedules that NASA had published for *Gemini*, required that *Soyuz* fly before the end of 1964. As the early months of 1964 passed, it became clear that *Soyuz* would not fly in the same time frame as *Gemini*. At this point, the history of the Soviet program becomes somewhat nebulous. Korolev reportedly told his First Deputy, Vasiliy Mishin, that Premier Khrushchev himself had directed Korolev to find some way of placing three men in orbit before the scheduled *Gemini* missions. However, Khrushchev's son writes that his father was not that aware of the technology aspects of *Gemini*. Sergei Khrushchev believed that it was Korolev himself who initiated the plan but did not want to be seen as advocating such a reckless and radical solution to their dilemma as was about to be produced.

Adding to the confusion, cosmonaut chief General Nikolay Kamanin reported in his memoirs that *"It was the first time I had seen Korolev in complete bewilderment… and could not see clear path on how to re-equip the ship [Vostok] for three[cosmonauts] in such a short period of time."* Thus, while it is unclear who was responsible for the decision, Korolev went forward with a re-layout of the basic *Vostok* to accommodate three men.

The project was given a new name, *Voskhod* (Sunrise) to disassociated it from the *Vostok* in every possible way in order to ensure that the West would perceive it as a true second generation effort. The official decree authorizing the project was signed in March 1964 with the expected first flight the following August. Simply flying the first three-man ship would not be enough. The announced plans of *Gemini* to introduce the EVA (a space walk) prompted a second part of this ad hoc program; a Soviet cosmonaut would also leave a *Voskhod* (probably the second flight) and venture out into the hostile environment protected only in his spacesuit (Figure 22).

Even the booster would have to be the upgraded *R-7* used for the Zenit satellites because of the projected 1,300 pound increase in the weight of the spacecraft. While the Soviets had often paced their program based on what the Americans had scheduled, this is the first real evidence that America was now driving the race and that their technology was being used as the measure of the Soviet's own progress. Perhaps the most important aspect of *Gemini*, the ability to rendezvous and dock, would not be present in *Voskhod*. With the flight of America's 1.3 million pound thrust *Saturn SA-6* in May of 1964, which carried an unmanned boilerplate of the *Apollo* spacecraft, Korolev may have become more than anxious.

As Korolev's engineers pondered the problem, it became clear to them that the project represented an almost suicidal venture. Because of the weight and space restrictions, there would be no room for the cosmonauts to wear pressure suits; it would be a true shirtsleeve environment. Perhaps the most precarious design change was that the ejection seat would also have to be deleted, so there would be no escape from the ship should there be a catastrophic failure. This also meant that during the return, unlike Vostok, the crew would have to ride the descent module to the ground. The impact would be softened by a solid-fuel rocket suspended from the parachute and fired just before touchdown. An escape tower being perfected for the *Soyuz* would not be ready in time for *Voskhod*.

Because of the tight budgets and timeframe, test articles were difficult to procure, but one test in particular was critical. Korolev authorized the use of Gherman Titov's *Vostok* II craft (which had been placed in a museum) to be equipped with the recovery parachutes and landing rocket to prove the design and technique. Unfortunately, the test did not go well, and the priceless artifact was destroyed. It was obvious that everything had to work perfectly or the three crew members would surely die.

The *Vostok* life support system, which would normally sustain one cosmonaut for 10 days, would only provide enough consumables for the expanded crew for about three days. If the retrorockets failed, the crew would most likely perish before the orbit would decay naturally to reentry, therefore a back-up retrorocket was added to the configuration. The Soviets could hide launch failures, but they definitely wanted to be able to announce a successful flight while the cosmonauts were in orbit; thus, the guaranteed safe return was pivotal. They had learned their lessons well from the public acclaim of Project *Mercury* and wanted to provide as much "openness" as would result in the maximum propaganda value.

As Korolev listened to his designers lament the possible consequences of the redesign, he offered a "carrot" to motivate their innovation: one of the three cosmonauts would be a member of the design team. While this might seem a bit odd in that the prospects for a successful flight were significantly narrowed by the improvisation that was taking place, the opportunity to fly in space was overwhelming. An experienced cosmonaut would command the mission, and the final slot would be allocated to a medical doctor.

The selection of the crew was a demanding and politically charged effort. Korolev had no authorization to make the flight offer to his staff. Other influential authorities had seen *Voskhod* as the opportunity to place a scientist in that third seat. After much bickering, several engineers were selected and underwent physicals to determine their suitability.

Konstantin Feoktistov, who had been in the forefront of the engineering work on *Vostok* and now was a key figure in its rebirth as *Voskhod*, was eager to get the opportunity to take his ride into space. He apparently completed the physical successfully (and actually was the only member of the engineering staff to do so). His selection would later be challenged with the comments that ulcers, poor vision, a deformed spine, and missing fingers were disqualifying. He was also not the most socially acceptable as he had a history of being difficult to communicate with and uncompromising. In support of Feoktistov, was the fact that he knew more about the workings of *Voskhod* than any other person.

The command position was reduced to three candidates, one of whom had a Jewish mother. This was deemed unacceptable to the Communist hierarchy. The candidate supported by Korolev was Vladimir Komarov. He had been grounded by a heart anomaly similar to that which was keeping the American astronaut "Deke" Slayton from a flight slot. Doctors finally relented to Korolev's insistence, and Komarov was cleared to be a candidate.

While it had been agreed that the second slot would go to a medical doctor, the Air Force had been insistent that he be a military man. Four candidates were selected for training, one of whom was not only the youngest (at 26) but also a civilian: Boris Yegorov. Compounding the crew selection and the final testing of the spacecraft and the *R-7* was the deteriorating physical and mental health of Chief Designer Korolev. His inability to control his emotional outbursts was costing him credibility and support of even those among his closest colleagues.

The actual training program was the shortest on record for either the Soviet or American space programs, lasting less than four months. In the end the final crew selection of Komarov, Feoktistov, and Yegorov was not made until a few weeks before the scheduled launch, which was postponed from mid-September to early October to allow for the re-test of the recovery system that had destroyed Titov's museum piece.

The planned duration for the first *Voskhod* flight was a full day (18 orbits), and there were a few improvements to the basic *Vostok* systems. The spacecraft orientation method used a new "ion" technique that allowed the ship to sense the horizon and maintain a selected attitude while in the night-time portion of its orbit. An improved video system transmitted 25 frames per second instead of 10 to smooth out the movements of the crewmembers to those viewing their activities from Earth.

The launch of a full dress rehearsal unmanned test satellite, announced as Kosmos 47 to mask its intent, occurred on October 6, and all the systems functioned properly. It was returned to Earth successfully after the full one-day mission.

There was considerable pressure on the entire launch crew as the countdown proceeded for *Voskhod* I on October 12, 1965. Korolev, in particular, was showing the strain, and he was observed to be visibly trembling during the initial launch phase as he had during Gagarin's flight. He knew that for the first 27 seconds of flight there would be no possible survival of the crew if there was a major malfunction during that time period. However, the *R-7* and its upgraded upper stage performed admirably, and the *Voskhod* was soon in orbit, with the surprise announcement to the world by way of the Soviet news agency TASS.

Despite the short training period, all three crewmembers had reasonably useful work to perform, although by the second and third orbit both Feoktistov and Yegorov were beginning to show signs of space sickness—that debilitating, nauseous feeling that was an extension of vertigo. Of significance is the fact that both were able to complete most of their scheduled assignments, and the flight ended successfully with a landing after slightly more than 24 hours in orbit.

Everyone involved with the flight that had knowledge of the actual planning and spacecraft configuration breathed a sigh of relief—everyone except Soviet Premier Nikita Khrushchev, who, during the course of the flight, had been deposed from power.

Khrushchev, who began his leadership by "de-Stalinizing" the Kremlin, introduced many reforms into the Soviet industrial and agricultural system. He had been in the forefront of an aggressive campaign to remove religion as well as dissident poets and writers from Soviet society. He had introduced early elements of détente with the West, including the "hotline" between the United States and the Kremlin, all of which brought criticism from his Chinese colleagues and the eventual distancing between Soviet Communism and its Chinese counterpart. His use of the Soviet space program (and Korolev in particular) had given him immense power in dealing with world politics. While he might have been at the pinnacle of his power in some respects, he had alienated too many within his own power structure and two men in particular. Aleksey Kosygin and Leonid Brezhnev had orchestrated a coup that toppled Khrushchev after his reign of ten years. As he returned from a vacation on the Crimea, he discovered he was *persona non-grate* and under house arrest. The centralized power between the Communist Party and the government that Khrushchev had declared was about to be diffused by collective leadership between Kosygin and Brezhnev.

The change in power also saw a change in the fortunes of the various design bureaus and the stature of the various Chief Designers. Each of the primary figures—Korolev, Glushko, Yangel, and Chelomey—sought to move closer to the new power structure, and change was inevitable. While *Voskhod* was given high marks in the world press and more aspersions were cast upon the "failing" Americans, the flight itself did not achieve the preeminence it might have, had the headlines of the world not been dominated by Khrushchev's relatively peaceful but undemocratic removal.

Voskhod II: A Walk in Space

As for the second part of the *Voskhod* objectives that provided for a space walk (code named Vykhod or exit), they were still in full swing. Korolev had hoped to launch *Voskhod* II in time for the November celebration of the 1917 revolution, but there were too many problems yet to resolve. Essentially the spacecraft would be similar to *Voskhod* I but would have only a two-man crew. The third couch was deleted; that space was needed for the astronaut to position himself for entry into an expandable, double-walled rubber airlock that was attached to one of the access hatches.

Unlike *Gemini*, in which the entire spacecraft was depressurized for the extra-vehicular activity (EVA), the *Voskhod* cabin could not be exposed to the full vacuum of space—its instrumentation and marginal life support systems were not designed for that environment. Thus, a collapsible, thermally-insulated airlock with a hatch at both ends was devised. Folded to only 28 inches, tight against the spacecraft sphere during the ascent, the airlock would unfold like an accordion to its full length of eight feet once in space and be pressurized to about 40 percent of sea level (Figure 22). The cabin pressure would also be lowered to this value, and as such, it would be necessary for both crewmen to wear pressure suits and breathe 100 percent oxygen.

Crew selection was simplified over the first *Voskhod* in that both cosmonauts would be members of the final group of four who had not yet flown (and were still physically qualified). The training program was more intense than for previous flights, as it was understood that the unknowns of a space walk could present significantly more exertion than simply riding the spacecraft into orbit and back. It was also recognized that space sickness could be life threatening during EVA if the cosmonaut had to vomit—it would obscure the faceplate and clog the environmental system, possibly asphyxiating the space walker. All four candidates underwent extensive zero-G training in a TU-104 transport aircraft (similar to the Boeing 707 "Vomit Comet" that was used by NASA to acclimate astronauts to weightlessness).

A test of all spacecraft systems was to be accomplished with the launch of Kosmos-57 on February 22, 1965. It was an identically equipped, unmanned ship launched under the nebulous banner of the Kosmos series to disguise its role. The mission was fully automated. Once in orbit the airlock was extended and the hatches activated in the same sequence as would be required for the manned mission. At the completion of the simulated EVA, the airlock was jettisoned. However, the spacecraft was not positioned properly for retro-fire and it entered a different orbit. This was sensed by the automatic destruct mechanism, and the ship was destroyed. Coming after a series of failures that had delayed the program, the KGB's presence was soon obvious, as they looked intently for a "subversive" among the test personnel. It was later revealed that erroneous commands from the ground caused the problem, and it was felt that, with one exception, the flight fully qualified the modifications.

The exception was the revised hatch ring (onto which the airlock was fitted) that remained on the spherical descent module. Because Kosmos-57 had not returned to Earth, it could not be examined to ensure the hatch would not present any problems during reentry. The workaround to verify the hatch however, proved relatively simple. Since the Zenit reconnaissance satellites used the same descent module, authorization was given to attach the hatch ring to the next Zenit flight (identified to the West as Kosmos-59), which occurred on March 7. Recovery of the descent sphere was made on March 15th and showed no problems.

The EVA *Voskhod* mission was set for March 18, and its scheduling was none too soon. The first *Gemini* was set to launch on March 23 (although the first *Gemini* EVA would not be accomplished until the second manned flight in June). Soviet mission planners reported in later years that they had felt pressured by the pending *Gemini* program.

Pavel Belyayev and Aleksey Leonov were chosen as the prime crew for the flight that was now officially designated as *Voskhod* II (assuming it achieved orbit). Belyayev, who would command the flight, was the oldest cosmonaut of the original 20 (then being 39 years of age) and the only one with a college education. His flight opportunity had been delayed by an injury received during a parachute exercise. Leonov, age 30, would perform the EVA. Perhaps one of the most articulate of the original cosmonauts, he was also an accomplished painter and would capture his EVA experiences in that form following the flight. He would also author several books in his later years and, after the fall of the Soviet Union, provide more insight into the Soviet space history.

The launch of the one-day *Voskhod* II flight began on the cold morning of March 18, fortunately without mishap as again there was no way of escaping should the booster malfunction. Korolev was his usual apprehensive self. He had written to his wife only a few weeks earlier about the importance of the safety of the crew: *"God grant us the strength and the wisdom to always live up to this motto and to never experience the opposite."* It was an interesting reference to the deity officially banned from Soviet culture by the Communist Party.

Once in an orbit that ranged from 108 miles to 311 miles, Belyayev activated the airlock expansion while Leonov prepared his suit. The pressure in the cabin was lowered to the value that had been established and verified in the airlock, about 300 millimeters of *Mercury* (Hg)—40 percent of the 763 mm sea level pressure. Leonov entered the airlock and connected the 16-foot tether to the suit as Belyayev closed the hatchway behind him. It was a tight fit for the cosmonaut and his bulky spacesuit with its life-support backpack. Although the airlock had an internal diameter of 40 inches, the hatchway was only 25 inches wide. All of the actions involving the extension of the airlock and the opening and closing of both the inner and outer hatches were under the control of the Belyayev. The pressure in the airlock was lowered to about 200 mm as Leonov verified the integrity of his suit. With the airlock finally depressurized to zero, the outer hatch was opened at about 90 minutes after liftoff as the spacecraft passed over Soviet territory. Several 16 mm film cameras provided for the historical footage, while a video camera recorded images transmitted to the ground and to a monitor for Belyayev, so he could be aware of Leonov's relative position and situation.

Leonov slowly emerged from the airlock into the expanse of outer space and tentatively held on to the edge of the hatch. His first observation was that he could *"see the Caucuses,"* the range of mountains that run across the heart of eastern Russia. He then let go of the spacecraft and the film shows him facing the camera and drifting slowly away. He would later recall, *"It was an extraordinary sensation. I had never felt like it before. I was free above the planet Earth and I saw it—saw it was rotating majestically below me."*

Science fiction films of the 1950s, such as *Destination Moon,* had predicted and depicted the space walk. Later films, such as *2001*, would add more drama to the event. But here for the first time man was floating above the Earth. Traveling almost 18,000 mph with nothing between him and a most hostile environment but the thin veneer of the spacesuit, Leonov waved to the camera. He made an effort to take a picture of the spacecraft from his unique vantage point but was unable to activate the shutter. As the spacecraft and Leonov continued their orbital track, the Pacific soon appeared under them, and it was time to end the ten-minute excursion.

As Leonov started to reposition himself for a feet first entry back into the airlock, he realized for the first time that the spacesuit had expanded to the point where both his feet and hands no longer extended into the boots and gloves of the suit. It was as though he had no hands and feet, just blunt limbs that were almost useless. Even the limited digital dexterity the suit had afforded was now gone. To make the situation even more critical, the suit had ballooned so that it was too wide to fit back through the hatch. He was beginning to feel the physical effect of the forced exertion as his heart rate now reached 143 beats per minute, and his respiration was twice normal, while his body temperature exceeded 100 degree F.

Leonov recognized the solution was to decrease the pressure in the suit to the lowest level available, about 200 mm. This done, he was able to enter the airlock headfirst and then turn himself around so that he would return to the cabin in the required feet first position. Given the dimensions of the airlock, this was an almost impossible task. He was near exhaustion when the increasing sweat and moisture condensed on his faceplate obscuring his vision. He elected to open the faceplate before pressure in the airlock was stabilized (against mission rules). He rested for a minute before opening the inner hatch. The EVA had lasted just 24 minutes.

Subsequent EVAs would encounter similar situations in both the American and Soviet programs. Movement in space was difficult and had to be carefully planned, and footholds and handholds had to be provided; otherwise there was nothing to provide leverage for moving the body.

For the new leaders in the Kremlin, Kosygin and Brezhnev, being aware of the launch and watching the video of the EVA "live" was a new and exciting experience. With Leonov now safely back in the spacecraft, the inner hatch was closed and the air lock released. The two cosmonauts breathed a sigh of relief—however, more danger lay ahead. Almost immediately, it was discovered that the inner hatch had not sealed properly. With the airlock gone, the cosmonauts could not reopen the hatch and attempt to remedy the problem. The life sustaining atmosphere of the spacecraft was slowly venting into space, and the life support system was replacing the air with an enriched oxygen content that reached almost 50 percent, increasing the risk of fire. Compounding the problem was a rotation of about two revolutions per minute imparted to the spacecraft when the air lock was jettisoned. It was decided not to address the rotation as the limited attitude-thruster fuel needed to be conserved for the re-entry positioning.

By the thirteenth orbit the cabin air supply had been depleted to about one-third of its capacity, but the pressure appeared to stabilize at about two-thirds of the sea-level value (500 mm), although the oxygen content still presented an explosive mixture. The cosmonauts proceeded with some simple experiments as they waited for the re-entry procedure to begin on the seventeenth orbit. However, at the appointed time, Belyayev reported the retro-fire did not take place. Apparently, the attitude had not been correct, and the automatic sequencer had detected it and prevented the engine from firing.

It was reported that there was much confusion among the ground control crew with Korolev taking command and assessing the situation. It was apparent that a manual retro-fire would have to take place on the next orbit. The procedure, although relatively simple in concept, required Belyayev to lie across both seats of the spacecraft. This allowed him to get a visual alignment through the porthole of an etched line that would permit the craft to be established in the correct attitude for retro-fire. Leonov had to lie under the couch, out of the way, yet holding Belyayev from floating around, while this was taking place. Then both cosmonauts would have to return to their seats before retro-fire commenced so that the center of gravity was in the correct position. The reseating took the better part of a minute and caused retro-fire to occur late. As had been the case with most of the *Vostok* missions, the instrument section failed to fully separate from the spherical descent module. This caused a higher G loading, but the two pieces eventually separated.

Because of the delay, an overshoot of almost 2,000 miles put *Voskhod* II down in dense forest and several feet of snow, and the report of a safe landing was delayed. It was several hours before a rescue helicopter spotted the red parachute but was unable to find a clearing to land. Warm clothes and food were dropped to the pair, and they were forced to spend a cold night in the spacecraft while wolves prowled its perimeter. The next day a rescue team was able to ski into the area, but because of confusion over who had jurisdiction for the recovery effort, Belyayev and Leonov, along with the rescue team, spent the second night in the wilderness before finally being airlifted out after having chopped a clearing for a helicopter to set down.

The harrowing ordeal, both in space and on the ground, was not revealed for many years. In fact, Leonov, on several occasions during speaking tours over the few weeks that followed the flight, noted how pleasant and easy his space walk had been, and Belyayev indicated that the *Voskhod* had the capability to change orbit (it could not).

Although there were plans for at least two more *Voskhod* flights, poor planning and indecision would result in the Soviets not orbiting a man for more than two years. The Soviets touted their 500+ hours in space compared to the Americans' 54 hours and the fact that they had now sent 11 cosmonauts into orbit compared to America's four (they discounted Shepard and Grissom's flight).

The *Voskhod* effort, designed solely to upstage *Gemini*, delayed the introduction of the *Soyuz* by at least a year, perhaps sealing the fate of the Soviet program and ensuring that, baring any significant problems to the American effort, the Soviets would not be the first to the Moon. First Deputy Mishin, in a comment made some 25 years later, indicated that the *Voskhod* program had not contributed to Soviet space advances and in fact had been *"simply a waste of time."*

Gemini III: Space Twins in Orbit

Attempts to launch the second unmanned *Gemini Titan* 2 (GT-2) during the first week of December 1964 were frustrated by a series of problems. The launch was recycled (a term meaning that the countdown had to be restarted), and it finally left the pad on January 19, 1965. The flight was a ballistic, sub-orbital trajectory to subject the heat shield to a calculated overload to ensure it would survive all of the likely reentry profiles. After just two *Gemini Titan* tests, the booster and its spacecraft were set for the start of the manned flight program.

However, it was far more than simply a new spacecraft and booster. In the months following the last *Mercury* flight, a new NASA facility had sprung up near Houston, Texas, and this hub of the Manned Spacecraft Center was now operational. Control of the manned missions would be transferred from Cape Kennedy to "Houston" (as it was blandly referred to on the communications channels) when the rocket cleared the umbilical tower. The Cape had undergone its own name change from Canaveral to Kennedy following the assassination of the president in 1963, and the NASA facilities there were known as the Kennedy Space Center. The Cape itself would revert back to its historic name of Canaveral in the years to come, to preserve that landmark.

Control of a manned space flight could only be effective if the data that reflected the condition of the spacecraft could be quickly and accurately transmitted to the Goddard Space Flight Center in Maryland, which remained the nerve center of the electronics network. Here, information from a greatly expanded telemetry system that now included 89 ground stations and ships around the world would be filtered and routed, and some of it processed by a set of new IBM 7094 computers working in tandem, before being sent on to Houston. Data rates increased dramatically, and the format went from hodgepodge of analog signals to exclusively digital. Each spacecraft would be sending information regarding 270 aspects of various temperatures, pressures, and states of electronic and mechanical relationships (compared to the 90 channels of data for *Mercury*).

A third group of 20 new astronauts was selected in October 1963 with the emphasis moving to engineering and scientific skills, although each still was required to be a pilot. To this point, only John Glenn had retired from the astronaut corps to pursue his political ambitions. Some, like "Deke" Slayton and Alan Shepard, were grounded because of medical conditions.

Virgil "Gus" Grissom had been named to command the first manned GT-3 flight, perhaps as a result of his personal rededication resulting from the "bad press" he received when his "Liberty Bell 7" *Mercury* capsule sank during the recovery operation. No one could deny that his engineering expertise and drive set the standard for ensuring that the first *Gemini* flight would succeed. Because of its tight confines and the fact that Grissom, being the smallest of the astronauts, was the only one who really fit into it, the spacecraft was occasionally referred to by the other astronauts as the "Gusmobile." As for a spacecraft name, Grissom selected "Molly Brown." It was from the title of a Broadway show—*The Unsinkable Molly Brown*—which gave reference to his unfortunate episode with his *Mercury* capsule. NASA did not like it, but Grissom somewhat lightheartedly suggested an alternate name of "*Titanic*." As a result of the situation, Molly Brown would be the last spacecraft to carry a name other than its official designation for the next four years.

Grissom's right seat "pilot" for the mission was the rookie John Young. Neither was particularly talkative, avoided the press, and tended to business. They suited each other and made a good team for the tight confines of *Gemini*. The astronauts had pressed NASA management to make the flight an open-ended affair with a target of perhaps four days. However, NASA was still in the conservative mode, and three orbits were sufficient to prove the basic systems. One aspect that was a strong indicator of how smoothly things were starting to fall into place in the American space program was the move of the flight from April forward into March.

Thus, it was just three days after Belyayev and Leonov had emerged from the dense forests of northern Russia that the 340,000 pound GT-3 rocket came to life on the morning of March 23, 1965. At T+151 seconds the first stage had exhausted its propellants and was shed as the second stage continued to accelerate the space ship until T+335 seconds when the orbital velocity of 17,600 mph was reached. Twenty seconds later, explosive bolts separated Molly Brown from the spent second stage, and Grissom fired the Orbital Attitude and Maneuvering System (OAMS—pronounced ohms) unit for the first time to move the spacecraft comfortably ahead of the now tumbling second stage which was also in orbit.

At the end of the first orbit, Grissom again fired the OAMS for 74 seconds to slow the spacecraft and lower its apogee by 34 miles. The historic event caused a few smiles across the faces of the controllers hunched over their consoles at Mission Control (MC). America had achieved a major 'first' in space—a manned satellite had changed its own orbit. Two more maneuvers were accomplished with the last one lowering the perigee so that *Gemini* 3 would reenter the atmosphere even if the retrorockets failed to fire. The OAMS unit was then jettisoned, and the retro-rocket sequence began. Both astronauts later reported that the "kick" from the 1,200 pound thrust units was much more than they had expected. The retrorocket adapter unit was likewise jettisoned, and *Gemini* 3 began its plunge back into the atmosphere.

As the spacecraft began to heat from the atmospheric friction, an ionized sheath of plasma enveloped it, cutting off communications with the ground. Grissom "flew" the spacecraft using computerized directions for pitch and roll. This in turn used the lift from the offset center of gravity to provide for some control over where it would land. It became immediately apparent, as the plot boards at Mission Control showed, that the spacecraft was not generating the predicted lift and would land some 50 miles short of its designated landing area.

The next surprise came when the spacecraft, descending on its parachute, transitioned from its vertical orientation to the "almost" horizontal position. The pitch forward was much more intense than either astronaut expected, and, perhaps because they had not "snugged" the restraining belts, they were thrown forward into the windows. The helmet faceplate of Grissom's was badly broken while Young's was deeply scratched.

During the mission review, virtually every one was satisfied with the performance of the spacecraft and its systems. Only the fuel cells, which continued to exhibit problems in their development, were not aboard. GT-3 flew with only batteries, as would GT-4 because its expected duration of four days was within the capabilities of the battery systems.

The technology embodied within *Gemini* was indeed a generation ahead of its Soviet contemporary, *Soyuz*, which would not fly for another two years. Moreover, when it did fly, it would require several years of upgrading to perform to the level of Gemini. Despite the weight lifting disparity of the Titan booster (10,000 pounds as opposed to the R-7's 14,000 pounds), miniaturization of spacecraft systems and its advanced engineering allowed the Americans to fly a truly capable machine. The only negative aspect of the flight was a corned beef sandwich smuggled aboard by Wally Schirra as another running gag among the astronauts. It brought the ire of management and a stern reprimand that "unauthorized" equipment or food would not be tolerated.

Gemini IV: America's First Walk in Space

Planning for GT-4 was in its advanced stages when GT-3 was hoisted aboard the aircraft carrier Intrepid. But the tentative EVA, however, was not a sure thing. There were several items that still needed to be qualified. The suit itself (called the G4C) would be a modified version of G3C used on GT-3. It would provide for more air circulation, and have two nylon micrometeoroid protection layers and two additional faceplates to reduce the intensity of the sun by 88 percent. Even politics were playing a role. There was some reluctance to perform the EVA at this early date as it might seem that the U.S. was reacting to Leonov's space walk from *Voskhod* II, when in fact the reverse was the case. All of the required tests and validations were completed by mid-May, and the EVA remained in the flight plan.

The biggest difference in the American space walk was that astronaut Ed White, who would fly right seat in GT-4, would have a hand held maneuvering unit to help him move in space. The small device, powered by two small oxygen cylinders, would allow measured thrust based on how hard White squeezed the trigger.

The biggest unknown in the GT-4 mission was the four-day duration itself—fully three times the flight of Cooper's 34 hour MA-6. What data the Soviets had published on the effects of long duration flights (*Vostok* 5 had the longest with just under five days) revealed that the body appeared to exhibit mineral loss in the bone as well as possible cardiac problems and the unpredictable disorientation of space sickness. The medical teams would keep a close watch on the biosensors as well as reports from the astronauts themselves. To help mitigate the symptoms, each astronaut would use a bungee to exercise frequently (as much as could be done in the tight confines of the cockpit) during the flight to stimulate the metabolic actions of the body.

The duration of GT-4 would also require the first real meal planning that consisted of reconstituted freeze dried and dehydrated foods that would provide about 2,500 calories a day. Of course, what goes in must come out, and GT-4 would be the first U.S. flight that had to deal with solid human waste. Plastic bags, which contained a biological neutralizing agent and an adhesive opening, would be placed on the buttocks for defecation. After the event, the bags would be sealed and the content "kneaded" to mix the neutralizer with the feces before storage. How easy this whole process would be in the weightlessness of space and the small volume of the spacecraft was yet to be discovered.

After its launch from the Cape, GT-4 would be the first flight where control of the mission would be passed to the Manned Spacecraft Center (MSC) in Houston. It would also be the first time that a scheduled round-the-clock Mission Control would be manned by three different sets of controllers known as the "red, white and blue" teams. Likewise, a Flight Director, whose prominence would become more visible to the public as the missions advanced in complexity and duration over the course of *Gemini* and then *Apollo* programs, would head each team. Many Americans would come to know the names Chris Kraft, Eugene Krantz and John Hodge, by way of the media reporting of Jules Bergman, Walter Cronkite, and Roy Neal. GT-4 would be the first launch telecast to Europe "live" by way of the *Early Bird* communication satellite that had been placed in orbit the preceding year. Moreover, the "voice of *Gemini* Control," was now assumed by NASA's Public Affairs Officer, Paul Haney, while Lt. Col. John "Shorty" Powers, who had provided the "voice of Mercury Control" for Project Mercury, faded into obscurity.

Two Air Force rookies were selected for the crew of GT-4; Captain James McDivitt would command the mission while Captain Ed White would fly right seat and perform the EVA. The countdown and liftoff on June 3, 1965 was flawless, and within minutes the astronauts were in orbit and observing the second stage tumbling slowly behind them at a distance estimated to be about 200 yards. As noted on the previous flight, it appeared that it might be an easy task to move closer to the Titan's second stage to exercise the maneuverability of the OAMS system, and this became the first objective of their schedule. However, the relative movement of two objects in slightly different orbits was much more complex than it intuitively appeared, and, with each firing of the OAMS, the elusive second stage actually seemed to move away from them. It was determined that, even at these close distances, they would have to use the computer to make the appropriate orbital changes, and the chase ended because too much OAMS fuel was being used. It was time to move on to the EVA.

Ed White began to prepare his suit by going through the checklist that included attaching the 15-foot tether that also provided the oxygen supply. McDivitt established the proper attitude and began the depressurization of the spacecraft. This first increment was to reduce the cabin's ambient pressure of 250 mm/Hg (one third of sea level pressure) to about 195 mm/Hg. If there were any leaks in White's suit, it would cause the cabin pressure to increase. There was no change. The next step was the complete spacecraft depressurization and opening of the hatch, which took place as the crew came into communication with the Hawaii tracking station.

White mounted a 16 mm camera to record the activities (there was no room or weight allocation for a TV camera). Twelve minutes after the hatch opened, White stated, *"O.K. I'm separating from the spacecraft."* As he proceeded to float out, he aligned the maneuvering unit with his center of gravity and pulled the trigger. The response was immediate and positive. He experimented with various power settings and thrust durations for the next few minutes. In the 1950 science fiction movie *Destination Moon*, a crew member used a tank of compressed gas to maneuver in space to rescue another crew member; science fiction had become reality in the short space of ten years.

With the maneuvering unit fuel exhausted, White began to describe the sensations. *"CAPCOM it was very easy to maneuver with the gun... this is the greatest experience...There is absolutely no disorientation associated with it."* As White started to move back to the spacecraft, the situation changed. Now he had to use the tether to move himself, and he found that his tugs and pulls sent him off at odd angles. Management at the Manned Spacecraft Center in Houston (MSC), began to get a little uneasy, and CAPCOM (Capsule Communicator) relayed that anxiety with the message: *"The Flight Director says get back in!"* There was some thought that perhaps White had become mesmerized by the experience, but that was not the case; it was just difficult to get himself back to the hatch.

White began the hatch closure process but quickly determined that he could not exert any pressure on the hatch lever in the weightless environment. McDivitt had to hold White down, and it took three hands to complete the closure. In the process, White's heart rate, which had risen to 155 beats per minutes during the EVA, jumped up to 180. Now the Americans had learned what the Soviets had come to understand—movement in space required planning and handholds.

The two astronauts proceeded to move through the reminder of the mission without any further problems, although it became obvious that staggered sleep periods did not work out very well since any communications or thruster activity tended to wake the sleeping partner.

A failure of the on-board computer required that the spacecraft fly a ballistic path during its reentry, resulting in slightly higher G loads. Other than some loss of bone mass and depleted blood plasma volume, the two astronauts came through the mission in good physical condition.

Gemini V: Fuel Cell Problems

Planning moved forward for *Gemini V*, the first spacecraft to use the troublesome fuel cells needed to generate electricity for the eight-day flight. The duration was determined to coincide with the time it would take to fly to the Moon, spend a day on the surface, and return. The significance was an important milestone in the upcoming *Apollo* program. Its schedule also represented an interesting departure from the previous routine. In the past, it was the hardware that had determined the schedule. With the *Gemini* and *Titan* coming together with much smoother precision and timing than the former *Mercury Atlas*, now the astronauts would slow the schedule. Because of the inability of the *Gemini* V crew to get sufficient training in the simulators, Chief Astronaut "Deke" Slayton had requested a two-week delay to the standard two-month launch cycle.

Also of some significance was the announcement of a fourth group of six astronauts, a group in which the primary consideration had moved from being a pilot and engineer to being a scientist. It was the first group in which being a pilot was not even a consideration. The selection was made from a list of 400 candidates supplied by the National Academy of Sciences.

In addition to the long duration flight, *Gemini* V astronauts Gordon Cooper and rookie Pete Conrad would practice the first attempts at orbital rendezvous using a Radar Evaluation Pod (REP) carried along in the equipment bay and released in orbit. In an effort to reduce the number of *Agena*s (and their associated cost), the REP would simulate the *Agena*'s radar response. *Gemini* would change its orbit and return to the REP to simulate the rendezvous technique using its on-board radar and a new computer. The range and rate data from the radar would be analyzed by the computer (which held the relative positional data of the spacecraft) and would then provide the appropriate spacecraft pitch and yaw attitude and the OAMS burn time to accomplish a rendezvous. Of course, a series of other scientific experiments and observations were scheduled that would keep the duo occupied for the eight-day mission.

A problem with the fuel cells and then a thunderstorm canceled the launch at the T-10 minute mark on August 19, 1965. Recycled to Saturday August 21, the launch occurred precisely at the scheduled 9:00 A.M. time. The ride itself was a bit disturbing to the two astronauts as the *Titan* exhibited some significant pogo effects (lateral oscillations). It was quickly determined that an incorrect control parameter in the guidance system had allowed the oscillations.

Within a half hour of achieving orbit, the pressure in the fuel cell oxygen tank began to drop. At first, it was not critical since the fuel cells actually operated better at the lower limits. When that lower limit was exceeded, Conrad attempted to increase the pressure by electrically heating the oxygen to get it to produce more gas. There was no response. The heating unit had failed, and the pressure continued to drop. Without the proper pressure, the cell would not develop its rated electrical power.

The planned release of the REP took place on the second orbit, and the *Gemini* radar successfully interrogated its radar transponder. However, before any rendezvous maneuvers could take place, the oxygen pressure fell to the point where it was producing little electrical power. At that point, activities with the REP were canceled, and there was talk of returning the astronauts early. There was no real concern for safety because the re-entry battery power, which had not yet been activated, was sufficient for their return should that be necessary, and the batteries themselves would provide power for approximately 13 hours.

An effort was now made to save the mission. Some believed that the fuel cells might continue to operate at the low pressures and, with enough oxygen "boil-off," could return to normal operation without the heaters. The mission controllers began to work through the electrical systems to turn off all unnecessary equipment—at the very least to reach the sixth orbit contingency recovery area.

As the oxygen pressure continued to fall, it reached a level from which it was not expected to make any recovery, and Cooper and Conrad began to prepare for a reentry. Mission Control was still determined to save the long duration aspect of the mission, and when the pressure appeared to stabilize, *Gemini* V was given a go for at least completing a day in orbit. The fuel cell was the critical part to subsequent *Gemini* and *Apollo* missions. NASA was determined to find a solution to the problem if at all possible—especially since the fuel cells were located in a part of the spacecraft that would not be returned for examination.

Over the next several orbits, ambient heat provided a minute but perceptible increase in the oxygen pressure, and, by the beginning of the second day, the mood had changed to one of guarded optimism. The REP module's batteries ran down, and that very important aspect of the mission was no longer possible. The Soviet news agency TASS accused the United States of jeopardizing the lives of the astronauts for the sake of beating the Soviet's long duration five-day space flight.

Politics aside, astronaut "Buzz" Aldrin (who had yet to fly) had done his doctorial thesis on orbital rendezvous, and he worked with Mission Control to establish a series of OAMS maneuvers to intercept an imaginary satellite, now that the REP module was no longer available. These exercises worked very effectively but a "live" rendezvous had not actually been accomplished. The fuel cells continued their slow rebound, and the mission continued on a day-to-day basis with the pressure returning to about one third of its intended value and consuming about one third of its oxygen. The thrill of flying in space wore thin for the astronauts in an environment of empty wrappers and equipment floating around in the small cockpit, accompanied by unpleasant odors that were too intense to be removed by the activated-charcoal filters.

When the last day of the mission finally "dawned" (a figure of speech when dealing with a satellite in low-Earth orbit), both astronauts were eager for the return to Earth. The medical team was very apprehensive about their ability to withstand the high G forces after such a long period in space. A series of data-entry errors again resulted in the spacecraft missing the recovery area by an embarrassing 100 miles. However, within 90 minutes Cooper and Conrad were plucked from the ocean, and a quick physical showed no more than the expected relative low levels of metabolic deterioration.

Gemini V represented another figurative turning point in the space race. The long duration (finally exceeding the Soviet record), coupled with the rendezvous capability and the use of the temperamental fuel cells, set a new level of technology. It would be almost a decade before the Soviets would exceed the Americans in space flight duration, and by then it would no longer be of major consequence.

Spirit of 76: Courage on Pad 19

Astronauts Wally Schirra and rookie Tom Stafford were assigned to *Gemini* VI; it was to be the first to rendezvous with a live *Agena*. As such, it would be a short, two-day flight and did not have the fuel cells aboard. As the two lay on their backs during the GT-6 countdown on October 25, 1965, the *Atlas-Agena* roared off a nearby pad. After the *Agena* came around the Earth following its first orbit, GT-6 would launch and rendezvous with it using the coelliptic method of the Hohmann Transfer. All went well until five minutes into the flight when the *Agena* D separated and initiated its firing sequence. Then all telemetry was lost. It is believed the *Agena* was destroyed in an explosion. The GT-6 flight was postponed. Not only would it be some time before a review board could determine the cause of the failure but there was also no *Agena* immediately available for a back up.

NASA was in a quandary as to how to proceed with the *Gemini* program. Then a surprising suggestion was made: unstack *Gemini* VI from Pad 19 and launch *Gemini* VII on its planned two-week flight, immediately restack *Gemini* VI and launch while *Gemini* VII is still in orbit, and have the two rendezvous. The plan was called "Spirit of 76"—reflecting the reversed launch order as well as a play on the patriotic aspect of the slogan.. It would require a change in how data from the spacecraft was returned to Houston. Although the Soviets had flown multiple manned missions, NASA had not contemplated having two manned spacecraft in space at the same time until *Apollo*, when the Command Module and the Lunar Module would present a similar problem.

With the alternative being a two month delay in proving the ability to effect orbital rendezvous, and the possibility of being upstaged once again by the Soviets, the pressure was on to see if the dual flight plan could really be accomplished. One of the few hardware changes would be the installation of the *Agena* radar transponder in *Gemini* VII. When it was established that the concept was sound and doable, the change in the program was sent to President Lyndon Johnson to allow him to make the plan public, which he did on October 28th. *Gemini* VI was unstacked the next day.

The launch of *Gemini* VII occurred on December 4, 1965—its objective was to acquire valuable data on the effect of much longer flights on the human physiology. The spacesuit is uncomfortable to wear, even in the weightless condition of space where there are no "pressure points" on the body. To ease the burden of discomfort, rookies Frank Borman and Jim Lovell wore new light-weight spacesuits that were capable of being removed, even in the tight confines of the *Gemini* cockpit—but mission rules required that one astronaut must be suited at all times.

Following orbital insertion, *Gemini* VII was able to perform "station keeping" with the expended second stage, using newly developed procedures that allowed the spacecraft and the second stage to move to within 50 feet of each other. Following that exercise, the astronauts raised their orbit and began a more leisurely pace of scientific and engineering observation and experimentation for the next 14 days. Activity on the ground, however, was much more intense, and Pad 19 was ready to receive GT-6 the day after the GT-7 launch.

The erection of GT-6 was smooth and without any significant problems, and the countdown proceeded to a launch on December 12. As the count reached the magic "zero" mark, the turbo-pump began its characteristic start-up whine, and the flame emerged from beneath the rocket. Almost immediately, the two *Titan* engines shut down. Mission Control made the first call indicating the situation: *"We have a shutdown, Gemini VI."* As there had apparently been some motion in the booster, a call from the spacecraft indicated, *"My clock has started."* Flight Director Chris Kraft countered with *"No lift-off, no lift-off."* The astronauts themselves could not see what was happening, and there was anticipation that one or both would pull the "D" ring that would eject them clear of what threatened to be a fireball. Their discipline held, and Schirra later commented that his distrust of the ejection system was such that it would have taken more than a premature shutdown before he would commit himself to using the "escape of last resort."

The astronauts were now sitting on over 300,000 pounds of volatile toxic chemicals, and some programmed sequential post launch events had begun with the start of the mission clock aboard the spacecraft. Mission Control and the blockhouse team immediately moved into action to "safe" the rocket. Gunter Wendt, the NASA technician who had secured all the previous astronauts in their spacecraft, positioned himself and his team to begin a hasty egress of Schirra and Stafford as soon as the booster had been declared "safe." More than an hour and a half would elapse before they would descend from the spacecraft. The thoughtful and somewhat courageous actions of the astronauts in not ejecting essentially saved the mission. Had they ejected, the spacecraft would have effectively been destroyed.

First indications were that one or more sensors had detected a slight decay in thrust after engine start, but it took a full day of searching before a dust cap was discovered in a virtually inaccessible area of the turbo-pump gas generators. It should have been removed during the booster assembly process at the factory. Was there time to salvage the mission? Following an assessment of the situation, it was determined that GT-6 could be recycled for three days hence—December 15.

The launch occurred on the designated day at the specified time. *Gemini* VI was in orbit some 1,250 miles behind *Gemini* VII, but in a slightly lower orbit (100 X 170 Miles) that would allow the astronauts to catch their target. The first maneuver, a slight phase change to raise the perigee, occurred 2 hours and 18 minutes into the flight when the two spacecraft were about 450 miles apart. This was followed by a plane adjustment of .007 degrees by yawing the spacecraft 90 degrees to the orbital track and firing the OAMS for 40 seconds.

Several small corrections now put the two spacecraft in a position where the *Gemini* VI radar could lock onto the transponder carried in *Gemini* VII, which it did at 271 miles. The computer called for another burn at 3 hours and 47 minutes into the flight, using the aft thrusters for 54 seconds. The computer now switched to the final rendezvous mode. As they crossed the African continent, the final burn was made while both spacecraft were still in darkness. The radar called the distance as less than a quarter of a mile and closing at about 6 feet per second. The spacecraft continued to move closer until they were separated by only 100 feet; it was 5 hours and 56 minutes into the flight. Schirra cancelled the closing rate with another small OAMS burn, and the two spacecraft began station keeping only yards apart. The first rendezvous in space had been achieved-the Moon was another step closer.

There was a lot of celebrating in Mission Control as the astronauts traded comments about the long beards on the occupants of *Gemini* VII. The crews took turns maneuvering about one another—station keeping. They noted on the rear of each craft, the trailing adhesive tape that had been used to seal the separation area between the spacecraft and the booster to keep dust and debris from entering the small opening while it sat on the launch pad. Following the maneuvers, *Gemini* VI used its OAMS to separate the two orbits to ensure they would not inadvertently run in to each other during the sleep period.

The following day, as *Gemini* VI prepared for reentry, astronaut Stafford gave a report that had Flight Director Chris Craft sitting upright: *"This is Gemini VI. We have an object, looks like a satellite moving from north to south, up in a polar orbit. He's in a very low trajectory... looks like he may be going to reenter very soon."* By now, all of Mission Control had their attention riveted on the report. *"Stand by... it looks like he is trying to signal us."* At this point Schirra played "Jingle Bells" on his harmonica, and Stafford added the bells—it was little more than a week until Christmas. Another "gotcha" by the playful astronauts!

The reentry went well. With the corrected lift-to-drag ratio now programmed into the computer, the spacecraft landed within 8 miles of the recovery forces. It had been a short and very successful flight for the crew of *Gemini* VI. But the spirits of the *Gemini* VII crew, temporarily buoyed by the rendezvous and banter with *Gemini* VI, sank a bit as they had to return to their tedious and somewhat uncomfortable mission that had two days to completion. It was obvious that long duration space flights would be as much a psychological as physical problem, when living in tightly confined spaces was factored into the equation. Toward the end of day thirteen, the task of the astronauts became one of cleaning up the spacecraft, putting their spacesuits back on once again, and preparing for their return to Earth, which was uneventful.

1965 was a year that saw the end of Soviet dominance in space. While it could not be recognized at the time, the overall capabilities exhibited by the two sparing nations began to show significant differences. With the successful lunar and planetary probes, as well as flying the new second-generation two-man spacecraft, America's vast technological infrastructure was steadily showing its supremacy. The initial five manned *Gemini* spacecraft, launched over a period of just eight months, had effectively achieved two of the three primary objectives (EVA and rendezvous). The third and critical docking exercise was yet to be achieved, along with the testing of different rendezvous modes and EVA equipment. These would be demonstrated in the coming year, while the Soviet's second-generation spacecraft, *Soyuz*, was still two years from its first flight.

The Untimely Demise of the Chief Designer

If the early successes of the Soviet space program infused life into Korolev and his team, the long string of failures in the lunar and planetary programs in the years between 1962 and 1965 created an air of despair for this talented group of people. However, it was undoubtedly the inability to fly the second-generation *Soyuz* in a time frame to keep the Soviet program ahead of the American's that seemed to have the most negative affect. As for Korolev himself, the emotional impact of these years had taken a significant toll on the man. He had lost his ability to work effectively with many members of his own team, and his physical health, impaired by the severe years of Gulag incarceration, was now aggravated by his incessant long hours of work and the intense conflict among the other Chief Designers.

Low blood pressure and a serious heart condition continued to take its toll, and he realized that he was in grave trouble. This was evidenced by a comment in a letter to his wife in late 1965 in which he stated, *"I am in a constant state of utter exhaustion and stress... I am holding myself together using all the strength at my command."* These words are from a man who was only a few weeks shy of his 59th birthday.

In mid-December, he underwent several medical tests which indicated a bleeding polyp in the intestines. The simple surgery was scheduled for January 14, 1966, and he expected only a few days in the hospital. During the operation it was discovered that he had a large malignant tumor which was removed with great difficulty and with much bleeding. Following a series of complications, aggravated by the lack of a competent medical staff, Korolev's heart failed, and the great man succumbed.

While some in the Soviet space program, such as his rival Glushko, were not saddened to see his demise, most of the Soviet scientific community who understood the role he had played in the advancement of Soviet rocketry, felt a keen loss. Soviet Party First Secretary Leonid Brezhnev, in approving a state funeral, also allowed Korolev's name to be revealed as "the Chief Designer," although most in the West still had little or no idea of the true significance of the role he had played.

Korolev's legacy was the *R-7*. It was created and perfected in a remarkably short time, and its effect on mankind's move into space, as well as the relationship between the two most powerful countries the world had ever seen, was profound. The fact that it remained the only manned space booster in the Soviet inventory for over 50 years says much about its engineering as well as the inability of the Soviet system to provide a viable replacement for either the man or the machine.

A top priority for the Soviet Space program was to find a replacement for Korolev. In reality, there was no one who had the intricate understanding of the workings of his design bureau. No one had his set of relationships (some of which were admittedly in disarray), and certainly no one had the ability to generate a vision with which others could readily identify. His staff was leery of outsiders and promptly requested that Vasiliy Mishin, Korolev's deputy, be appointed. His lack of people skills and political connections caused some to oppose him. However, his engineering abilities and knowledge of the existing program structure and technology base provided assurance of a degree of continuity for Korolev's work in progress, and Mishin received the appointment, but a critical five months elapsed. As with the other Chief Designers, Mishin's status would remain a state secret for years to come.

As one of his first priorities, Mishin undertook a review of the manned space program and *Voskhod* in particular. Three more flights had been planned at that point and, in fact, had been delayed from February 1966 by a series of failures in various subsystems. The twenty-two day flight in March of Kosmos-110 (a *Voskhod* type spacecraft) carrying two dogs was supposed to pave the way for *Voskhod*

Sergey Korolev, Chief Designer of the Soviet Space Program

III, but the poor condition of the dogs and problems with several life support systems caused the manned flight to be postponed.

America's success with *Gemini* loomed large, and many in the *Voskhod* program were reluctant to proceed with any more flights but wanted to move on to *Soyuz*—even if it meant a delay of 6-8 months. The only real milestone that *Voskhod* might salvage was a long duration flight of perhaps three weeks and higher apogees (altitudes) of 600 miles. The world now recognized America's ability to rendezvous, and adding longer duration flights would not impress even the media, which was becoming more knowledgeable of the various technologies. As a result, further *Voskhod* flights were canceled—additional evidence that the Soviets were now reacting to America's lead.

As 1966 drew to a close, the Soviets had finally settled on three primary projects designed to retain the lead in space technology and to continue their efforts to convince the uncommitted nations that Communism was the wave of the future. These projects consisted of the three-man *Soyuz* spacecraft with the ability to rendezvous and dock with another *Soyuz* (7K-OK), the *UR-500* circumlunar flight (*UR-500K-L1*), and the *N1* for manned lunar landing—possibly using both EOR and LOR techniques (*N1-L3*). Rather than develop a new spacecraft for the lunar expeditions, a modified *Soyuz* (less its living compartment) would be used for both the *UR-500* circumlunar flight and the *N1* lunar landing. The latter required the development of a lunar lander.

Gemini VIII: Emergency in Space

Gemini VIII, with rookies Neil Armstrong and David Scott, sought to achieve the elusive docking milestone and test a sophisticated manned maneuvering unit with an ambitious EVA. It was scheduled for a mid-March 1966 launch. The various simulators were shared with the crew for the following *Gemini* IX mission. Astronauts Elliot See and Charles Bassett flew up from Houston to St. Louis to begin their work with the docking simulator since their mission would involve the first restart of the *Agena*. As their T-38 approached Lambert Field where the McDonnell plant was located, poor weather dictated an instrument approach to minimums. The T-38, a small, two-seat, twin-engine, high-performance-jet trainer, was used by NASA for the personal transportation and flight proficiency of the astronauts.

As their aircraft broke out of the low overcast, See recognized that they were not in a good position to continue the approach to touchdown. Inexplicably, he executed a sharp right turn at low altitude apparently to perform a circle-to-land maneuver instead of climbing straight out as the missed approach procedure called for. With the afterburners roaring, an indication of See's recognition of the critical situation, the aircraft clipped the building in which their spacecraft was completing its final assembly and crashed, killing both astronauts—the only fatalities of the accident. Astronauts Tom Stafford and Eugene Cernan, who had accompanied See and Bassett in a second T-38, landed safely and would eventually be named as the prime crew for *Gemini* IX. The hazards of being an astronaut were not limited to the space missions as would be demonstrated again within a year.

The *Gemini* VIII *Atlas Agena* target successfully achieved orbit on March 16 with astronauts Armstrong and Scott following 101 minutes later on GT-8. Using the basic coelliptic rendezvous technique, the *Gemini* twins (Armstrong and Scott) were station keeping a few yards from the *Agena* six hours later and preparing to perform the historic first docking. The Target Docking Adapter (TDA), located in the nose of the *Agena*, consisted of a cone into which the nose of the *Gemini* would be inserted.

Following an intense scrutiny of the *Agena* systems to verify that all was well, Armstrong lined up with the TDA and began to move *Gemini* into it at a speed of about 8 inches per second. The spacecraft were still on the night side of the Earth, with a variety of position and status lights illuminated on the *Agena*, when the two spacecraft made contact, and the mooring latches engaged. A set of hydraulic dampers acted as shock absorbers to soak up the impact of the *Gemini*. Small electric motors came to life and pulled the *Gemini* into what is referred to as a "hard dock" configuration. This allowed the two spacecraft to achieve a relatively rigid assembly. Armstrong announced *"Flight, we are docked"* to confirm the telemetry indications that Mission Control was seeing on the ground. The last of the three critical aspects of rendezvous had been achieved—rendezvous, station keeping, and now docking. The ability to link two independent spacecraft (docking) was the last unproven aspect that was the cornerstone of the Lunar Orbit Rendezvous (LOR) technique that America had gambled would take them to the Moon *"before this decade is out"*.

The various systems of the two spacecraft were now electrically integrated to allow the astronauts to issue a variety of commands. Either the *Gemini* spacecraft or the *Agena* could control the attitude of the docked assemblies. Looking out the windows, the astronauts could see a set of illuminated instruments positioned on the *Agena* to provide direct readings of various conditions. All seemed to go smoothly.

Within 30 minutes of the docking, a slow yawing (left turning) of the assembly was noted. Scott decided to switch off the *Agena*'s attitude control only to find that the yaw and roll moments to the vehicle increased. For some reason both astronauts felt that the anomaly was being input by the *Agena* and thought that their efforts to disengage the *Agena* control had failed. Soon the movement was so severe that they realized they had to disconnect from the *Agena* for fear the forces being experienced would tear the two assemblies apart. They were fortunate that the undocking process was quickly and effectively achieved without ramming the two together. However, with the *Gemini* spacecraft now freed from the *Agena*, its roll and yaw rate increased rapidly, making it obvious that the problem was with their spacecraft and not the *Agena*. The roll rate of the *Gemini* and its occupants reached 360 degrees per second. As a last resort, they systematically began to shut down as much of the OAMS system as possible and activated the reentry attitude control. Their wild ride was finally brought under control.

It was conceivable that the motion rates Armstrong and Scott had been subjected to may have significantly impaired their ability to cope with the problem. They were fortunate to have recovered from an apparent electrical short in the number eight thruster as soon as they had. As the event had occurred when the spacecraft was out of direct communication with mission control, the appearance of two separate craft to the tracking ship *Coastal Sentry Quebec*, on station in the Pacific, presented a puzzle that discussions with the crew soon resolved. With the activation of the reentry control system, the mission rules dictated that the flight be terminated at the earliest opportunity. The crew then began preparing for reentry.

Over the next several weeks, a review of the incident caused the recovered spacecraft (and those yet to be launched) to be carefully examined. To virtually everyone's surprise, there were several areas where electrical shorts were possible. A further inspection of the yet-to-be-launched *Gemini* IX also revealed many areas that could present problems. It was obvious to NASA and McDonnell Aircraft that the quality of workmanship and design may have been sacrificed to meet deadlines, and quality had to be improved to avoid a future disaster. Within the year, similar poor design and workmanship by yet another contractor would have devastating consequences.

Gemini IX: The Angry Alligator

The *Gemini* IX mission with the backup crew of Tom Stafford and Eugene Cernan (who replaced See and Bassett) prepared for their launch on May 17, 1965. However, for the second time the target *Agena* failed to achieve orbit (the *Gemini* VI target had been the first). On this occasion the problem was with the *Atlas*. The GT-9 launch was scrubbed, and an alternative target for rendezvous was prepared.

Several months earlier, McDonnell Aircraft was tasked by NASA to prepare a simple docking collar that would be launched by the *Atlas*, as it was recognized that the *Agena* could be a problem. This inexpensive stand-in (6 million dollars as opposed to 9 million dollars) would not provide any functions other than attitude control and the ability for the *Gemini* spacecraft to dock with the assembly. The satellite was called the Augmented Target Docking Adapter (ATDA), and NASA decided to turn to it to allow the remainder of the *Gemini* IX experiments, particularly a complex EVA, to be flown. There were still three more missions, and *Gemini* IX did not really need the ability to fire-up the *Agena*.

The back-up plan was put into action, and an *Atlas* with the 2,200 pound ATDA was launched successfully on June 1st. Now another problem clouded the mission. Telemetry indicated a problem with the protective shroud (nose cone). Consequently, it was decided to launch *Gemini*, as there was no way of confirming the specific problem.

Stafford and Cernan followed the launch of the ATDA two days later on June 3rd. Stafford, having flown on *Gemini* VI, became the first person to have a second Gemini space flight. The rendezvous used a slightly different technique that required only three orbits (termed M=3) instead of four. On joining up with the ATDA, it was obvious that there was a problem with the shroud; the rear stainless steel band was still in place although the forward portion had separated. Stafford characterized the appearance to Mission Control as looking like an "angry alligator." There was some thought of dislodging it by bumping it with *Gemini*, but it was decided that the risk was too high.

Three rendezvous maneuvers were accomplished during the first day, and then *Gemini* moved to another orbit to ensure there was no conflict with the ATDA during the sleep period. The second day was devoted to the planned EVA, the second in the U.S. program and following Ed White's space walk by one year. The primary objective was to test an Astronaut Maneuvering Unit (AMU) that would allow Cernan to become a separate satellite himself. The first part of the EVA went well since it was comprised of relatively simple tasks that did not involve the AMU, which was stored in the back of the equipment adapter.

Following an initial 30-minute work period, Cernan took a planned rest and then proceeded to the equipment bay to get into the AMU. He quickly discovered that the plan for strapping on the backpack AMU was flawed. A new material called Velcro had been used extensively to provide the astronaut with the ability to get some leverage, but this proved insufficient. Without adequate handholds, it was almost impossible to prepare the AMU or position himself to attach to it. As his pulse soared to

180 beats per minute and respiration accelerated, the small chest pack that provided the spacesuit environment was unable to keep the water vapor and temperature within the established tolerances, and his faceplate began to fog over.

At about 90 minutes into the EVA, Cernan and Stafford decided that it was futile to continue the exercise. Cernan returned to the spacecraft in what appeared to be an almost completely flawed set of tasks. Following another sleep period, *Gemini* IX was brought back to Earth. If there was any redeeming value to the mission, it was that the new reentry program put the spacecraft down within a mile of the recovery ships. Examination of the photos taken of the "angry alligator" ATDA showed that the two electrical leads that should have been connected to activate the separation of the shroud were taped neatly to the side of the ATDA. The technician who was responsible to connect them had been suddenly called away to attend to his wife who was giving birth, and the person assigned to close out the pre-launch procedures didn't know what to do with them.

Gemini X: Firing up the Agena

Sometimes failure breeds success. With the knowledge Cernan gained about what would be of value to help in movement during an EVA, the next missions would be much better prepared—*Gemini* X would be extraordinary.

On July 16, 1966, the next scheduled *Atlas Agena* successfully achieved orbit, followed by *Gemini* X with John Young and Mike Collins 101 minutes later. A late afternoon schedule was dictated by a backup plan should the *Gemini Agena Target Vehicle* (GATV) 5005 not achieve orbit. *Agena* 5003, from the aborted *Gemini* VIII the preceding March, had been moved to a higher orbit and was still available should it be needed. The initial rendezvous (M=4) with *Agena* 5005 did not go well due to a variety of factors including the inability to get good navigation sightings, but fortunately the spacecraft had 50 percent more OAMS fuel than previous flights. Performance enhancements to the *Titan* and the availability of space in the equipment adapter allowed for considerable growth potential. The two spacecraft were docked within six hours of liftoff and ready to perform the last major objective of the *Gemini* program—a burn of the *Agena* engine to boost the astronauts into a higher orbit.

With the *Gemini* firmly latched into the docking collar on the front end of the *Agena*, the astronauts were facing backward. It would be the first time this configuration had ever been attempted because of the normal constraints of G forces. Facing backward, Collins and Young would experience only about one negative G (sometimes called "eyeballs out") because the weight of the combined spacecraft was about 14,000 pounds and the Primary Propulsion System (PPS) of the *Agena* was 15,000 pounds of thrust.

Although their view of the *Agena* was somewhat obscured by the docking collar, John Young recalled vividly the events: *"At first the sensation I got was that there was a "pop" and then a big explosion…We were thrown forward* [restrained by their harnesses]. *Fire and sparks started coming out of the back of that rascal. The light was something fierce."* Their new orbit took them almost twice as high as any human had ever ventured, 476 miles above the Earth, where its curvature was more pronounced.

Following a nine-hour sleep period, the PPS fired again, lowering the orbit to permit a later rendezvous with *Gemini* VIII's *Agena* 5003. Again, Young was impressed with the burn. *"It may be only one G but it's the biggest one G we ever saw."* Mike Collins performed a "Stand-up EVA" which did not involve actually leaving the spacecraft; he simply opened the hatch and floated only a foot or so above his seat to perform some experiments. The EVA was cut short when both astronauts experienced a burning in their eyes that was later traced to the lithium hydroxide that removed the carbon dioxide from their environment. The problem immediately cleared when the hatch was closed and they turned off the spacesuit environment system.

The next objective was to rendezvous with *Agena* 5003. They undocked from their *Agena* and executed several burns of the OAMS that soon put them station keeping with the now inert target. Collins then proceeded to perform a major space walk in which he maneuvered over to the *Agena* and retrieved a micrometeorite package that had now been in space for three months. A hand-held maneuvering unit, similar to what Ed White had used, aided him. He still experienced considerable physical effort to complete the 72-minute EVA. Following a final sleep period, *Gemini* X prepared for a reentry that dropped them within 3 miles of their primary recovery ship.

At this point in America's program, preparation for the first *Apollo* flight (which had now slipped into early 1967) was consuming more of the time of both the astronauts and Mission Control. There was some talk of canceling the last two *Gemini* flights, but again reason prevailed. These last two flights would explore several areas that *Apollo* would have to spend time on; one in particular was an M=1 rendezvous. This scenario would simulate the Lunar Module lift-off from the Moon and the one orbit rendezvous with the *Apollo* Command and Service Module (CSM). Any techniques that could be accomplished and proved with *Gemini* would shorten the time to the lunar landing—the last two flights were approved.

Gemini XI: M=1 Rendezvous

Lift-off of *Gemini* XI with Pete Conrad and Dick Gordon occurred on September 12, 1966, for a first orbit "M=1" rendezvous with its GATV *Agena* 5306. The launch had to occur within seconds of its scheduled time to perform the maneuver successfully. One hour and thirty-four minutes later, *Gemini* XI was docked with its *Agena*. The ability to rendezvous and dock in such a short period of time surprised even a few in NASA, who had their doubts about this most difficult rendezvous. Launch, phasing, plane, apogee and Terminal Phase Initiation tasks all had to be accomplished within a 100-minute span.

The flight was a combination of scientific experiments and observations, several EVA activities, and the use of the *Agena* engine to achieve a higher apogee (853 miles). The new apogee was positioned to avoid the Van Allen radiation belt. Conrad and Gordon exhibited the same thrill with the *Agena* ignition and the view from high above the Earth. *"You can't believe it,"* exclaimed Conrad. *"I can see all the way from one end, around the top... about 150 degrees... the water really stands out and everything looks blue... there is no loss of color and details are extremely good."* At one point, when a rest period occurred during an EVA, both astronauts nodded off while Gordon floated in space.

Like the previous EVAs, however, Gordon also experienced the inability to position himself effectively for the various activities despite more handholds, and one EVA session was ended early due to his fatigued condition. Several important experiments were conducted using a 100-foot tether between the two spacecraft following an undocking.

Positioning the heavier *Agena* between the Earth and spacecraft, the effect of the gravity gradient (as had already been demonstrated with some unmanned satellites) confirmed that the heavier object would remain closer to the Earth, stabilizing the assembly. A second task involved rotating the two around each other to create artificial gravity. Although the rotation was only once every six minutes and the astronauts indicated they could feel no gravity force, unsecured items in the cabin floated away from the rotational center. The reentry on the 44th orbit was conducted solely by the autopilot for the first time and resulted in a splashdown only 3 miles from the recovery force.

Gemini XII: All Objectives Accomplished

The flight plan for *Gemini* XII, which had been constantly revised since astronauts Jim Lovell and "Buzz" Aldrin began training more than five months before lift-off, was still in a state of flux when *Gemini* XI returned to Earth. While all of the primary objectives of the *Gemini* program had been achieved, the inability to perform an effective EVA still eluded NASA.

The Air Force had yet to strap an astronaut into its AMU and was anxious to fly it again. However, NASA was adamant that the basic ability of the crew to do effective work during a space walk had to be addressed. To accomplish this, it was decided to make the EVA the focus of an experiment using a set of "busy boxes" that Aldrin would use to perform a variety of mechanical tasks that included threading and applying torque to bolts, moving switches, and plugging electrical connectors. These types of tasks would be critical to "building" structures in space, as visualized by such luminaries as von Braun, Clarke, and Korolev. To aid in preparing the astronaut, NASA had installed a large swimming pool in Houston designated the "neutral buoyancy facility." By immersing the suited astronaut under water, his movements would closely replicate the weightless environment of space. This proved an invaluable tool for decades to come.

What emerged from the water training simulations was no less than 44 handholds, foot restrains, and tethers that could be positioned to hold an astronaut in a specific place while activities were being performed, without having to use his hands to maintain that position. On November 11, 1966, as Lovell and Aldrin walked to the *Gemini* spacecraft from the elevator that had brought them to the top of the giant *Titan* rocket, each wore a sign on his back with a single word that expressed the finality of the mission—"The" on one and "End" on the other.

The launch of both the *Atlas Agena* and the *Titan* were flawless, and *Gemini* XII began an "M=3," co-elliptical rendezvous. The radar failed when the two ships were still 70 miles apart. This proved not to be a problem since "Buzz" Aldrin (often called Dr. Rendezvous by his cohorts at NASA) used a sextant and navigation charts to complete the task.

An anomaly in the thrust-chamber pressure of the *Agena* PPS during its initial burn into orbit indicated a possible problem, and it was decided not to restart the *Agena* for a planned orbital change to a higher apogee. This cancelled activity allowed another event (which had been previously considered) to take place: the observance of a solar eclipse from space. Little more than seven hours into the mission, Lovell positioned the spacecraft into a slightly modified orbit so that it would fly through the shadow of the Moon as it moved between the Earth and the sun. It was a short, seven-second event that was captured on film for later analysis by astronomers on Earth. A two-hour stand-up EVA was then conducted, followed by a sleep period.

The next day, problems with a fuel cell and an OAMS thruster caused some concern, but these had workarounds, and the long EVA with "busy box" tasks began. Using frequent rest periods, Aldrin's respiration remained around 20 per minute and his heart rate around 100. A repeat of the tether experiment of *Gemini* XI was conducted with a 100-foot tether linking the two spacecraft. The EVA ended after two hours—man could effectively work in space. *Gemini* XII returned after 94 hours which, when added to the previous manned flight, gave the United States a commanding 1,993 hours in space compared to the Soviet's 507.

The *Gemini* program was a resounding success. It had demonstrated man's ability to fly reliably for periods of up to two weeks in the hostile environment of space and to perform useful work there. It had allowed the various techniques of rendezvous, station keeping, and docking to be performed. Moreover, *Gemini* had shown that, with redundant systems and flexible flight plans, equipment failure could have a high degree of tolerance. Finally, *Gemini* had revealed that the technological depth of the American program had overtaken the Soviets. But it had also engendered, in the United States, a degree of complacency with respect to the high risk aspect of their spacecraft design—the pure-oxygen environment. Within months, this would almost cripple the American lunar program.

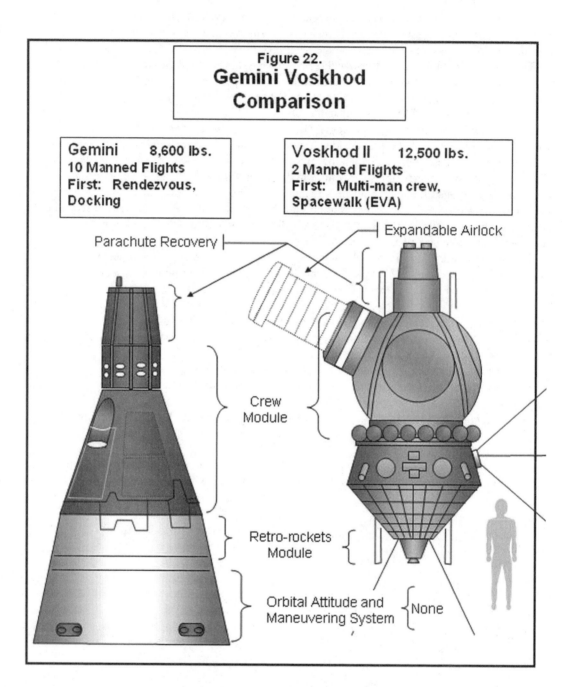

Figure 22.
Gemini Voskhod
Comparison

Gemini   8,600 lbs.
10 Manned Flights
First: Rendezvous,
Docking

Voskhod II   12,500 lbs.
2 Manned Flights
First: Multi-man crew,
Spacewalk (EVA)

Expandable Airlock

Parachute Recovery

Crew
Module

Retro-rockets
Module

Orbital Attitude and   None
Maneuvering System

January 1967 arrived with the United States enjoying the recent conclusion of its highly successful two-man *Gemini* project and the imminent launch of the first three-man *Apollo* spacecraft. Success with the *Mariner* planetary probes and the *Surveyor* lunar lander gave strong indications that a change in leadership of the space race may have taken place. The Soviets had encountered both technical and political bottlenecks that had brought their manned space program almost to a standstill. However, the *Apollo* spacecraft was entering its eighth year of gestation and facing some significant problems in moving to its first manned flight. The competition between the two superpowers for the minds of the uncommitted nations was entering its tenth year since the Soviets had surprised the West with their *Sputniks*. Nevertheless, 1967 would prove to be a fateful year for both countries.

Project Apollo

As it was envisioned in 1959, *Apollo* was a three-man spacecraft initially developed as a follow-on to Project *Mercury*. *Mercury* had been constrained by the lifting ability of the *Atlas* and the compressed time element to compete with the Soviets. Although *Apollo* did not have these limitations, it had no booster until the *Saturn* IB became available in 1966. With a 13-foot diameter, the 10.5-foot high cone-shaped craft was considerably wider than the diameter of either the *Atlas* or *Titan*. The basic thinking was that *Apollo* would serve as the basis for whatever missions followed *Mercury*. It would use an attached Service Module that could be configured to support specific assignments (for durations of up to two weeks) and added structures for possible lunar landing stages as well. The Service Module, similar in concept to the *Gemini* adapters, contained the electrical power, attitude control, and an orbital maneuvering engine. It would be jettisoned before reentry and would not return to Earth. (This configuration philosophy would be resurrected by the United States 40 years later).

When President Kennedy established the lunar landing goal in May of 1961, *Apollo* became the primary spacecraft for that mission. In that role the reentry protection for the return to Earth required considerably more coverage around the entire spacecraft—not just the heat shield. The spacecraft would dip into the upper layers of the atmosphere on its return to dissipate the high energy levels generated by the 25,000 mph speed using an offset center of gravity to achieve a degree of lift to skip like a flat stone across a lake. To some degree, this would mean that the upper portions of the cone would be exposed to the heat pulse.

The heat shield, a phenolic epoxy resin ablative covering would vary in thickness from 2.5 inches at the base to one-half inch on the upper portion of the conical cabin. Special provisions were made for openings for the thrusters, antenna, hatches, and four windows—which were a special high-temperature resistant glass. As a result of the need to envelop virtually the entire outer portion in ablative covering, equipment access could not be achieved through external openings for maintenance and modification prior to launch, as had been done with *Gemini*, but would have to revert to internal access, similar to *Mercury*. It was this configuration that would, as it had with *Mercury*, delay the ability to install, test, modify, or replace components.

The first design called for a two-gas environment for the cabin (nitrogen and oxygen) at one-half atmosphere (362 mm Hg) because of the unknown physiological effect of breathing pure oxygen for durations of up to two weeks. Subsequent ground based experiments showed no significant problems; and, because the two-gas system would weigh at least 35 pounds more, the pure oxygen environment of *Mercury* and *Gemini* would be adopted (at one-third atmospheric of pressure—258 mm Hg).

Because the basic design of the spacecraft was conceived before the Lunar Orbit Rendezvous (LOR) mode had been established, there had been many changes along the way. However, the basic cone-shaped reentry vehicle remained a fundamental part of the design. With LOR established as the mode to the Moon and the *Saturn* V as the vehicle, the weight of the Command Module was set at 12,000 pounds, the Service Module at 52,000 pounds, and the lander 22,000 pounds. This total had to be kept within the *Saturn* V's ability to accelerate 88,000 pounds to escape velocity. All through the development period, weight would be a contentious factor as trade-offs became critical. The basic structure would be aluminum honeycomb bonded between aluminum sheets.

The diameter of the cylindrical Service Module matched the outer 13-foot cone of the spacecraft and was 15 feet long plus an additional six feet for the Service Propulsion System (SPS) engine's protruding nozzle. The SPS had a thrust of 20,000 pounds and could be restarted as many as 50 times for periods as short as one-half second or as long as twelve minutes. Its job was to provide mid-course corrections to the lunar trajectory, slow the spacecraft so that it would enter lunar orbit, and accelerate the spacecraft out of lunar orbit and back toward the Earth. Each of these tasks was mission critical.

Failure to restart would put the astronauts' lives in peril. The engine had to fire every time it was commanded to perform. The choice of hypergolic propellants (those which ignite on contact) was for obvious reasons, one being the length of the mission—typically eight days. Even though cryogenic propellants would offer superior performance, their still uncertain nature, added to their propensity to "boil-off," virtually eliminated any thought of their use. Thus, 50 percent unsymmetrical dimethylhydrazine diluted with 50 percent hydrazine was the fuel, and nitrogen tetroxide was the oxidizer. Each propellant tank contained a sump tank at the outlet with a retention screen to ensure that there was always an initial flow to start the engine, and then the acceleration would settle the fluids within the tanks for continued supply in the weightless condition.

The Reaction Control System (RCS) for attitude control supplied 28 pounds of thrust for as little as 12 milliseconds and also used hypergolic propellants. The problem of settling the RCS propellants in their tanks prior to firing was resolved in an innovative manner with a Teflon bladder in each tank surrounded by helium gas, which provided positive pressure to feed the propellants into the engine. Again similar to *Gemini*, there were two completely independent systems, one in the Service Module for mission activities, and one in the Command Module for reentry.

As with both *Mercury* and *Gemini*, *Apollo* would use parachutes for the final descent back to Earth. This took some engineering, since the spacecraft weighed 12,000 pounds and fell at a speed of over 300 mph following reentry. This would be reduced to less than 200 mph by a series of drogue chutes and then pilot chutes. With the limited space and weight available, there would be no back-up system; thus, three chutes were used, of which any two would be sufficient to arrest the descent to less than 20 mph. The total package weighed less than 500 pounds.

The abort system was again similar to *Mercury* with a "tractor" arrangement of an escape rocket to pull the Command Module clear of an exploding *Saturn* V. Weighing over 8,000 pounds itself, the four-nozzle solid-fuel rocket provided 150,000 pounds of thrust which could carry the Command Module to an altitude of almost one mile (and off to one side) should an abort happen on the launch pad.

The decision to award the contract for building *Apollo* to North American Aviation (NAA) in 1962 was somewhat complex. Martin Aircraft submitted a technically superior bid and scored the highest points in virtually all areas of the selection criteria. North American Aviation displaced the Martin bid when the point allocation was unexpectedly redistributed to place a higher emphasis on cost. However, when LOR mode was selected, North American Aviation realized that the Lunar Module would not be a part of the contract. It would not be a North American Aviation product that would actually take the astronauts to the surface of the Moon. There was some political maneuvering to return to a mode that would allow North American Aviation to build the entire assembly. The influence extended into the Kennedy White House through the efforts of Presidential Science Advisor Jerome Weisner. Had it not been for Kennedy's preoccupation with the Cuban Missile Crises, the lunar mode decision might have been revisited.

The LM contract ultimately went to the Grumman Aerospace Engineering Company and would be the first spacecraft designed purely to operate in the environment of outer space and on the lunar surface in one-sixth Earth's gravity. There were no provisions for the craft to be able to reenter the Earth's atmosphere, as that would be the job of the *Apollo* spacecraft. The LM was a two-stage rocket having a first stage composed of the descent engine and associated propellant tanks along with the landing gear. This stage would remain on the lunar surface and act as a launch pad for the second stage. The second stage contained the ascent engine and its propellant tanks coupled to a small pressurized cabin for the two astronauts who would descend to the lunar surface.

Apollo 1: "Fire in the Cockpit!"

The *Apollo* spacecraft was built in three versions—boilerplate (simply the structural elements used to prove structural and heat shield integrity) and two manned iterations called "Blocks." Block I was the initial production version that was designed to prove the basic systems, but it did not have the requisite weight or advanced systems to make it lunar capable. The Block I articles were constantly being subjected to revision and redesign of various systems as NASA and North American Aviation sought to bring as much improvement (and weight reduction) to the spacecraft as possible. The change process became quite chaotic, and by late 1965, Major General Sam Phillips, who had been recruited from the Air Force to serve as *Apollo* Program Director, performed a detailed audit of NAA's *Apollo* effort. His findings were very critical of company management, and he stated that they were *"not giving sufficient attention to the details of the direction and execution of these contracts."* He recommended that more attention be paid to the *"details and their problems."* Subsequent changes were instituted that moved NAA's quality control procedures and schedule in a more positive direction, but never beyond "average" in overall performance.

Initial flight-testing, which involved von Braun's *Saturn* I and then IB, was going reasonably well with no significant failures, although the schedule for manned flight continued to slip. Tests of the escape system, which had been such a problem for *Mercury*, achieved a higher level of success using a similar solid-fuel rocket concept, called Little Joe II, and boilerplate spacecraft to simulate the various abort scenarios.

The first unmanned sub-orbital flight of a Block I Command and Service Module (CSM-009) aboard the first *Saturn* IB (AS-201) in February of 1966 was a success. Another suborbital flight, AS-202 (with spacecraft CSM-011) in August of 1966, cleared the way for the next flight to be manned.

Astronaut "Gus" Grissom was again selected to "shake down" the new spacecraft in the same way he had with the first manned *Gemini*. Accompanying him would be America's first space walker, Ed White from *Gemini* IV, and a rookie, Roger Chaffee. It had been hoped that this first manned flight, AS-204, would occur before the end of 1966, but there were just too many flaws in the various systems of spacecraft CSM-012. Unlike the conservative three-orbit flight of Grissom's *Gemini* III, AS-204 was open-ended with a target of 14 days. The aggressive plan to fly a manned mission after only three flights of the *Saturn* IB was a part of the Manned Space Flight Director George Mueller's new policy of "all-up-testing."

CSM-012 arrived at Kennedy Space Center in August and moved through a series of test programs. Among these was an Overall Space Vehicle Plugs-In Test, simulating the launch and ascent phases with the space vehicle remaining connected to all the umbilical connections. The second, the Plugs-Out Test, simulated the launch to the point where all the electrical connections were ejected, and the CSM operated on its own internal power.

A basic design flaw was discovered in the Environmental Control System's (ECS) oxygen regulator, and ultimately the entire ECS was replaced in late October. The flight slipped into the first part of 1967. Multiple leaks in the water/glycol coolant system caused the fluid to flow over electrical wiring, but by December this problem appeared resolved. On January 18, 1967, the Plugs-In Test was completed, but discrepancies required that the test be repeated a week later. The Plugs-Out Test was scheduled for January 27 when the spacecraft would be switched to internal power during a simulated launch. Whenever the spacecraft was not directly involved in one of these tests, the technicians from North American Aviation swarmed over it in an attempt to keep up with the more than 623 changes scheduled for implementation since CSM-012 arrived from California. The ability to make changes to many of the systems was hampered by the fact that some of the more than 15 miles of wiring in the craft had been routed and rerouted so that a specific wire that might have to be spliced could not easily be found. Standards of workmanship and quality control procedures were occasionally inadequate.

Some felt that this situation with the spacecraft was leading to a very high-risk condition. Thomas Baron, a quality control inspector (who had a reputation for what some considered "extreme" diligence) spread his alarm to the very highest levels. However, his reasoning and evidence were not compelling to upper management, and he was fired from North American Aviation in early January of 1967.

For the Plugs-Out Test, the astronauts were in their space suits and entered CSM-012 where the inner pressure hatch was then installed. This required the astronauts to use a special ratchet tool to remove the six bolts that held it in place. Next, the heat shield material hatch followed; it was closed by a simple latch mechanism. Finally, a fiberglass section was put in place as a part of a cover for the entire spacecraft to shield it from the exhaust gasses of the escape tower jettison-rocket motor. In all, it would take about 90 seconds to egress the spacecraft should a problem require a rapid exit. A quick release hatch using explosive bolts had originally been designed but was replaced with the more cumbersome method when Grissom's MR-4 hatch prematurely blew, and the spacecraft was lost. Nevertheless, the new design was criticized as not only requiring more time to open but not readily allowing EVA. A redesign was in process for the Block II spacecraft. Grissom was concerned about the awkward exit process, and he had scheduled an emergency egress exercise at the end of this Plugs-Out Test.

As the entire pre-launch process was being simulated, the spacecraft was filled with 100 percent oxygen at slightly higher than standard atmospheric pressure to assure there would be no movement of external gasses into the spacecraft. Despite a cabin filled with oxygen, the test was not considered hazardous (no propellants were involved), and therefore the fire and medical personnel were on "standby" rather than "alert".

The test was interrupted by several problems but nothing of a serious nature. Grissom complained about the poor communications between the spacecraft and the blockhouse several times, once commenting, *"If we can't talk between a few buildings here at the Cape how are we going to talk on the Moon?"* The activity wore into the early evening and was nearing completion when, at 6:30 P.M. a small power surge was noticed on AC Bus 2. Ten seconds later Grissom made the fateful discovery: *"Fire! We've got a fire in the cockpit!"*

It was understood by everyone, including the astronauts, that a fire in these circumstances would be fatal. The flames spread rapidly, feeding on the flammable materials that had crept into the spacecraft over the past few years. White had immediately dropped down from his couch and had engaged the ratchet into the first bolt. Chaffee stayed at his position and simply relayed what everyone on the outside knew. *"We've got a bad fire—let's get out... we're burning up in here!"*

Within 15 seconds, the fire had consumed most of what it could with the available oxygen and had created a significant overpressure that caused the spacecraft pressure vessel to rupture. This allowed the flames to move towards the split and to be sucked over the three men.

Chaffee's last words repeated his first plea. *"We're burning up!"* In less than 30 seconds, it was all over. The fire had burned itself out, the smoke had rendered the astronauts unconscious, and the flame had imparted severe burns. As those on the outside raced to rescue the crew, they were initially driven back by the dense toxic smoke that spewed from the ruptured craft. The large solid-fuel abort rocket motor positioned just above the spacecraft was ignored by those intent on saving the three crewmen. By the time the hatches could be opened, almost five minutes had elapsed, and there was no hope for a miracle.

The nation mourned the loss and, to some extent, was in shock that something like this could happen. The possibility of death accompanied each launch, but a ground test should not have had the prospect of becoming lethal. An outpouring from most of the free world expressed sympathy to the widows and support for the effort of the United States.

As expected, the Soviet Union's response sought to make as much political propaganda as possible. Expressing sympathy for the deceased astronauts and their families, the Soviet press stated that the astronauts were *"victims of the space race created by the American space program chiefs"* based on *"hate"* for the peoples of the USSR. While their use of the word "hate" would certainly overstate the basic premise of the United States towards the people of the Soviet Union, the essential argument was correct. In its haste to retake the technological lead, NASA and its contractors had taken many short cuts that did increase the risk. The Soviet leadership might have been more charitable had they known what was in store for their own program just a few months into the future.

The spacecraft was unmated from its *Saturn* IB and taken apart piece by piece. What was discovered had already been known—there were many areas in the electrical wiring that presented a hazard, primarily because of the constant access for changes and troubleshooting problems. Although the specific cause of the fire was burned beyond the ability to identify, an electrical arc in the vicinity of the lithium hydroxide panel came closest to being the point of origin. As a part of the investigation as to what caused the tragedy and how similar catastrophes could be avoided, the pure oxygen atmosphere was again examined.

A number of investigation boards were established, and a congressional inquiry was undertaken. While there was no criminal culpability proved, the lack of management oversight caused several people in NASA and North American Aviation to lose their jobs. Of the triple constraints of project management, cost and performance had generally not been compromised—but time was the factor that always appeared to be the culprit.

The Block I spacecraft were basically discarded and the Block II were revamped with new wiring harnesses built with three dimensional wiring harness jigs to avoid the bending that had been a major cause of stress. The wiring codes and quality practices were upgraded. The new hatch and a series of panels to isolate flame propagation, to allow the astronauts more time to escape or combat a fire, were installed. Had *Apollo* been designed and constructed with the awareness that originally surrounded the use of a pure oxygen environment, these improvements would probably have been a part of the original design. However, the success that NASA had enjoyed over the five years of flying manned spacecraft had taken its toll on caution. This situation would arise again twenty years into the future.

It would be more than 18 months before the first manned *Apollo* would launch, and the possibility of beating the Russians to the Moon was now only a glimmer. The delay in *Apollo* for design changes and quality improvements would allow other aspects of the program that were lagging to catch up.

Soyuz 1: "An Internal Matter"

The development of the three-man *Soyuz*, like *Apollo*, played a pivotal role in Soviet space progress. It was their vehicle for perfecting Earth Orbit Rendezvous, and it would provide for circumlunar flights projected for 1967. It would also be the "mother ship" for the LOR lunar lander that was scheduled to fly in 1968.

The first flight-qualified *Soyuz* spacecraft had gone through a troubled period of testing since its delivery in May of 1966 with more than 2,000 discrepancies and changes being recorded; they caused its scheduled 30 day check-out period to extend to four months. Several of the systems had likewise taken an extensive period of development, and some of the less sophisticated segments, such as the parachute recovery system, had proved troublesome.

For the first flight, it was decided to launch two unmanned *Soyuz* and have them rendezvous and dock. It was an ambitious approach, but the success of the American *Gemini* program had to be addressed. The first spacecraft designated Kosmos-133 (to avoid revealing its true nature) was launched on November 28, 1966. The 113 mile by 145 mile orbit was less than anticipated as the new *R-7* upper stage (being used for the first time) did not provide the expected performance. Within an hour serious problems with the "mooring and orientation" engine (the equivalent of the *Gemini* OAMS) had virtually depleted all of its propellant, and the craft was in a slow roll. Recognizing that the situation did not allow for the planned rendezvous and docking, the launch of the second unmanned *Soyuz* was canceled, and attempts were made to immediately return Kosmos-133. After an extended effort to provide the correct orientation, the spacecraft overshot its intended recovery area, and the automatic destruct mechanism activated.

It was decided to launch the second unmanned *Soyuz* as a solo mission and test its systems before committing to another rendezvous. On December 14, 1966, when the count reached "zero," the four boosters of the R-7 ignited, but the central core did not. This caused the automatic system to shut down the boosters. The launch crew immediately moved in to safe the giant *R-7*. They had just begun to raise the various service structures into place when the solid-fuel escape rocket on top of the *Soyuz* fired, some 27 minutes after the first stage had shut down. The *R-7* was destroyed, but, by a stroke of luck, only one of the launch crew perished although many were severely injured.

There have been two reasons for the mishap that were published 30 years after the event. The first is that one of the service towers was raised out of sequence and, coupled with some gusty winds, caused the *R-7* to tilt. The gyroscopic guidance system (which had not yet been shutdown) sensed an angle that exceeded the seven-degree tolerance for the first vertical phase of the launch sequence and signaled an abort. The second reason again relates to the gyroscopes but ascribes the out-of-tolerance condition to the rotation of the Earth during the 27 minutes following the aborted ignition sequence. The gyroscopes maintain a "rigidly" in their position and as the Earth continued to rotate on its axis, the gyros sensed the "tilt." The failure not only destroyed the launch vehicle but also badly damaged the launch pad.

The next *Soyuz* launched on February 7, 1967, and was reported as Kosmos-140. It again experience problems with its attitude control system and sun sensor. The spacecraft was bought down early and landed short on the frozen Aral Sea, eventually breaking through the ice and sinking in shallow water, but was later recovered. It was yet another set back for the Soviet program.

The first L1 (lunar) version of *Soyuz* flew an unmanned flight on March 10, 1967, six weeks after the *Apollo* fire and four weeks after the failure of the third unmanned *Soyuz*. The L1 did not have the "living quarters" module. This was to keep its weight to 11,500 pounds; the upper limit of the now four-stage *UR-500* Proton that boosted the intended circumlunar mission. The new trans-lunar injection stage (called the Blok D) put the L1 in a parking orbit. At the appropriate point, the Blok D fired a second time sending the spacecraft into a highly elliptical orbit that would have reached the Moon had that been the plan. However, this was an engineering test, and there was no desire to involve the Moon at this point, so its timing ensured that the Moon was not in the vicinity when the spacecraft (named Kosmos-146 to conceal its intent) reached its apogee, some 220,000 miles out. On its return, the craft burned up in the atmosphere, as there was reportedly no goal of recovery, although other sources indicate that it was to survive. The soviet team was ecstatic at the success, as most of the systems performed well and those that did not presented relatively minor problems (which tend to discount the survivability issue).

The second L1 spacecraft (Kosmos-154) was launched a month later on April 6, 1967, but this test required the Blok D to remain in its parking orbit for 24 hours before its second firing to send it on to the Moon. The Blok D failed to fire the second time due to a problem that was traced to an incorrect switch setting. The third flight (which was scheduled to be the first manned flight around the Moon) had been tentatively set for July. This schedule was unrealistic, as the first manned Earth orbiting *Soyuz* was set to fly within a few weeks, and there were many unknowns yet to be discovered. Since American's *Apollo* effort was grounded for at least a year as a result of the tragic fire, the chances were now excellent that the first man around the Moon would be a Soviet citizen.

In spite of the series of problems that plagued the *Soyuz* program, it was decided to proceed with the manned Earth orbital flight. It was felt that all of the difficulties encountered could be remedied or at least be under the control of the pilot—although *Soyuz* was designed to be flown as an automated spacecraft with the crew as "passengers." This had been the Soviet philosophy from the beginning. As had been previously demonstrated, the presence of the pilot had saved several missions, although the lack of true integration of manned and unmanned systems had revealed the designs of the *Vostok* and *Voskhod* to be inadequate; *Soyuz* sought to remedy this problem.

The inaugural flight of the manned *Soyuz* would be another Soviet spectacular. Not one but two manned spacecraft would rendezvous and dock. Two of the three cosmonauts from the second ship would space walk to the first to join its lone occupant and return in it. It was believed that the prestige lost during the *Gemini* program could be regained in one flight that demonstrated most of what the Americans had accomplished with ten *Gemini* flights.

The first manned *Soyuz*, enclosed in an aerodynamic shroud similar to *Vostok*, launched on April 23, 1967, with its lone occupant, Vladimir Komarov (who had commanded the first *Voskhod* flight in October of 1964). Yuri Gagarin had lobbied hard for being placed back on flight status but was passed over for several possible reasons. The first was that he was a national hero, and it was important to keep him as a living icon of Soviet progress. The second reason was that Gagarin had not stayed in good physical condition nor had he remained as technically competent as the others had. Both reasons were, in part, caused by the need for him to spend much of his time as a good will ambassador for the Soviet space program.

Komarov did not wear a spacesuit in the shirtsleeve environment of nitrogen and oxygen at normal atmospheric pressures. With the Living Module on the front end, isolated from the Control Module during EVA, the Soviets believed there was no need for the astronauts to wear spacesuits during normal operations which included launch and reentry. Following the five-minute powered flight and arrival in orbit, the mission was immediately announced by the TASS news agency as *Soyuz* I. Many rumors that a second craft would rendezvous had preceded the launch.

As with the American program that had established a Manned Spacecraft Center for Mission Control in Houston, Texas, the new *Soyuz* flight inaugurated a new Chief Operations and Control Group called the Scientific-Measurement Point No. 16, located in the Crimea. It didn't take long before the Control Group recognized that the left solar panel had not deployed, the back-up telemetry antenna was not functioning, and the attitude control sensor had apparently become fogged over by exhaust gases.

Komarov worked professionally to address each of the problems and effectively communicated his status to the Control Group. The loss of 50 percent of the electrical power and the problem with the attitude control dimmed hopes for a rendezvous and docking, the second *Soyuz* launch was cancelled as Komarov attempted to manually orient his craft.

Although a series of problems precluded the use of the ionic sensors to help establish the proper attitude, Komarov was able to complete the retro-fire sequence after a full day in orbit, and he was on his way back to Earth after jettisoning the living and instrument compartments. However, his trajectory would be a basic ballistic path, causing him to endure 10 Gs as opposed to 5 Gs had he been able to use the lift capability of the craft.

No further communication with *Soyuz* I was received after the expected reentry radio blackout period, seemingly this was not a major concern to the Control Group. A recovery helicopter got a fleeting visual on the descending craft and then spotted it lying on its side on the ground next to the parachute. As they observed the scene, the soft-landing rockets, which should already have fired to cushion the descent under the parachute, were seen to ignite, and the craft caught fire and was completely destroyed before a rescue could be attempted. Arriving sooner would not have helped Komarov, as the primary chute had failed to fully deploy, and the back-up became fouled in the primary. The ship hit the ground at over 100 mph, and Komarov was killed by the impact.

Announcing the failure and the death of Komarov to the world was a bitter pill for the Soviets to swallow, especially after their accusations only two months earlier regarding the *Apollo* fire. They refused to allow American astronauts Gordon Cooper and Frank Borman to attend the state funeral services, citing that it was "an internal matter."

As with the *Apollo* fire, there would be a lengthy investigation and fault-finding exercise. Several areas were uncovered in the parachute folding and ejection process that contributed to the malfunction. As for the problems in flight, it was determined that the spacecraft systems had not been sufficiently tested before committing to a manned flight. Indeed, the desire to catch up with the Americans had caused the Soviets to fly *Soyuz* before it was ready.

Others Who Succumbed

The fire in the isolation chamber that killed cosmonaut-trainee Valentin V. Bonderenko in 1961 and the *Apollo* fire in January 1967 that took the lives of Grissom, White and Chaffee demonstrated that the risks of being an astronaut were not reserved for actual space flight. America had lost it first astronaut, rookie Theodore Freeman, in a T-38 accident in October 1964. This was followed by the T-38 crash of Elliot See and Charles Bassett in St. Louis in 1966. Edward Givens was killed in a car crash in June 1967, and Clifton Williams was killed in another T-38 crash in October of that year. The demise of Komarov in the April *Soyuz* flight was the only one of the 10 deaths that had occurred during an actual mission.

It was no wonder, then, that the Soviets were reluctant to allow a second space flight for the world's first astronaut Yuri Gagarin. Gagarin was far more useful to the Communist regime as a good-will ambassador, and some were not displeased to see his physical conditioning and technical competence wane. Gagarin himself recognized that his situation was personally humiliating, and he decided to make a comeback and dropped some unneeded weight. By late 1967 he had completed his dissertation for a graduate degree and had been promised another flight with the completion of his academic work. Cosmonaut Chief General Kamanin was concerned with Gagarin's newfound fervor; but, until the time came to assign a flight, there was no need to confront the popular cosmonaut.

Gagarin completed a new medical and was cleared for assignment to another space flight but was not permitted to fly high performance jets solo. He was required to use the two-seat UTI-MiG-15 and a senior safety pilot. On March 27, 1968, Gagarin and Colonel Vladimir Seregin took off in the MiG from an airfield on the outskirts of Moscow for a one-hour proficiency flight. What actually occurred during the fateful last few minutes of the flight remains obscured, but the smoking wreckage of the jet was discovered several hours later after it dove into the ground almost vertically at 500 mph.

The initial official inquiry cited "pilot error," but a reinvestigation twenty years later sought to assign some of the blame to procedures of the Soviet airspace system. The most likely cause was a high-speed stall from which the jet was too low to recover. This situation may have been exacerbated by several layers of low clouds that contributed to disorientation. Gagarin, a man who would forever be linked with the conquest of space exploration, was dead, and the Soviet people mourned his loss as though he were their own relative. The Americans and other nations were allowed to attend the official state funeral.

The crew of Apollo 1, Edward White, Virgil Grissom and Roger Chaffee

The pilot of Soyuz 1, Vladimir Komarov

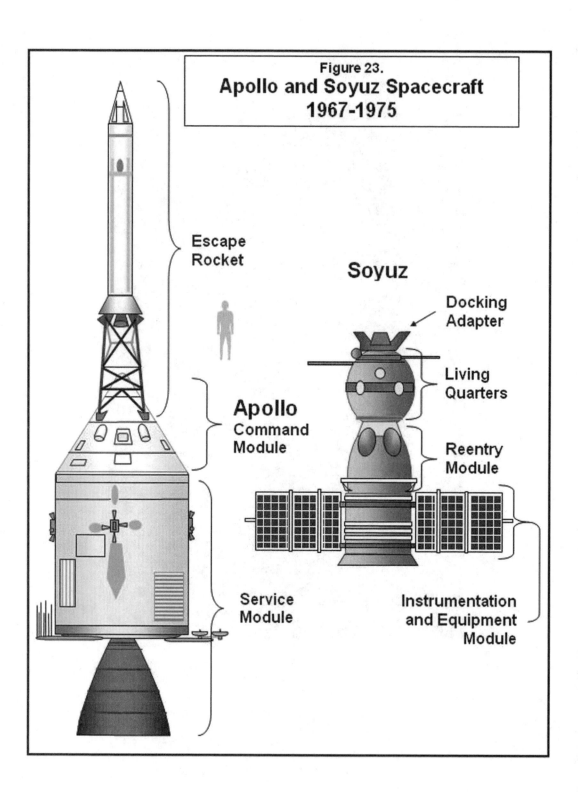

Figure 23.
Apollo and Soyuz Spacecraft
1967-1975

Escape Rocket

Soyuz

Docking Adapter

Living Quarters

Apollo
Command Module

Reentry Module

Service Module

Instrumentation and Equipment Module

Soyuz: Struggling for Success

The effort to get the basic *Soyuz* spacecraft cleared for another manned attempt followed completion of the critical recommendations that were made after Komarov's fateful flight. Two more unmanned spacecraft attempted to rendezvous and dock. The first launched on October 27, 1967, with the designation of Kosmos-186, again precluding any association with the Soviet manned space effort. It would be the "active" participant in the rendezvous attempt that was to occur. Following three days in orbit to test and remedy some problems that arose, the second *Soyuz* designated Kosmos-188 left the launch pad on October 30th.

The *Igla* radar system on Kosmos-186, the active satellite, and the computer designed to interpret the range and rate information performed flawlessly, and the two vehicles docked together in less than one orbit as called for by the technique they employed. Televised views of the impressive event were broadcast on the Soviet news, but the angle and clarity of the transmission revealed little of the configuration of the two craft.

It had been more than six years since Yuri Gagarin's flight, and the Soviets had yet to release pictures of either his *Vostok* or the *R-7*, and they weren't about to unveil their new spaceship. They were justly proud that they had accomplished yet another "first" with the completely automated rendezvous and docking of two satellites. While it was not a manned circumlunar flight, Kosmos-186/188 was the best the Soviets could provide for their people to celebrate on the 50th anniversary of the revolution.

All was not well with the event (although the world would not know that for another 30 years). The spacecraft were unable to achieve a complete latching into a "hard-dock" configuration. After several orbits, the two ships separated, and Kosmos-186 returned to Earth on October 31, using a ballistic path when its attitude control was unable to provide a less G-intensive "guided" reentry; however, it touched down safely. Kosmos-188 also had to execute a ballistic return, but its path threatened to fall short of Soviet territory, and it was intentionally destroyed.

Almost one year after the death of Komarov, a revised and better tested, but still unmanned, *Soyuz* rose from its launch pad at Baykonur (Tyura-tam) on April 14, 1968. Named Kosmos-212, it was followed the next day by the unmanned Kosmos-213. The launch parameters were so accurately timed that the second satellite was less than three miles away from the first at orbital insertion. The automated docking was followed closely by the ground controllers on live television. The two spacecraft undocked after about four hours, and then each proceeded to spend five days in space conducting tests of the various systems. Kosmos-212 de-orbited and performed a guided reentry to a successful landing, followed the next day by Kosmos 213 which performed equally as well.

In spite of the very successful flights, the political hierarchy was reluctant to commit to the next mission being manned. The bad press from *Soyuz* I was still fresh in everyone's mind. As a result, Kosmos-238 was a solo flight launched on August 28, 1968, and returned on September 1, finally clearing the way for a manned flight.

Soyuz 2 successfully launched on October 25, 1968, but it was unmanned as there was still reluctance to trust the *Soyuz* systems. *Soyuz* 3, with only cosmonaut Georgiy Beregovoy aboard (the first manned flight since Komarov's death), launched the following day. The automated rendezvous apparatus brought the two spacecraft to within 500 feet and Beregovoy attempted to perform the docking manually. The lack of coordination between the automated and the manual systems thwarted his attempts.

Although the Soviets publicly claimed that docking had not been one of the objectives, privately most of the blame for the failure to dock was placed on Beregovoy for his inability to properly control the ship. In the process, he almost exhausted the attitude control fuel, and the docking was cancelled. (In the post flight debriefing, he would complain that the attitude control system was too sensitive and not easily managed in space.)

He spent the remaining time performing experiments and providing three TV sessions, which were broadcast live on Soviet television, before being brought back safely on October 30. The new "openness" of the Soviet program to television (which still did not reveal the launch vehicle nor the space-craft configuration) was a direct result of the *Apollo* 7 flight which had occurred a week earlier.

Saturn V Flies

The American plan to fly the modified Block II *Apollo*—delayed by the emphasis on improved wiring, the quick opening hatch, and a reduction of flammables in the cabin—would not see another manned attempt until the fall of 1968. In the meantime, the effort to fly the first *Saturn* V took on new meaning. It was now imperative that the big booster perform its first flight virtually flawlessly if the Kennedy commitment was to be satisfied. This was asking a lot of a rocket that was four times as powerful as the *UR-500*.

The designations of the *Saturn* V vehicles began with AS-501 as the first flight article. An inert facilities test assembly (designated *Saturn* 500-F) was used to verify all of the electrical and mechanical connections. A renumbering of the *Apollo* spacecraft also occurred at this time. They were retroactively assigned numbers that recognized the deceased crew of Grissom, White and Chaffee; it was designated *Apollo* 1. *Apollo* 2 and 3 were the boilerplate versions that flew on the *Saturn* IB AS-201 and AS-203. *Apollo* 4 sat high atop AS-501. It was a true Block II spacecraft with virtually all the systems active.

On August 26, 1967, when the first flight-ready *Saturn* V was moved to KSC Pad 39A, it was a media event that caused almost as much excitement as a launch itself. The massive assembly was truly an awesome sight to watch as it moved ponderously along the five miles to the flame pit, while thousands of spectators gathered to witness the event. It would be another two months before the launch as more delays crept in to the schedule.

On Thursday morning November 9, 1967, the fully fueled rocket weighed almost six million pounds. As the final seconds of the weeklong countdown ticked by, a computerized progression of events removed the actual firing sequence from human hands at T-3 minutes. The command to start the S-IC engines of the first stage was sent at T-8.9 seconds when the turbo pumps began their high-speed whine to deliver the propellants to the five huge 17-foot F-1 thrust chambers. Within a few more seconds, smoke erupted into the flame pit at the base of the rocket. The six umbilical causeways ejected their connections to the rocket and retracted. Then slowly, ever so slowly, the giant began to lift.

To this point, there had been no sound, as the nearest official spectators were more than 2 miles from the rocket. When the acoustical impact of the shock wave traveling at over 1,000 feet per second hit the gathered viewers, it was not just a physical impalement of force, but an emotional, and to some, a spiritual experience. The sound literally shook the surrounding ground, structures, and rib cages. Even to the old hands who had witnessed many launches at the Cape, this was different.

Almost immediately the *Saturn* V responded to a very subtle command to yaw slightly to provide more clearance to the umbilical tower. At T+11.9 seconds a slight pitch and roll maneuver began that aligned the rocket with its intended inclination to the equator and initiated the ballistic trajectory to an orbital path. It took a full ten seconds for the mammoth rocket to clear the 400-foot umbilical tower while the 7.5 million pounds of thrust devoured almost 3,400 gallons of propellant each second.

As the rocket rose higher into the atmosphere and gained speed, the characteristic "frozen lightening" appeared, and the exhaust plumes of the five F-1 engines began to widen as the atmospheric pressure lessened. Mach I (the speed of sound) was passed at T+68 seconds, and this was followed shortly by maximum dynamic pressure as evidenced by a vapor cloud that briefly surrounded the mid section of the rocket. At T+135 seconds into the flight, the center engine shut down as planned to lessen the forces on the rocket when the first stage burnout occurred at T+150 seconds: 40 miles high and 6,000 mph. The first stage separated, and eight small solid-fuel rockets fired opposite to the direction of flight to slow it down. This ensured that it would not impact the remainder of the rocket which was now experiencing the effect of eight small ullage rockets that were designed to provide a slight acceleration to settle the propellants in the second stage tanks.

Now for the first time in flight, the five LH2 J-2 engines came to life providing one million pounds of thrust. By incorporating two major milestones into a single flight, all-up-testing had saved its first $135 million dollars (the cost of a *Saturn* V launch). Thirty seconds into the burn, the corrugated interstage adapter was released to reduce weight. The super efficient LH2 cluster burned for 360 seconds and accelerated the upper stages to an altitude of 120 miles and a velocity of 16,500 mph, parallel to the Earth's surface.

The second stage separated, aided by four forward firing solid-fuel rockets to again ensure the spent stage would not run into the remaining rocket assembly. The single J-2 engine in the S-IVB ignited, and its 200,000 pounds of thrust continued to accelerate the *Apollo* CSM for 145 seconds to an orbital velocity of 17,500 mph where it would "park" for a little more than 3 hours.

The objective of the flight now called for re-ignition of the S-IVB to simulate the translunar injection burn. Rather than a full burn, the S-IVB would push the *Apollo* up to a speed of "only" 20,000 mph, which would result in an apogee of just over 10,000 miles. The *Apollo* CSM was then separated from the S-IVB, and its own Service Propulsion System (SPS) fired to raise the apogee another 600 miles and to establish a reentry path of 8.75 degrees, simulating a return from the Moon (All-up-testing had now saved yet another $135 million dollars).

At eight hours and ten minutes into the flight, the SPS fired again for 270 seconds to accelerate the now descending spacecraft to a full lunar return velocity of 25,000 mph. The *Apollo* 4 Command Module separated from the Service Module and proceeded to execute a "double dip" into the atmosphere. It bled off speed (energy) on its first encounter (exposing the spacecraft to about eight Gs) and used the resulting lift to skip back to a higher altitude before performing the second and final phase of the re-entry (resulting in four Gs). External temperatures rose to 5,000 degrees F while the interior temperatures showed an increase of but 10 degrees. The craft was retrieved by the recovery forces in the Pacific.

The first *Saturn* V test was successful beyond anyone's expectation. In one inaugural flight, it had proved the first and second stages and had fully qualified all of the spacecraft systems on a simulated lunar voyage. While a second planned unmanned flight remained on the schedule, the third *Saturn* V flight would be manned—if the first manned *Apollo Saturn* IB flight, still almost a year away, was successful.

The second test of the *Saturn* V, AS-502, occurred on April 4, 1968, sending *Apollo* 6 on a similar flight profile to that of *Apollo* 4 the previous November. This flight revealed several problems beginning at T+133 seconds when the vehicle experience longitudinal oscillations (approximately five cycles per second for the last ten seconds of burn), referred to as the "pogo effect." The phenomenon had occurred with the *Gemini-Titan* vehicle, but instrumentation on AS-501 had not shown any significant problems. The forces exerted on AS-502 caused portions of the spacecraft adapter to break away, but the huge rocket continued its flight.

During the S-II second stage thrusting, engine two shut down prematurely, followed by engine three. The guidance system was able to control the rocket, and the remaining three engines were commanded to burn longer (using the fuel allocated to the inoperative engines) to compensate for the lack of thrust. However, the energy was insufficient to totally make up for the second stage, only producing 60 percent of its rated thrust, and the S-IVB had to fire 29 seconds longer to put the *Apollo* CSM in the proper parking orbit. At the appointed time for the second burn of the S-IVB, for the simulated translunar injection (TLI), it failed to restart. A contingency was available for the mission, and the CSM Service Propulsion System provided a long seven-minute 22-second burn to send the CSM 14,000 miles into space. It then returned to Earth and was recovered from the Pacific.

The problems with AS-502 were immediately addressed. The pogo oscillations would be dampened by adding helium to shock absorbing cavities in the LOX line pre-valve assembly to change its natural resonance. The pogo had not occurred during the first flight, apparently because of its lighter weight. A leaking LH2 igniter fuel line caused the premature shutdown of engine two in the S-II stage. The anomaly was detected from thermocouple temperature readings in the tail section of the S-II. These readings indicated that the problem was caused by the flow of cold gas (the leak) resulting in the lowering thrust from the engine. The signal to shut down the engine was sent to the fuel valve of engine two, and to the LOX valve of engine number three as a result of a wiring error—thus the loss of two engines.

Why did the fuel line develop a leak? Exhaustive tests were carried out with special attention to the accordion-like bellows section that was protected by a metal braid to allow the line to flex. Using higher flow rates, pressures, and temperatures, none of the test items failed. Turning to the possible implication of the conditions in space, eight test items were placed in a vacuum chamber and exposed to vibration while LH2 was pumped through the lines. All eight test items failed within 100 seconds. When tested in normal atmospheric conditions, the LH2 caused the air around the lines to liquefy and this liquid was held between the bellows and the braiding, dampening the vibrations—so failures did not occur. Only in the vacuum condition, where there was no air to liquefy, did the vibrations set up resonances that caused the malfunction. The failure of the S-IVB to fire for a second time was due to a failed igniter. That the *Saturn* experienced these failures and was able to continue its flight was particularly gratifying to the entire MSFC team—especially von Braun.

The next question, then, was when to fly AS-503. If it was simply to verify the corrections being made, it could have flown as early as July of 1968, but with the concept of all-up-testing, there had to be more to the mission. NASA management decided not to repeat another unmanned *Saturn* V flight. As the schedule now appeared, all of the remaining flights would be manned. If AS-503 were to be flown manned, the current schedule called for it to provide an Earth orbit test of the Lunar Module, LM-3. This flight would not occur until January 1969 because of delays in installing changes. In the interim, *Saturn* IB AS-205 would fly in October of 1968 as the inaugural two week manned *Apollo* flight. The question that faced NASA management was what to do with AS-503.

Soviet Manned Circumlunar Efforts

That the Soviets were still committed to a lunar program seemed to be validated just a month after the *Soyuz* tragedy. Two American astronauts, David Scott and Michael Collins, had an opportunity to speak with cosmonauts Pavel Belyayev and Konstantin Feoktistov at the 1967 Paris Air Show. Collins related that Belyayev spoke openly of Soviet manned circumlunar flights in the near future. On his return to the USSR, Belyayev was soundly reproached for his comments, and the task of toning down his rhetoric was apparently given to Academy Sciences' Leonid Sedov a few months later when, during a press conference, he made a point of saying that *"manned flight to the Moon is not in the forefront of Soviet astronautics."*

While the Komarov accident dimmed the prospects for a manned circumlunar flight in time for the November celebration of the 50th anniversary of the Russian revolution, the Soviets pressed forward with perfecting the *UR-500*/7K-L1 combination. The first opportunity to test the vehicle occurred on September 28, 1967, when the two-million-pound thrust rocket left the pad at Tyura-Tam. Observers noted that the huge rocket appeared to rise slower, while those in the blockhouse knew why: only five of the six engines had fired. The rocket's guidance system was able to sustain a near normal trajectory until 61 seconds into the flight when the abort sensors recognized that the rocket was unable to maintain the desired path and triggered the escape mechanism for the spacecraft. The engine problem turned out to be a rubber plug in a propellant line that had not been removed during manufacture (similar to the first attempt to launch *Gemini* VI). The failure removed any possibility of a manned circumlunar flight in 1967. There would be no space spectacular for the 50th anniversary of the Bolshevik revolution.

Another attempt was made on November 22, 1967, and this time it was the second stage that experienced a failure when one of its four engines failed to fire. Again, the abort sensors detected the malfunction and the spacecraft was rocketed clear of the booster. During the landing the solid-fuel rocket motors in the base of the *Soyuz* fired too early, and the craft experienced a "hard" landing, so it was fortunate that there were no occupants.

With the 50th anniversary behind them, there were no more "political" deadlines, and Korolev's old team, now headed by Chief Designer Vasiliy Mishin, had at least a year to complete their goal of a manned circumlunar flight before the Americans could possibly make a similar attempt. Nevertheless, the confidence of the engineers as well as the cosmonauts themselves had been shaken by the series of failures.

The next flight of the *UR-500*/7K-L1 on March 2, 1968, was made without the Moon as its target in order to expedite the flight without having to wait for the tight time constraints of a lunar launch window. The unmanned spacecraft successfully achieved a high apogee orbit that sent it out to a distance of 220,000 miles. This time the TASS announcement referred to the ship as Zond 4 (Russian for "probe"), a moniker that had previously been used for much smaller interplanetary probes a few years earlier.

Two days into the flight, a mid-course maneuver was attempted. However, due to a problem with the attitude orientation system (which failed to identify correctly the navigation star Sirius when employing a low-density filter), it was not executed. The burn was not critical, and a second attempt the following day using the high-density filter was also unsuccessful. A third attempt on March 7 finally succeeded, using a medium-density filter. The trajectory for reentry back into the Earth's atmosphere appeared nominal, and the spacecraft was separated from the equipment module for the fiery plunge.

But the projected path, which was to take the descent module to within 30 miles of the surface before skipping back out to almost 100 miles, did not occur. Instead, it continued into the atmosphere, exposing the spacecraft to extremely high temperatures and up to 20 Gs. Had a crew been on-board, they probably would have survived but may have experienced significant injuries. When it became apparent that the "skip" would not take place and the spacecraft would actually land near the west coast of Africa, the decision was made to destroy it for fear of its falling into American hands.

It was again believed that the star-tracker opticals had been contaminated by the engine exhaust gasses, and provisions were made for subsequent craft to provide a sensor enclosure that would be disposed of after the final burn of the Blok D engine on the next attempt on April 22, 1968. The star tracker never had a chance to operate on that flight because the abort detection system malfunctioned during the first stage flight, and the emergency escape system was activated.

Frustration was apparent, not only with Chief Designers Mishin (for the 7K-L1 spacecraft) and Chelomey (for the *UR-500*) but with their entire teams. The political hierarchy was also showing signs of irritation at the inability of the circumlunar project to show success. Even the tight security that surrounded each launch and subsequent failure was not immune from some leaks through a variety of paths to the American CIA. But the western press was still captivated by the Soviet mystique and continually predicted an imminent spectacular.

As the next vehicle was being ground tested for the August 1968 lunar launch window, an oxygen tank on the Blok D stage exploded and nearly caused the loss of the entire rocket as well as the 7K-L1 spacecraft. It was only by heroic efforts that more damage was not sustained, although one life was lost.

The next *UR-500*/7K-L1 spacecraft, launched on September 15, 1968. This flight contained live tortoises among the many biological specimens aboard. The booster performed flawlessly, and the upper stage Blok D sent the spacecraft towards the Moon with its Trans-Lunar Injection (TLI) burn. The mission was designated Zond 5 to the world, and attitude control once again threatened its success.

A constant stream of work-around commands allowed Zond 5 to complete its loop around the Moon (coming to within 1,200 miles of its surface). It then proceeded back to Earth after taking the first high quality pictures of the Earth, showing it as a complete blue disk suspended in the black void of space. (America's Lunar Orbiter had taken similar pictures, but the resolution of its TV scanner could not compare with the direct image photos returned by Zond 5.)

Once again, the attitude control was critical for the reentry, and, following another series of difficult maneuvers, the 4,500 pound Zond 5 splashed down in the Indian Ocean just 65 miles from a Soviet recovery ship. Although the reentry was ballistic (there was no attempt to "skip" off the upper layers of the atmosphere to lessen the G load), it would have been survivable for a cosmonaut. The flight was a milestone in that it was the first spaceship to be recovered after traveling to the vicinity of the Moon. Soviet space planning was again becoming more solidified.

Zond 6 launched on November 10, 1968, on yet another unmanned circumlunar mission, and it carried a biological payload similar to its predecessor. It took high-quality pictures of the far side of the Moon and returned to Earth on November 17, using the more sophisticated atmospheric "skip reentry" that would be necessary for a manned flight. Its trajectory allowed it to land in the Soviet Union—only 10 miles from the launch site. However, the spacecraft experienced a loss of pressurization that killed most of the biological specimens and caused the parachute recovery system to activate early and to release before the craft had landed. This caused a high-speed impact with the Earth that effectively destroyed Zond 6. Amazingly, the camera film was recovered intact.

Technologically, there were no more hurdles preventing the Soviets from launching cosmonauts on the next flight. They had the equipment necessary, but reliability remained questionable. The next lunar launch window for the Soviets was the second week of December 1968, and there was great speculation in the West that this flight would be manned. But, the Soviet's realized that they would have to skip that window and perhaps look to January, 1969. Cosmonaut Chief Kamanin wrote, *"I have to admit that we are haunted by U.S. intentions to send three astronauts around the Moon in December. Three of our un-piloted L1 spacecraft have returned to Earth at the second cosmic velocity, two of them having flown around the Moon. We know everything about the Earth-Moon route, but we still don't think it is possible to send people on that route."*

Surprising Apollo Schedule Change

The desire of both nations to complete the first manned circumlunar flight was high on their priority list. For the Soviets, who were still struggling to achieve the same rendezvous, docking, long duration, and high-apogee flights that *Gemini* demonstrated, circling the Moon before the Americans was critical to maintain the appearance of space technology leadership. As for the Americans, who were finally getting it all together after so many years of playing catch-up, they were not about to lose the race to circle the Moon.

During the summer of 1968, NASA Administrator James Webb was kept informed by the CIA of the pending prospects for a Soviet circumlunar flight. He had been made aware of the Soviet project to build a *Saturn* V equivalent (the *N1*) and had alluded to it in congressional testimony. American reconnaissance satellites had regularly photographed Baykonur and had seen the ground-test mock-up of the *N1* on the launch pad in the spring of 1968. Because Webb could not reveal his sources (and the available information was rather skimpy), many thought his use of the mythical giant rocket was simply a ruse to pressure congress into releasing more money for NASA (whose funding was now on the decline). The phantom Soviet rocket was often referred to derisively as "Webb's Giant."

Associate Administrator for Manned Space Flight George Mueller had decided that AS-503 would be a manned flight following the resolution of the problems with AS-502; the next *Saturn* V would carry a full crew on only its third flight. With the first flight-qualified Lunar Module (LM-3) not expected to be ready until early 1969 for a test in Earth orbit, what should be the mission of *Apollo* 8— the first manned *Saturn* V? To wait for LM-3 would lose a full *Saturn* V launch cycle of 10 weeks, and it was critical that a full five launch cycles be available in the last calendar year of the decade to have a chance of meeting the Kennedy goal.

NASA Associate Administrator George Low started to think "outside the box." With a possible Soviet circumlunar flight close at hand (judging from the appearance of Zond 4), why not use AS-503 to send *Apollo* 8 (CSM-103) around the Moon. Although it would be a change in the flight test progression that had been established, it would be a valid test of the translunar trajectory, tracking, and communications and might just beat the Russians to this important milestone.

During the first week of August 1968, Low contacted several members of the launch team to explore the possibility of flying a circumlunar mission with *Apollo* 8. A thorough review of all the various systems and ground support elements was made, and it was determined that there were no serious obstacles to such a mission at the next launch opportunity. To ensure that it satisfied the "all-up" criteria, it would be launched to reproduce as closely as possible a manned lunar landing profile—only the LM would be missing. This meant that it would not just go around the Moon and return, but that it would fire the CSM's SPS engine to place it into an orbit around the Moon. The timing would be such that it would encounter the same lighting conditions as a lander, thus dictating a late December launch.

By August 12, a basic flight plan had been formulated, and Low started its circulation through the various NASA organizational structures as a confidential internal document. There were more than a few who felt the proposal was carrying "all-up-testing" beyond reasonable limits, that the risk of a lunar flight on only the second manned *Apollo* flight was too great. It is interesting to note that when the *Apollo* schedule was initially defined, there was no explicit circumlunar flight—all were designed to move towards the goal of a lunar landing. In the mean time, Arthur Rudolph, the man who had headed V-2 production at the Mittelwerk in the Harz Mountains of southern Germany and was now the manager of the *Saturn* V Program Office, was directed to assemble AS-503 in the Vehicle Assembly Building.

Despite all of the quiet planning that was taking place to possibly send AS-503 on a lunar mission, it was emphasized to the NASA staff that *Apollo* 7 had to fly a flawless first manned mission on AS-205.

Astronaut chief Deke Slayton conferred with James McDivitt, who was scheduled to command the original *Apollo* 8 mission with LM-3. McDivitt declined the switch in missions as his crew had already spent considerable time preparing for the LM-3 mission profile. That decision moved the *Apollo* 9 crew, headed by Frank Borman, into the *Apollo* 8 slot, and, since his mission had been planned as a high-apogee flight, it more closely suited the initial training of his crew, although Borman would get the CSM-103 spacecraft.

When the proposal reached NASA Director James Webb (who had been out of the country), he was not pleased with the idea nor that his subordinates had made so much progress in planning such a dramatic step without contacting him earlier. As he was filled in on the tentative plan, he quickly saw both the technical and political logic.

He approved and the planning continued. However, no commitment would be made until after the *Apollo* 7 flight and all of the test results were in, relative to changes generated by AS-502. Moreover, the plan would remain an internal document, not to be disclosed to the press.

Apollo 7: Disgruntled Astronauts

By the time the redesigned *Apollo* was ready to fly in October 1968, 23 months had elapsed since the last successful American manned flight—*Gemini* XII. Four astronauts had been killed (Grissom, Chaffee, White, and Komarov), and, for America, the situation on Earth had deteriorated with the escalation of the war in Vietnam and the riots in the streets of several major cities.

The expensive space race seemed to have lost it appeal to many who now considered the resolution of racial strife and achieving peace to be far more pressing national goals than the Moon. President Lyndon Johnson had assumed the presidency following Kennedy's assassination and had won the presidential election of 1964 over Republican Barry Goldwater. Johnson's presidential banner was "The Great Society," through which he sought to end poverty and provide "Negroes" with equal opportunity.

However, his unbridled willingness to continue not only an unpopular war but also one in which the tactics proved difficult for the American military to effectively handle became his downfall. Faced with opposition, primarily from within his own Democratic Party, Johnson announced in March 1968 that he would not seek re-election the following November. Then, just days later, the civil rights leader Martin Luther King was assassinated. This was followed two months later by yet another assassination—this time former President Kennedy's brother Robert, who was running in the presidential primaries.

Even the funding for *Apollo*, which had hit its peak in 1966, was now on the down slope, and thousands of young engineers and technicians who had staked their careers on the space program were being laid-off as the Moon program moved closer to its goal. Congress showed little interest in a follow-on manned mission to Mars or a permanent space station or lunar base program. The dynamic head of NASA, James Webb, had resigned only weeks before the scheduled *Apollo* 7 flight and was replaced with Thomas O. Paine

Thus, by the time astronauts Wally Schirra, Walter Cunningham, and Donn Eisele were strapped into the *Apollo* 7 spacecraft on October 11, 1968, America was a divided and depressed nation. Many of its youth, using the Vietnam war as an excuse, had turned to drugs and abandoned the ethical and moral standards that had made the nation great. One of the many counter-culture movements professed that "God is dead."

Schirra, who had flown a perfect six-orbit mission in MA-6 and demonstrated the first rendezvous with *Gemini* VI, was tasked to test fly the first manned *Apollo*. The primary objectives of the eleven-day flight were the qualification of all the *Apollo* systems, crew coordination and performance, rendezvous techniques, as well as compatibility with ground support facilities.

The *Apollo* 7 crew had followed CSM-101 through its redesign, fabrication, and test. Now they sat in it, atop AS-205, the first manned *Saturn* IB, on pad 34. The Block II ship was the result of over 1,800 changes since the fatal fire, and it had moved through the ground testing phases relatively smoothly, as the contractor, NAA, had made noticeable improvements in its quality assurance and fabrication methodologies. *Apollo* 7 would be flying atop only the fifth flight of the new *Saturn* IB booster; which had never taken men into space. This was in stark contrast to more than 100 *Atlas* firings and 35 *Titan* II launches that qualified those rockets for a manned flight.

At T-3 seconds the eight H-1 engines of the first stage came to life and quickly built to 1.6 million pounds of thrust. The launch was flawless, and the huge rocket blazed a trail through the sky—a sight that had become familiar to space fans for the previous ten years. At T+143 seconds the four inboard engines cut off, followed by the four outboards four seconds later. The S-IVB immediately separated, and its single 200,000-pound-thrust LH2 J-2 engine ignited. The escape tower was jettisoned 20 seconds later, and the S-IVB burned for a total of 470 seconds, placing the 36,000-pound CSM in an orbit that ranged from 142 miles to 178 miles. The CSM then separated from the S-IVB, and the astronauts proceeded to exercise the Reaction Control System and SPS propulsion systems to perform a series of rendezvous maneuvers with the spent S-IVB over the next few days.

Unlike previous missions, all three astronauts were able to remove their space suits after the launch and spent the mission in light blue jump suits that were much more comfortable. The *Apollo* was much roomier than *Gemini* and allowed the American astronauts to un-belt and float around the cabin for the first time. In addition, Americans were now able to visit the astronauts through the miracle of television.

Almost from the first, the crew developed severe head colds that were aggravated by the weightless condition which allowed their sinuses to expand but didn't allow them to drain naturally under the influence of gravity. This condition caused considerable discomfort and made the astronauts less agreeable as the eleven days wore on.

Schirra, in particular, was constantly annoyed by changes to the schedule and new tasks that were communicated up to the crew. He criticized many of the activities that they were asked to perform, as did the other two astronauts. At one point, he cancelled the first scheduled TV period and remarked to Mission Control, *"The show is off! The television is delayed without further discussion. We've not eaten. I've got a cold, and I refuse to foul up my time."* Later, when the crew was awakened from a sleep period earlier than planned, Schirra again complained that his request for an extra hour of sleep had been ignored. There were many other heated exchanges, but when subsequent TV periods were scheduled, Schirra conducted them without any hint that he had been angered by various situations, and the American public was excited to be able to see the astronauts at work on live television for the first time. Although Schirra had indicated that this would be his last flight, neither of the other two astronauts ever flew again—more than likely because of repercussions from their "disagreeable" nature.

Apollo 7 was the first and only use of the *Saturn* IB as a manned launcher during this period. The accelerated NASA schedule immediately began to use its larger brother, *Saturn* V for subsequent *Apollo* flights. It was never used for any other program despite its efficient payload carrying capability. It was called upon to support the *Skylab* space station with three manned launches of the *Apollo* spacecraft and the single Apollo-Soyuz Test Program of the mid-1970s.

If the head colds and the added tasks impeded the working conditions and the rapport between Mission Control and the crew, the outcome of the flight could not have been better. All of the assigned objectives were met or exceeded. The booster and the spacecraft worked to perfection. It was time to take a bold step forward.

Apollo 8: The Christmas Flight Around the Moon

Following the successful flight of *Apollo* 7, a review board met on November 7, 1968, and declared that all of the elements necessary for the tentatively planned circumlunar flight were complete. In an unusual Sunday meeting with key NASA and contractor representatives on November 10, it was unanimously agreed that *Apollo* 8 would go to the Moon. It is not known if the launch of Zond 6, that same day, had any influence on the final decision. One contractor objected to the CSM entering lunar orbit, suggesting that it merely loop around to avoid the possibility of its being stranded should the CSM's SPS engine not restart. However, redundancy of the engine systems along with the fact that the spacecraft carried three fuel cells (and could operate with only one), coupled with two independent oxygen systems, gave solid assurance that the risks were being minimized.

The following day, the new NASA administrator, Thomas Paine, approved the mission scheduled for December 21, 1968. If the Soviets were to attempt a circumlunar flight, their launch window was the second week of December. Even with its bold plan to launch Americans to the Moon on the first manned launch of a *Saturn* V (and only the third launch of that rocket), America might still lose another opportunity to be first.

Paine met with the press the following day to make the momentous announcement: *"After a careful and thorough examination of all the systems and risks involved, we have concluded that we are now ready to fly the most advanced mission for our Apollo 8 launch in December, the orbit around the Moon."*

While the revelation was met with excitement and anticipation by most, it was also the subject of some critical analysis. Among the detractors was Sir Bernard Lovell, Director of Britain's Jodrell Bank Radio Observatory, who noted that, within a few years, missions to the Moon could be done by robotic spacecraft, avoiding the risk to a human crew. He also commented that the mission represented *"a dangerous element of deadline beating,"* an obvious reference to the Kennedy commitment and the race with the Soviets. Some critics cited the lack of an extensive "track record" for either the *Saturn* V or the *Apollo* spacecraft, and there were a few who predicted outright failure.

If there was a dominant theme in why America should turn its back on the Moon, it was often the cost factor and the reference to the millions of people who lived in poverty in the United States. To this argument the proponents of a vigorous space program would provide statistics that showed the space program budget for 1968 stood at 4.7 billon dollars (down from its peak of 5.9 billion in 1966), which represented about 2.5 percent of federal expenditures. This was less than Americans spent on cigarettes. President Johnson's war on poverty had escalated to more than 20 billion dollars that year— more than four times the NASA budget. The duel in space between the Americans and the USSR was a high stakes affair that had political, economic, and military implications that demanded both nations offer their best efforts. In the final analysis, most felt that it was man's destiny to explore the unknown, even if robots could eventually accomplish the task. Man needed to satisfy that inner desire "to go where no man has gone before."

Preparation continued for the launch. It appeared to be another situation where one nation would perhaps win by a week or two, not enough to make a significant technological statement. However, when the Soviet tracking ships returned to port early in December, it seemed obvious that there would be no lunar attempt on their part for a month or more.

The *Apollo* 8 crew entered their spacecraft in the pre-dawn hours of December 21, and the countdown proceeded without any major problems. The excitement of the day was highlighted because of the Saturday launch date and the Christmas school break that allowed millions of Americans to witness the event through the electronic media of television.

Air Force Lt. Col. Frank Borman was a veteran of the *Gemini* VII flight and commander of the *Apollo* 8 mission. Navy Commander James Lovell, who shared Borman's *Gemini* VII flight and who had flown a second time on *Gemini* XII, would be the second man to go into space three times. It would be the first trip into space for William Anders.

The automatic sequencer assumed control of the *Saturn* V countdown at T-187 seconds. At T-10 seconds, the flame pit was deluged with tons of water that served to cool the launch pad and dampen the acoustical effects of the 7.5 million pounds of thrust. Ignition sequence started at T-8.9 seconds with each of the engines starting at a slightly staggered interval to avoid the stress of the explosive-like impact of the volatile liquids flaming into life. It took almost six seconds for the engines to stabilize and for the computerized diagnostics to verify their health before the four steel arms that held the rocket to the pad were given approval to move back and allow the rocket to begin its ascent.

The eight umbilical arms that had, until seconds earlier, exchanged electrical signals with the rocket now moved out of the way and were themselves drenched in a cascade of water to minimize the damage from the rocket's exhaust.

The astronauts, as well as the millions of spectators who lined the beaches and roads of central Florida, experienced a soaring degree of excitement as the rocket rose higher into the sky. Still more tens of millions viewing on television had higher levels of heart rate in the vicarious experience of man's first attempt to break fully his bonds with the Earth.

Except for some pogo oscillations in the S-II stage, the first and second stages performed to specification, and the S-IVB finished the job of inserting *Apollo* 8 into a parking orbit 180 miles above the

Earth, eleven and one-half minutes after lift-off. The crew worked with Mission Control to verify the proper functioning of all of the systems before committing the spacecraft to its journey to the Moon.

Soon the call was communicated to the crew from Mission Control: *"Apollo 8, you are 'go' for TLI, over."* TLI (translunar injection) was the final burn of the S-IVB that would provide the needed 24,500 mph escape velocity. Lovell responded, *"Roger, understand we are 'go' for TLI."* For those who understood the significance of the call, it was a milestone in mankind's exploration of his universe (Figure 24).

The S-IVB lit up, and the CSM began accelerating the final 7,000 mph increment over a period of 5 minutes and 18 seconds. Following shutdown, Borman separated the spacecraft from the S-IVB and turned it around to face the spent third stage. Had a lunar module been aboard the S-IVB, this was the point at which the *Apollo* CSM would have docked with the LM and pulled it free. But the maneuver on this day was just to simulate the activity, and soon another burn of the RCS provided separation from the S-IVB which would pass by the Moon and enter an orbit around the sun.

The crew removed their space suits and began to perform a series of post launch trans-lunar navigation sightings using the Guidance and Navigation (G&N) system while floating within the cabin. Within hours all three began to experience space sickness, possibly induced by the psychological aspect of floating free. The *Mercury* and *Gemini* astronauts had not had a significant problem because of the small confines of those spacecraft. Borman also contracted intestinal flu-like symptoms, but none of these problems appeared to inhibit crew performance or represent a hazardous situation.

The G&N system was critical in determining the path of the spacecraft and the mid-course maneuvers that would be necessary to assure the Moon was intercepted at the appropriate time in space. The digital computer in the unit contained 38,864 bytes of read-only memory and 2,048 bytes of read/write memory. The computer had a small keyboard with 19 keys that represented the ten digits and 9 function keys. Another element of the G&N system was an optical telescope that served as a sextant.

At eleven hours into the flight, the spacecraft was now traveling at a mere 5,600 mph, having been slowed by the pull of the Earth's gravity. The first mid-course firing of the large SPS engine occurred at that point with a short burn of 2.5 seconds, slowing the CSM by 14 fps. It would take 69 hours to reach the Moon.

Sunday, December 22, *Apollo* 8 broadcast its first TV session, 31 hours into the mission (and almost halfway to the Moon). It included a view of the relatively small disk of the Earth. The black and white camera lacked the ability to display the brilliant blue that the astronauts described. *"It's a beautiful, beautiful view, with predominantly blue background and just huge covers of white clouds,"* said Borman,—who by that time was mostly recovered from his bout with the flu.

At 55 hours into the mission, a historic milestone was reached in the annals of mankind when the three crewmen passed into the gravitational field of the Moon. Two scheduled mid-course maneuvers had not been needed, and even now, only a slight velocity correction by the RCS was required. Now, traveling at a mere 2,700 mph, they would begin to accelerate towards the Moon. The trajectory they were on was called a "free-return" because, if they did nothing to their speed, they would simply loop around the Moon, because of the lunar gravity, and return back to the Earth. However, the mission called for them to fire their SPS engine on the far side of the Moon to slow their velocity and allow them to be captured by the lunar gravity into an orbit around it.

As the CSM neared the Moon, all the systems and navigation were again verified, and the call *"Go for LOI"* (Lunar Orbit Insertion) was sent up to the astronauts. As the spacecraft disappeared around the far side, it was now out of communication. If the burn was successful, *Apollo* 8 would emerge after a silence (Loss of Signal) of 33 minutes. If the burn failed to take place, the signal would be acquired early.

When the signal was not acquired early, everyone following the mission began to sense the anticipation that indeed the ship was now in orbit around the Moon. Precisely on schedule, the telemetry from *Apollo* 8 came streaming across the quarter million mile distance, and a cheer went up from Mission Control. The SPS had slowed the spacecraft by 2,000 mph to 4,500 mph providing for a perilune (closest approach to the Moon) of 70 miles and an apolune of 195 miles.

From their viewpoint, the astronauts could see the details of the Moon as no man had ever seen it, not even through the most powerful telescopes. When asked what the Moon looked like, Lovell responded, *"O.K., Houston. The Moon is essentially Gray, no color... we can see quite a bit of detail...*

The craters are all rounded off, there's quite a few of them... the walls of the craters are terraced, about six or seven different terraces all the way down." They were viewing the surface at a passing rate similar to what is experienced on Earth when traveling at high altitude on a jet liner.

On Christmas Eve, *Apollo* 8 sent its last TV transmission from the Moon. One more revolution and they would begin heading back to Earth. Borman began by saying, *"The Moon is different to each one of us...I know my own impression is that it's a vast, lonely and foreboding type existence...."* Lovell then commented, *"The vast loneliness up here... is awe-inspiring... The Earth is a grand oasis in the big vastness of space."* Anders added about the lunar "terminator" (the area where the sun is rising or setting on the Moon), "...[it] *brings out the stark nature of the terrain with the long shadows...."*

Anders began the closeout of the TV session, "...*the crew of Apollo 8 has a message that we would like to send to you. 'In the beginning, God created the heavens and the Earth...'."* Starting at Genesis 1:1 (the Genesis Factors that are the foundation of rocket science) the three astronauts took turns reading from the Bible the first ten verses of the Book of Genesis as the rugged and desolate terrain of the lunar surface moved slowly across the TV screen. It was an emotional occasion for the millions who witnessed it.

It was time to return home as the spacecraft passed behind the Moon for the last time and Mission Control experienced LOS (loss of signal). If the SPS failed to fire, *Apollo* 8 would spend eternity locked in its orbit around the Moon. But right on queue, 37 minutes after disappearing from view, the voice of Jim Lovell confirmed what the telemetry was telling Houston, *"Houston, Apollo 8, over."* Houston replied, *"Hello, Apollo 8 loud and clear."* Lovell responded, *"Roger, please be informed that there is a Santa Claus."* The reference was to a successful SPS burn. They had spent 20 hours orbiting the Moon.

The 63-hour return trip was uneventful, and 35 minutes before re-entry the Command Module switched to its own battery supply and separated from the Service Module 15 minutes later. The spacecraft then made its initial flight into the upper fringes of the atmosphere at an angle of 5.5 degrees, dropping down to 35 miles from the surface before skipping back up another five miles for its final descent. The 5,000-degree temperature that ionized the atmosphere around the spacecraft caused the expected five-minute communications blackout period. *Apollo* 8 landed only a few miles from the recovery task force.

America received worldwide acclaim. Even the Soviet's often-critical response to America's achievements offered its congratulations and added, *"The world now stands on the brink of entirely new experiences in interplanetary exploration."* The effort to downplay the Soviet's own aspirations were voiced by Leonid Sedov, who was the primary spokesman for their space program: *"There does not exist at present a similar project in our program. In the near future we will not send a man around the Moon."* He noted the Soviets preferred to use unmanned robots, playing on the comments of Sir Bernard Lovell a few weeks earlier. In fact, the giant *N1* Moon rocket was nearing its first critical flight test. Although the Soviets had lost the vital opportunity of circumlunar flight, there was still much that could go wrong in the American program. Kamanin wrote in his diary, *"The flight of Apollo 8 is an event of worldwide and historic proportions. This is a time for festivities for everyone in the world. But for us, the holiday is darkened with the realization of lost opportunities and with sadness that today the men flying to the Moon are not named Valeriy Bykovskiy, Pavel Popovich, nor Aleksey Leonov..."*

1968 had been a difficult year for America with the Vietnam War, the racial strife, and the assassinations. Many felt that the successful circumlunar mission gave hope that, if man could overcome such highly technical problems that a flight to the Moon presented, the troubles here on Earth could be managed as well. However, engineering solutions are more easily mastered than the behavioral problems of mankind.

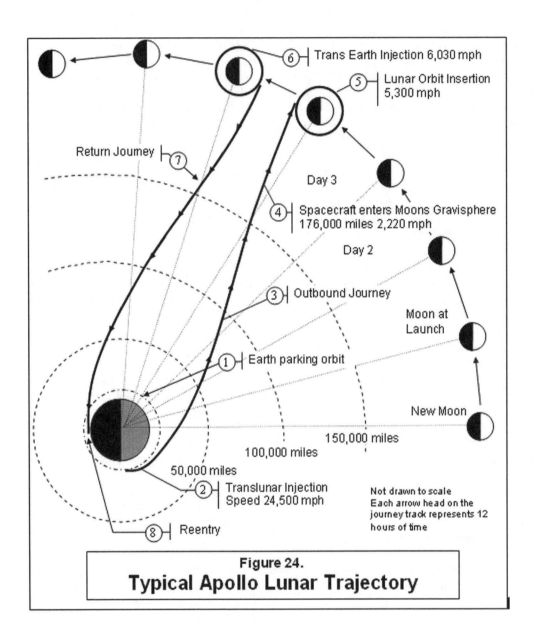

Figure 24.
Typical Apollo Lunar Trajectory

Chapter 27 — One Giant Leap for Mankind

Soviet Response to Apollo 8

At a meeting of senior Soviet space officials just two weeks after the return of *Apollo* 8, the agenda focused on how to handle the success of that flight. To this point the Soviets had used the weightlifting advantage of the *R-7* to accomplish an amazing string of firsts. However, *Sputnik* I (which many in the West equated with a technological Pearl Harbor) was now more than 10 years in the past, and the amazing accomplishments of *Gemini* and then *Apollo* had come without a comparable response from the Soviets.

The meeting sought to determine both long- and short-term goals that might counter the American success. As for their circumlunar bid that had come so close to success, it would be at least several months before a comparable Soviet effort could emulate *Apollo* 8. As it would provide no real increment in capability or technology, it was decided to effectively abandon their manned circumlunar effort. The remaining L-1 spacecraft would be used in an unmanned role with "scientific excellence" being touted. Because the *Zond* moniker had never been differentiated between the smaller early unmanned craft and the *Soyuz* flight test articles, the world would not know of the Soviet's failure in this area.

As for responding to the possibility of an American manned lunar landing, the Soviets took their queue from Sir Bernard Lovell; they would emphasize unmanned robots and make a point of showing that there was no need for man to travel to the Moon and that its exploration could readily be accomplished with less expensive and "no-risk" automated spacecraft. Over the past several years, they had been developing two unmanned lunar landers. The Ye-8 Lunokhod was a wheeled vehicle that would explore the Moon and perhaps provide a homing beacon for a manned lander. The Ye-8-5 automated laboratory was designed to return samples from the lunar surface. This could be accomplished with the *UR-500* booster, whose reliability was still questionable. The decision was made to accelerate the Ye-8-5 project in an attempt to beat the Americans in returning samples of the Moon. This in itself would be a major triumph.

However, the majority of those present still felt that a manned landing was possible in a time frame that would be competitive with the Americans; they would continue with their *N1/L-3* manned lunar program. However, its lack of lifting capability (no LH2 upper stages) and the overweight condition of their lunar lander would require that they use two launches to assemble the required lunar vehicle.

As the Soviets had not publicly declared the Moon as a national goal, they could still hide the program behind their wall of secrecy if they failed to achieve their objective. And it was unlikely that the American's could continue their unbelievable string of successes—it would take only one major failure to set back the *Apollo* landing program for a considerable time. What was the likelihood of the Americans achieving six consecutively successful *Saturn* V launches?

There was also some discussion on turning to near-Earth goals of a manned space station. This, too, would be a part of their deception, depending on how the year played out. Some thought was given towards an elaborate Mars program that would culminate in a manned landing in 1977. This, as well, would never be funded as the events of the year would take their toll on any advanced interplanetary planning. While money was still a major problem for the Soviet space program, the lack of decisiveness and effective management had likewise impaired engineering progress.

Soyuz 4/5

Almost two years after the death of *Soyuz* 1 cosmonaut Komarov, the Soviets were once again ready to try to dock two manned spacecraft. *Soyuz* 4 was launched on January 14, 1969, with 41-year-old Lt. Colonel Vladimir Shatalov as its lone occupant. Following a slight change in his orbit with the on-board maneuvering engine, he held a brief TV session in which he pointed out the two additional seats but made no commitment on how they were to be filled.

The following day, *Soyuz* 5 launched with a crew of three: Lt. Colonels Boris Volynoz and Yevgeniy Khrunov, and civilian Aleksey Yeliseyev. As with previous *Soyuz* missions, a single orbit rendezvous (M=1) had been planned, but for undisclosed reasons the process was extended over several orbits. The docking was accomplished manually, avoiding the problems that had plagued the previous mission. With a combined weight of over 28,000 pounds, the Soviets claimed "the world's first experimental space station." Unlike the *Apollo*/LM combination that had yet to be flown, the docking apparatus did not allow for an internal transfer of cosmonauts from one ship to the other—they had to exit into space from one ship and then enter the other ship. Khrunov, and Aleksey Yeliseyev moved into the living compartment of *Soyuz* 5 and donned space suits with the help of Volynoz, who then returned to the descent module. It was an awkward and difficult task in the limited confines and under weightless conditions.

Following a test of the integrity of the suits, the pressure was lowered in the living compartment, and the hatch was opened. Tethered to the craft, both men made their way over to *Soyuz* 4 and entered the hatch to its living compartment. The 37-minute excursion was only the second space walk for the Soviets since the historic first walk of Aleksey Leonov almost 4 years earlier. As had been the response from all space walkers, the two were amazed and excited by the experience.

Following the successful transfer of the two cosmonauts, the spacecraft were undocked and proceeded to complete independent experiments before *Soyuz* 4, now containing three crew members, returned to Earth on January 17. Volynoz, now alone in *Soyuz* 5, was about to experience the most harrowing re-entry yet. Following the retro-burn, the three segments of the spacecraft were to separate. However, as Volynoz could see from his porthole, only the living quarters had jettisoned. The service module was still in place as the antenna on the solar cell "wings" were plainly visible. Volynoz reported the problem to mission control. Several *Vostok* missions had experienced a similar problem in that they had partially separated, and some of the wiring had held the two components together. However, the two segments of *Soyuz* 5 appeared to be locked together mechanically, and this was essentially a fatal situation.

The reentry began with the two units tumbling end-over-end as the attitude control fuel was quickly exhausted in its attempt to stabilize a mass that was twice what it had been designed for with a center of gravity that was itself out of bounds. Only the heat shield had an ablative coating to survive reentry, and this was now embedded between the descent module and the service module. The atmospheric friction began to destroy both segments. The hydrogen peroxide tanks in the service module were the first to show the damaging effects of the reentry heat when they blew-up. Miraculously, the explosion accomplished the separation, and Volynoz realized for the first time that perhaps he might not die.

The center of gravity was now where it belonged, and, despite the lack of attitude control, the spacecraft assumed the proper aerodynamic position to allow the heat shield to do its job. When the parachute opened, the spinning spacecraft began to twist the risers, and he feared that the chute might collapse. Again, providence prevailed, and the chute was able to stop the rotation and unwind itself, allowing Volynoz to touchdown safely—some 400 miles from his intended recovery point. The TASS News Agency reported that the flight ended successfully, without any mention of Volynoz's unbelievable brush with death.

In an interesting postscript to the flight, as Volynoz and his crewmates were being driven to an awards ceremony a week later, they were fired on by a lone gunman at the gate to the Kremlin. The target of the assassin was Communist Party Secretary Leonid Brezhnev, and the gunman had simply assumed that anyone being driven into the Kremlin in a fancy limousine might well be Brezhnev. An army lieutenant named Ilyin was sentenced to 20 years in prison for the attempt.

On January 20, 1969, another *UR-500* Proton rocket with its unmanned L-1 spacecraft was launched, which would have been designated Zond 7 had not one of the engines in the fourth stage shut down 25 seconds prematurely. The flight program should have been able to accommodate the problems and continue, but confusion in the abort sensor caused the L-1 escape rocket to fire. This lifted the unmanned spacecraft clear from the launcher which was then destroyed and the spacecraft returned to Earth under parachute. On February 19, 1969, another *UR-500* booster with the first Ye-8 lunar lander was launched, but, as it passed through Max-Q (maximum aerodynamic pressure) at T+54 seconds, the new payload shroud failed and caused the destruction of the rocket. These were difficult failures for the Soviet program and the people who had worked so hard to prepare these missions.

The N1 Flies—Almost

The *N1* rolled out to the pad on February 3, 1969, and was about to take flight. Weighing over 6,000,000 pounds when fully fueled, it was slightly heavier than the *Saturn* V. It was moved to the launch pad on a railroad transporter that carried it horizontally in a cradle of steel supports. Its success would put the Soviet program back in competition with the United States. Its failure would perhaps doom the Soviet lunar program.

The firing sequence began on February 21, 1969, as 30 rocket engines, developing more than 10,000,000 pounds of thrust (one-third more than the *Saturn* V), came to life in an unbelievable display of sound and fire. Deputy Chief Designer Chertok described the event: *"Even if you have attended our Soyuz launches dozens of times, you can't help being excited. But the image of the N1 launch is quite incomparable. All of the surrounding area shakes, there is a storm of fire, and a person would have to be insensitive and immoral to be able to remain calm..."*

However, problems had begun even before lift-off. The Engine Operation Control System (KORD) had already sensed trouble with one of the first stage engines (number 12) and had shut it down along with the corresponding engine on the opposite side of the thrust ring to keep the vehicle in balance. As designed, the mammoth rocket continued to fly. However, more problems awaited the giant as it climbed steadily into the winter sky. At T+70 seconds all the engines shut down, and the abort system rocketed the L1S spacecraft clear of the booster.

Although the launch had not been a complete success, it had not been an outright failure. It proved many aspects of the most complex and powerful rocket ever launched by the Soviets—a system that had never even been test fired with all its engines operating.

As the telemetry results were studied, there were several problems revealed that had been working in the giant rocket during its brief flight. The failure of engine 12, detected by KORD just before launch commit, had been an erroneous indication—there had been no need to shut it down or its counterpart number 24. Acoustical vibrations, in particular, had played havoc with connections, and a fire started in the aft end, ultimately proving to be the final source of failure. It was clear that insufficient static testing played a major role in the failure. Although certainly disheartened, the "almost" aspect of the launch still gave hope for the beleaguered members of the space program.

Apollo 9: The Lunar Module

When NASA chose the Lunar Orbit Rendezvous (LOR) as the mode for sending men to the Moon, the Lunar Excursion Vehicle (LEV) became an independent space craft. *Apollo* became the "mother ship" that would accompany it to the Moon but would leave it there and return to Earth without it. This concept represented a complete departure from the traditional spacecraft in terms of the LEV's design elements, as the only environment that it would operate in was the vacuum of space and the 1/6 G lunar surface.

The Grumman Corporation, which had made its name building aircraft for the Navy, had worked hard to become a part of the new aerospace industry and won the contract to build the LEV in November 1962. The name of the vehicle was changed at that point to reflect the "modular" aspect of each of the three segments of the *Apollo* program: the Command Module, the Service Module and the Lunar Excursion Module, or LEM. The basic criteria of weight was paramount as Grumman set to work designing the LEM with the NASA people watching and advising virtually every engineering aspect. NASA engineer Maxime Faggot, who had played a pivotal role in the design of all US spacecraft, was again there to provide counsel.

Because the concept of LOR implied taking only what was needed as far as it was needed, the LEM started life as a two stage spacecraft. The first stage was the Descent Stage that contained a 10,000-pound thrust Descent Propulsion System (DPS) engine that would fire twice. The first time it would drop the craft out of lunar orbit towards the lunar surface, and the second time it would slow the LEM to a soft touchdown on the Moon.

This second firing would demand much from the engine. It would require a throttle to allow the computer, or the Lunar Module Pilot, to change the thrust setting. This would permit the craft to approach the surface at a slow rate of descent and to compensate for the constantly changing weight of the craft as it burned off fuel during the descent (it would consume more than half its weight during the

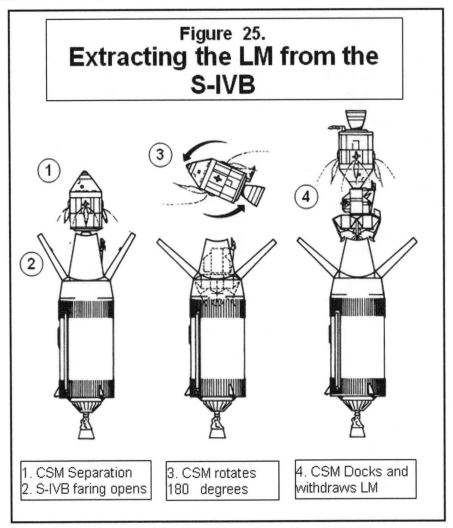

Figure 25.
Extracting the LM from the S-IVB

| 1. CSM Separation
2. S-IVB faring opens | 3. CSM rotates
180 degrees | 4. CSM Docks and
withdraws LM |

landing).

Initially, the Rocketdyne Division of North American Aircraft Corporation was considered as the contractor for the Descent Propulsion System because of their vast experience. However, the Space Technologies Laboratory (later renamed TRW, Inc.), who demonstrated a high degree of innovative and reliable engineering, ultimately received the contract. In addition to containing the descent engine, this stage also had to provide the legs on which the craft would rest on the lunar surface and provide a stable platform from which the ascent stage could be launched. The initial design called for five legs, but this was soon changed to four. The legs had to be folded up during launch from Earth so that the LEM would fit within the S-IVB forward area. The Command and Service Module (CSM) was situated atop the V-IVB.

Following the firing of the S-IVB, the CSM would separate and turn 180 degrees to point back towards it (Figure 25). It was critical that the S-IVB's attitude control system hold the assembly steady so that the Apollo spacecraft could approach and dock with the LEM that was still held fast to it. Following the accomplishment of a hard dock, the LEM would be released by the S-IVB as it was now a part of the set of modules (Apollo, CSM, and LEM) headed towards the Moon.

The pads on each leg were constructed with a crushable aluminum honeycomb material, and each leg had a shock absorbing capability. Even the exhaust nozzle of the DPS had to be frangible and could be "compressed" up to 30 inches because of the possibility of the LEM landing at high descent rates. This would cause the telescoping of the various leg and pad structures and might result in the DPS engine being shoved up into the ascent structure.

The LEM electrical power and environmental systems had to be capable of spending up to 24 hours on the surface. Initially, the fuel cell was considered for electrical power, but, as the problems with this

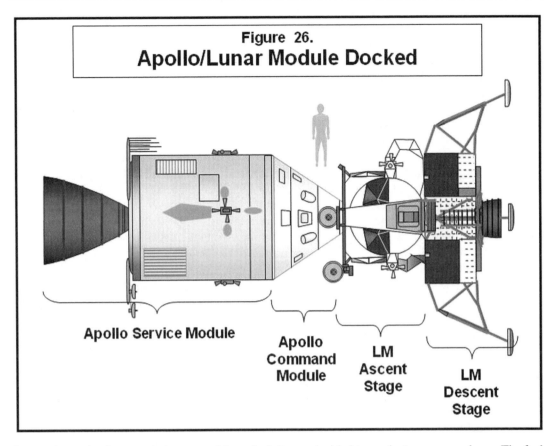

Figure 26.
Apollo/Lunar Module Docked

Apollo Service Module

Apollo Command Module

LM Ascent Stage

LM Descent Stage

innovative technology made its use problematical, it was decided to use battery power alone. The fuel cell would ultimately prove itself aboard *Gemini*, but the decision to use batteries had already been made.

The second stage, the Ascent portion, would launch the craft off the Moon and place it into orbit for rendezvous with the CSM that remained in orbit around the Moon. This scenario would never be fully tested until it would be used in a manned flight. The ability of the spacecraft to contain all of the necessary attributes—to provide its own launch pad, countdown sequencing, and guidance into orbit around the Moon—was a formidable task. Unlike the DPS, the Ascent Power System (APS), built by Bell Aerospace Corporation, employed a single thrust engine. Because of its single purpose role, it did not have to be gimbaled as the DPS required. The RCS provided the appropriate pitch, roll, and yaw during launch from the lunar surface into orbit around the Moon. It did have a slight bias to its mounting, being angled 1.5 degrees towards the forward leg—the leg that would be pointed to the azimuth of flight.

The compactness of the entire LEM is illustrated by fact that the APS engine compartment enclosure protruded up into the cabin area and provided a seat on which the astronauts could sit while on the Moon. The DPS, APS, and RCS engines all used nitrogen tetroxide as an oxidizer and a blend of unsymmetrical dimethylhydrazine as a fuel in separate tanks. The RCS propellants could be cross-fed from the APS tanks if needed.

Initially, two docking ports where considered, one on top and another on the front of the LEM. However, it was quickly realized that only a single overhead port was needed for docking, and the second one was converted to simply a hatch to allow the astronauts to exit onto the lunar surface. This also necessitated overhead windows through which the astronauts would have to be looking over their heads during the docking process. This would turn out to be a "pain-in-the-neck" process (Figure 26).

Weight was a critical consideration, and pounds were shaved from every possible system and structure. Even the astronauts had to give up their couches. They would be suspended by a harness in a standing position, as there would be no high-G powered phases in the 1/6 G lunar gravity. This also meant that they would spend their sleep period sitting on the floor of the cramped cockpit.

The initial specifications called for the LEM to weigh 24,000 pounds, but, by the time it flew, that figure had jumped to 32,000 pounds. The increase could be accommodated only because of the continual improvements to the *Saturn* V. Because the CSM configuration was frozen several years before the LEM, the majority of the increased weight lifting ability was allocated to the LEM. Likewise, cost increases pushed the LEM from an initial estimate of $400 million to a final figure of $2 billion (the CSM would end up at $3.5 billion).

The efforts of the *Ranger* and *Surveyor* lunar probes had been intended to support the design of the LEM. However, the accelerated time-line for the decisions that had to be made regarding its configuration was such that virtually no feedback from these projects was incorporated into the LEM. What was of value, however, was that *Surveyor* confirmed the assumptions made regarding the lunar surface.

Pushing the weight reduction program, while trying to accommodate all of the required complex propulsion, environment, and navigation systems, caused innumerable delays. It was fortunate that the other two modules, the CSM, could be tested without the LEM.

By 1966 the term "Excursion" was dropped from the LEM to avoid confusion in the minds of Congressmen who were then debating the merits of a follow-on to the *Apollo* program. The LM (as it was renamed) did not have the ability to perform excursions remote from the lander itself as some believed the name implied.

A *Saturn* IB (AS-204) flew unmanned on January 22, 1968, and placed *Apollo* 5 and the first iteration of the Grumman Lunar Module (LM) in Earth orbit. The *Apollo* was not docked to the LM, but instead the LM was released from its storage area in the S-IVB to perform a series of burns of both its ascent and descent engines. A computer glitch shut down the first burn of the Descent Propulsion System (DPS) after only four seconds. But a contingency plan provided for two short burns of the DPS while performing a series of throttling profiles, followed by the Ascent Propulsion System (APS) exercising a simulated descent abort. Yet another programming problem quickly depleted the attitude-control fuel. Because of the ability to cross feed the fuel from the APS to the RCS, a stabilized second burn of the APS was achieved until the cross feed was closed, and the ascent module proceeded to tumble. By then virtually all of the primary objectives had been satisfied. The mission was declared a success, and a second unmanned test of the LM was removed from the schedule.

In spite of the 18-month delay caused by the *Apollo* 1 fire, the entire program came together for the first flight of all three modules with the *Apollo* 9 mission on March 3, 1969. Jim McDivitt, Rusty Schweickart, and David Scott flew the second manned *Saturn* V and experienced some pogo effect in the second stage but encountered no other problems as they achieved an initial 120-mile Earth orbit. The plan called for the astronauts to exercise all three modules in a sequence that would emulate a lunar flight—while remaining in Earth orbit. This was the most demanding flight yet, but it did not receive as much publicity as Apollo 8 because it remained in Earth obit.

The CSM was turned around and docked with the LM three hours into the flight. The LM was then released from the S-IVB and McDivitt moved the CSM/LM combination a safe distance from the S-IVB which was then restarted twice and flown to escape velocity and into an orbit around the Sun. The astronauts removed their space suits and began a ten day series of tests, most of them directed towards proving the various LM systems.

With McDivitt and Schweickart aboard the LM, the two vehicles undocked. For the first time since *Gemini* 4 (Molly Brown), spacecraft were again given distinctive names so that communications between the two could be differentiated. The LM, because of its obvious wide legs, was "Spider" and the CSM was "Gumdrop" because of its silver cylindrical shape. A series of burns of the DPS and APS tested the integrity of both systems as well as the ability of either the CSM or LM to function as the "active" participant in the rendezvous process. These maneuvers were critical, as the LM had no way of returning to Earth should the final rendezvous fail. A brief stand-up EVA was also conducted to prove the ability of the astronauts to transfer from the LM to the CSM if a lunar return docking failed to be achieved. The lunar space suit was also tested on Schweickart during the brief (37 minute) EVA. After completing all of the LM tests, McDivitt and Schweickart returned to the CSM, and the LM was eventually jettisoned on the fourth day of the mission.

Apollo 9 continued with a series of experiments to complete the ten days in orbit before firing the SPS to return to Earth only three miles from the prime recovery ship. The way was now clear for a second trip to the Moon.

Apollo 10: Rehearsing the Moon Landing

Apollo 10 launched on May 18, 1969, with the intent of flying a lunar landing mission that would be complete in every detail except for the final burn of the DPS to the Moon's surface. Crewed by Tom Stafford, Eugene Cernan, and John Young, it was the most experienced crew to go into space—five *Gemini* flights between them… no rookies! It was also the roughest ride the *Saturn* V had yet given its occupants and prompted John Young to comment, *"Charlie, you sure we didn't lose Snoopy on the staging?"* The reference was to the CSM being named for the Charles Shultz comic character "Charlie Brown" and the LM named "Snoopy." It would be the last of the somewhat flippant names applied to spacecraft by their crews. Patriotism and science would dictate future communication monikers.

For the second time in the history of mankind the call went up to a spaceship; *"Go for TLI"* (Trans-lunar Injection). The S-IVB lit up for its second burn to accelerate the CSM/LM vehicle out of its earthly parking orbit and to escape velocity. Halfway into the six-minute burn a high-frequency vibration began to rattle the spacecraft. Stafford weighed the abort option as the vibrations continued to worsen to the point that he could hardly read the instrumentation. But, he had come this far and wasn't about to end a mission if all the numbers still indicated that the rocket was taking them to the Moon. (Later analysis revealed the problem to be with the pressure relief valves and was remedied for future missions). "Charlie Brown" then docked with "Snoopy" and pulled him from his "doghouse."

The trip to the Moon was uneventful with the crew using many rest opportunities to beam color TV sessions back to Earth and the millions of viewers around the world. The camera had broadcast quality resolution with 525 lines and 30 frames per second. On the fourth day, *Apollo* 10 slipped behind the Moon and executed the LOI burn of the SPS engine to place the combination into lunar orbit.

Stafford and Cernan entered the LM and completed all the preliminary checks prior to the undocking and the initiation of the new set of "Peanuts" call signs. A novel problem was encountered. During the docking three days earlier, there apparently had been a slight rolling motion that caused the docking latches to become slightly skewed. It was determined that the undocking could take place if the alignment was 6 degrees or less; the mis-alignment was 3.5 degrees. The only caveat was communicated from Houston: *"Charlie Brown, we're concerned about this yaw bias in the LM and apparent slippage of the docking ring. We'd like you to disable, and keep disabled, all roll jets until after undocking…"* The undocking was uneventful—another indication of the wide tolerance engineering built into the *Apollo* systems.

Snoopy then began a series of orbital maneuvers that would simulate the initial descent profile (DOI) to the Moon and at one point came within 50,000 feet of the surface. This never before seen perspective caused Cernan to again make a light-hearted comment as he made his initial contact with Earth as they came around the Moon. *"Houston, this is Snoopy. We is Go, and we's down among 'em Charlie."* The spacecraft was now within a few miles of some of the mountain peaks, and, at their velocity, it was a thrilling ride. He then added, *"Ah, Charlie, we just saw Earthrise and it's got to be magnificent."* Stafford interjected, *"There's enough boulders around here to fill up Galveston Bay, too."* Cernan emphasized the situation again when he said, *"We're low babe, we're low."*

The descent engine fired again to raise the orbit, and Snoopy prepared to separate the Ascent from the Descent stage. A switch (Automatic vs Attitude Hold) had been left in the wrong position and would cause the Ascent stage automatically to seek the CSM communication signal. Thus, when the stages separated, the Ascent stage with Stafford and Cernan aboard executed an unexpected 360-degree pitch causing Stafford to emit the expletive, *"Son of a Bitch!"* Despite the un-programmed maneuver, the astronauts were quickly back in control, and the remaining rendezvous and docking was accomplished without problems. Stafford reported, *"Snoopy and Charlie Brown are hugging each other."*

Speculation in the media suggested that perhaps Stafford and Cernan might just set the LM down—the opportunity to be the first on the Moon over-riding the mission profile. What the media neglected to inform its viewers was that the propellant in the Ascent stage had been limited to only 2/3 of its capacity to simulate the weight and balance of the final return to Charlie Brown. A second consideration was that the LM was the early version that did not have the weight reductions and could not have achieved the return flight even with full fuel. Had they attempted such a stunt, they would not have had enough propellant to complete the return to lunar orbit; so much for speculation.

The TEI (Trans Earth Injection) burn occurred on schedule, and as the spacecraft came from behind the Moon and into communications, Mission Control was treated to singer Dean Martin's rendition of "Goin' Back to Houston" from the on-board tape recorder. The last dress rehearsal for a manned lunar mission was over. The LM had passed all of its qualification tests. It was time for the real thing!

One Last Opportunity

The Soviet program had attempted to move forward during the first half of 1969, following the directives it had established after the *Apollo* 8 flight. A small core of eight cosmonauts continued to train for a manned lunar mission, but it was obvious to them that, even with total success of the next *N1*, they would probably not fly until 1970. A CIA assessment predicted that 1972 would be a more likely period. There had been many ambiguous statements by both cosmonauts and academicians regarding Soviet intentions. The cosmonauts continued to drop hints of their preparedness while the politically oriented science community followed the official line that not only was there no Moon race, but Soviet efforts were geared towards automated scientific methods of returning lunar samples.

The head of the cosmonaut office, Kamanin, had written in his memoirs during the *Apollo* 10 flight that he was disappointed by the *"unrestrained lying"* of Soviet officials regarding the intent of the space program. His thoughts perhaps reflected not so much the need for national security that lies often protect, but that the truth would ultimately prove to be a national humiliation. He then noted that *"We have come to the end to drink the bitter chalice of our failure and be witness to the distinguished triumph of the U.S.A. in the conquest of the Moon."* Chief Designer Vasiliy Mishin, who had taken over Korolev's responsibilities following his untimely death, was summoned to a meeting with Communist Party Secretary Brezhnev in April of 1969 to explain the current state of affairs as the American effort moved closer to success. His report simply reiterated what everyone in the space program knew: the organizational bureaucracy, coupled with a poorly coordinated program and a lagging technology infrastructure, had been severely impacted by lack of funding.

Five Ye-8-5 robotic soil-return probes had been prepared, and the first launched on June 14, 1969 in an effort to upstage the Americans. However, the fourth stage of the *UR-500* "Proton" booster malfunctioned, and the vehicle dropped back into the atmosphere to burn up. The *UR-500* had now failed five consecutive times and eight times in 14 attempts. Intense analysis appeared to point to quality control as the dominant factor. Random failure as opposed to systemic failure indicated that the basic design was sound. Now there was but one more lunar window, which would be coincidental with the scheduled *Apollo* 11 flight. It might just allow the Soviets to pull off the "scoop of the decade" by returning a lunar sample before the Americans.

The second test of the *N1* would precede that effort. The giant rocket came to life in the early hours of July 4, 1969. (Because it was just before midnight in Moscow, the official launch date is often given as July 3, 1969.) There was much apprehension that accompanied the frantic effort to prepare the rocket. Night launches are always more spectacular than those in the daylight as the intense lighting contrast highlights the unbelievable event. The *N1* rose slowly, but, just as it was clearing the lightning towers that surrounded the launch pad, the abort sensor fired the escape rocket, and the unmanned L3S lunar module was pulled free. A few seconds later, an explosion enveloped the vehicle, and the equivalent of a small nuclear bomb devastated launch complex 110. Safety precautions had prevailed, and there were no casualties other than the local wildlife.

Analysis of the telemetry revealed that five engines had failed at lift-off. KORD had detected the fatal anomaly and had attempted to shutdown all the engines at T+10 seconds. However, one engine continued to fire, and this caused the rocket to tip almost horizontal until it impacted the ground. There were several reasons for the failure, but again quality control loomed apparent. Of course, the failure would never be reported to the world.

Now there was only one opportunity left for the Soviets to recoup some of their sagging prestige. Another *UR-500* with a Ye-8-5 had been prepared in one last effort to upstage the Americans. It launched on July 13, 1969. The vehicle performed perfectly, and the payload, *Luna* 15, was on its way to the Moon at last. Twelve years after the first *Sputnik* and eight years after President Kennedy had made his commitment, the Moon race would, in fact, play out over a period of literally hours to the checkered flag—although one participant was a robot.

Apollo 11: " The Eagle Has Landed"

The crew for *Apollo* 11 had been announced on January 9, 1969: Neal Armstrong, Michael Collins, and Edwin "Buzz" Aldrin. All three were born in the same year—1930. Armstrong had piloted the X-15 before being selected as an astronaut in the second group following the original *Mercury* 7. Collins and Aldrin were from group three. All had *Gemini* flight experience—Armstrong's had been the first and only American flight brought down early because of the stuck thruster problem. He had also ejected from the Lunar Landing Training Vehicle (LLTV) the previous year. The LLTV was an ungainly looking device with four legs that used a jet engine positioned vertically to support 5/6ths of its weight to simulate the lunar gravity. Hydrogen peroxide thrusters then allowed the "pilot" a limited amount of maneuverability equal to the LM. Following an in-flight malfunction of the LLTV he narrowly escaped death.

At the time of their appointment, there was no way of really predicting that *Apollo* 11 would actually be the first manned lunar landing. But by late 1968 that is what the schedule called for. With the unerring success of *Apollo* 9 and 10, *Apollo* 11 was to carry out its assigned mission—land on the Moon. From the time that their crew assignment had been made public and as the date for the launch grew closer, all three recognized that it would be a long time before their life would be their own—if ever. Now they were public figures, and the situation would only get worse—if they were successful.

LM-5 and CSM-107 had been assigned to ride atop AS-506, the first of the fully operational *Saturn* V rockets that would not have to carry the complete set of engineering sensors. Only 1,348 measurements were provided as opposed to the 2,342 on the *Apollo* 10 flight. The crawl from the Vehicle Assembly Building to the flame pit of the completed assembly occurred on May 21, 1969, two months before the scheduled flight.

Initially, the first four landings were to be primarily engineering efforts to "prove" the path to the Moon. However, as costs soared and criticism continued to be leveled at the usefulness of the Moon program, it was decided to include more scientific aspects in the early flights. The Apollo Lunar Surface Experiment Package (ALSEP), designed to provide some of the science performed from the Moon's surface, would be deployed by the astronauts.

Collins, Aldrin, and Armstrong trained intensely during June and there was some concern that all might not be ready for the July launch date. A ninety-minute conference call on June 12 that included Slayton, von Braun, Chris Kraft, as well as the astronauts discussed the situation. In the final analysis, Armstrong summarized the training situation by saying that it would be tight, but they would be ready. With the final input on the booster, the spacecraft, the ground teams, and the crew, Apollo program director Sam Phillips made the decision to proceed with the launch on July 16.

Preparations to report the first landing on the Moon and to accommodate the spectators at Cape Canaveral for the *Apollo* 11 launch were greater than any other happening in history. Many believed this date would become as famous as Columbus's 1492 discovery of America. It would be celebrated in real-time as no event in the past had. The gaudy displays of the Super Bowl halftime had yet to be invented, and there was no basis for comparison. More than 18,000 official visitors were invited; many were representatives from around the world. Even within the restricted areas of the launch facility, more than 40,000 people gathered. The real crush of humanity was the hundreds of thousands who jammed the beaches and highways that surrounded the Cape in an attempt to see history in the making. Almost without exception, the assembled masses congregated without incident. During the two months leading up to the launch, even natural deaths declined noticeably; apparently the desire to witness the first man on the Moon was greater than the pull of the Grim Reaper.

Rev. Ralph Abernathy, who had succeeded the assassinated Dr. Martin Luther King as the head of the Southern Christian Leadership Conference, a group lobbying for Civil Rights legislation, arrived at the gate to the Kennedy Space Center with a group of poor people to protest the expensive lunar program. He was met by NASA Administrator Thomas Paine who compassionately, but realistically, told them that, *"…if it were possible to not push that button and solve the problems you are talking about, we would not push that button."*

Hermann Oberth, the man who had envisioned a flight to the Moon some 50 years previous, was present. Willy Ley, the prolific writer and advocate of space flight, unfortunately passed away just one month before the flight. The wife of Dr. Robert H. Goddard, the man criticized by the New York Times for prophesying that man could one day send rockets to the Moon, was also among the distinguished dignitaries.

Many of the Saturn V launch staff had been members of the von Braun team that arrived in the United States 24 years earlier as a part of Operation Paper Clip. However, the bulk of the staff in Mission Control were young "home grown" men in their late twenties and early thirties. Flight Director Gene Kranz, with his crew cut, white shirt, tie, and white vest, epitomized those in the front line of consoles referred to as "the trench." The vest had become a symbol—made special by Kranz's wife Marta for each launch. There were thousands of support people who were every bit a critical part of the mission—but only the three men in Apollo 11 had their lives on the line.

As first envisioned by NASA planners in the early 1960s, only one of the two astronauts would actually leave the LM to explore the lunar terrain and gather soil samples. The second crewmember would remain "safely" in the lunar module to avoid any hazards and to be ready to leave if problems arose. By 1968, as the planning began to take on more detail, it was realized that having both outside made the setting up of scientific equipment easier and provided the ability to help each other should trouble occur.

The question of who should exit first was also an interesting issue. Of course, it was realized that the first man on the Moon would be the one in history books. The initial planning called for the Lunar Module Pilot (Aldrin) to exit first. But, the final design of the LM positioned the men in such a way that it would have been almost impossible for Aldrin to squeeze by Armstrong to exit first. Although Armstrong was a quiet and introverted person, he left no doubt that, as Mission Commander, he would be the first to set foot on the hallowed soil. Aldrin, who was also somewhat of a loner, had missed being selected in astronaut selection group two because he was not a graduate of a test pilot school. He made his mark in the astronaut corps as an academic, immersed in the physics of orbital mechanics. He made some effort to reverse the decision as to who would exit the LM first but realized that it was futile.

The time had arrived—July 16, 1969. Aldrin recalled they were awakened early and had breakfast with Dr. Thomas Paine, the NASA Administrator. *"He told us that concern for our own safety must govern all our actions, and if anything looked wrong we were to abort the mission. He then made a most surprising and unprecedented statement: if we were forced to abort, we would be immediately recycled and assigned to the next landing attempt. What he said and how he said it was very reassuring."* Paine wanted the three to know that there was no reason for them to push a bad situation thinking it would be their only opportunity to be first to the Moon. *"We ... began to suit up—a rather laborious and detailed procedure involving many people."*

Then there was the 24 minute ride to the launch pad. The humid tropical heat of that July morning was beginning to build, and the sky was almost cloudless. Aldrin recalled, *"While Mike and Neil were going through the complicated business of being strapped in and connected to the spacecraft's life-support system, I waited near the elevator on the floor below. I waited alone for fifteen minutes in a sort of serene limbo. As far as I could see there were people and cars lining the beaches and highways. The surf was just beginning to rise out of an azure-blue ocean. I could see the massiveness of the Saturn V rocket below and the magnificent precision of Apollo above. I savored the wait and marked the minutes in my mind as something I would always want to remember."*

Collins recalled, *"I am everlastingly thankful that I have flown before, and that this period of waiting atop a rocket is nothing new. I am just as tense this time, but the tenseness comes mostly from an appreciation of the enormity of our undertaking rather than from the unfamiliarity of the situation. I am far from certain that we will be able to fly the mission as planned. I think we will escape with our skins, or at least I will escape with mine, but I wouldn't give better than even odds on a successful landing and return. There are just too many things that can go wrong."*

The countdown proceeded smoothly, and, almost before mankind was ready for it, the numbers dwindled—3, 2, 1, Zero! As if in character, the trio aboard the roaring beast had few comments as they were thrust into the atmosphere—part of 6,384,400 pounds of gross weight. Unlike AS-505, this *Saturn* V flight was smooth and presented no pogo effect—the engineering changes were working.

At T+140 seconds the inboard F-1 shut down as planned. Twenty-one seconds later, the S-IC first stage completed its burn, and the S-II began to convert hydrogen and oxygen into super efficient power. The escape tower jettisoned at T+166 seconds. The S-IVB began its job at T+549 seconds, and, at the end of its two and one-half minute burn (ten minutes and 26 seconds after leaving the ground), the 297,000 pounds of mass that remained was in its parking orbit 120 miles above the Earth.

Tracking data was now analyzed and systems checked to insure that nothing was left to chance. *"GO for TLI"* was issued, and 2 hours and 9 minutes into the flight the S-IVB ignited for the second time to propel the space travelers free of the Earth's gravity. The exact velocity of 24,394 mph was achieved at 5 minutes and 47 seconds into the burn. The Apollo Command & Service Module (CSM) separated from the S-IVB, turned 180 degrees, and docked successfully with the LM which was housed, legs folded in the S-IVB. The LM was then released from the S-IVB, and the CSM thrust the two vehicles clear of the spent third stage, which would proceed into an orbit around the Sun.

To control the extreme temperatures in space, the CSM/LM was turned broadside to the sun and put into a slow roll of one rotation every two minutes; it was called the "barbeque mode." At 25 hours into the flight, *Apollo* 11 passed the halfway point on its journey as its speed continued to diminish under the pull of the Earth's gravity; it was then traveling at only 6,000 mph. The trajectory was so precise that *Apollo* 11 would come within 208 miles of the lunar surface. This was adjusted by a short 2.9-second burn of the SPS engine to provide a miss of only 70 miles some two days later.

Despite the pending history-making event that they were now a part of, the astronauts ate and slept in relatively normal patterns. Their engineering discipline served them well as they were able to perform all the required equipment tests and experiments en route. Several TV broadcasts were made using the color camera (although most people viewing still had only black and white sets). When the LM was entered for inspection, it was found that, like *Apollo* 10, there had been some roll component during docking, but the two spacecraft were offset only 2.05 degrees. *Meanwhile...*

The Soviet Luna 15 successfully fired its descent engine to enter lunar orbit on July 17. The world now wondered, hyped by the media, if the Moon Race had actually come down to a virtual sprint to the finish line?

At 61 hours into the mission, with more than 40,000 miles to go and traveling at 2,050 mph, , *Apollo* 11 passed into the influence of the Moon's gravitational field. It then began to accelerate toward the waiting target 14 hours into the future. Mission Control brought the three travelers up-to-date on how the world was reporting their adventure. They were told that the Soviet newspaper Pravda referred to Armstrong as the "Czar of the ship," to which Collins quipped, *"The Czar is brushing his teeth so I'm filling in for him."*

As the Spacecraft prepared to pass behind the Moon, the array of flight controllers in Houston was polled for the status of their systems, and a ripple of "GO" responses was heard. With ten minutes to LOS (loss of signal), Mission Control sent up the call, *"Eleven, this is Houston. You are GO for LOI* [lunar orbit insertion].*"* Another milestone had been passed.

As the spacecraft looped around the far side, the SPS engine burned for six minutes and two seconds, and *Apollo* 11 was now in lunar orbit. The telemetry numbers received when the spacecraft reappeared from behind the Moon confirmed that all was well. As had the crews of the *Apollo* 8 and 10, Armstrong, Aldrin and Collins were mesmerized by the sight of the lunar landscape passing beneath the ship.

Luna 15 made an orbital correction on July 19 that put it into a 70-mile by 10-mile orbit. Its radar altimeter, however, showed no smooth areas beneath its orbital track at the perilune. Preparation for its landing continued.

Preparation of the LM began at T+81 hours and 25 minutes, followed by another night's sleep before the big day. This sleep period was unlike the previous ones in that none of the astronauts appeared to sleep soundly (according to the biomedical data being transmitted from the sensors on their bodies)—could any human have taken this occasion with normal responses?

The day of the landing had now arrived, and the astronauts wriggled into their spacesuits. Armstrong and Aldrin floated into the LM for the final preparations before separation. Collins remained in the CSM. The LM landing legs were extended and all systems given one final check. *"Go for undocking,"* was the advisory from Mission Control. Now a new set of names that had been chosen to distinguish the two separate spacecraft began to highlight the political aspect of the mission.

"Eagle, Houston. We're standing by, over." Eagle was the call sign of the LM. It was the symbolic bird of America, represented as a large eagle in flight on the decorative patch that adorned the space suits of the astronauts. *"The Eagle has wings,"* was the response from the LM as it undocked. Now

a series of numbers were communicated to update the onboard computer for the Descent Orbit Initiation (DOI) burn. *"GO for DOI"* was issued as the two, now independent, spacecraft disappeared behind the Moon.

Armstrong and Aldrin made the final preparations for executing the first burn of the DPS (Descent Propulsion System) engine. Eagle was facing away from its direction of flight and parallel to the lunar surface so that the thrust would reduce its 3,800 mph speed and drop it out of orbit. The astronauts were initially oriented so they faced the Moon and could identify various landmarks. At ignition the engine developed only 10 percent of its rated thrust so there was little detectable G-force. It was then throttled up to 40 percent for 30 seconds before shutting down. The spacecraft was then reoriented to face away from the Moon, allowing them to view only the star-studded blackness of space. Eagle now fell in a curved path toward the lunar surface. There were no wings and no parachute that could save a fatal impact should the DPS engine fail to fire for the PDI (Powered Descend Initiate), and only an emergency abort using the APS engine was an alternative to an unsuccessful landing. It is also interesting to note that this would be the very first powered "landing" of a manned rocket. The first time it had ever been attempted!

"Columbia, Houston, Over," was the call from Mission Control to Collins as the CSM emerged from behind the Moon. The word "Columbia," derived from Christopher Columbus's name, had for almost two centuries personified the identity of the United States. It was also a symbolic reference to Jules Verne's fictional "Columbiad" that had carried three travelers to the Moon 100 years earlier. These two call signs, Eagle and Columbia, left no doubt about the intended meaning of the mission— America was winging its way to the lunar surface in response to a deadly serious challenge perceived by its Commander-in-Chief some eight years earlier.

As Eagle appeared from behind the Moon, it was just 17 minutes from PDI, and there were some problems establishing communications with it and with Columbia. The high-gain antennas on both vehicles had to be pointed at the Earth for the signals to be successfully transmitted. With the anticipation that the situation would be quickly resolved, the directive *"Go for Powered Descent"* was passed to Eagle by Collins. In Columbia, he had a direct communication path with the LM.

The PDI was a twelve-minute burn of the DPS—all the way to the surface. The engine again came to life at 10 percent, and after a short period throttled up to 100 percent. Now feeling the effects of positive Gs, the two astronauts were supported by a set of harnesses that positioned them in front of two triangular windows, which still looked out into space. It was not until Eagle began to pitch forward (relative to the direction of flight) that the astronauts were finally able to view the surface of the Moon. This motion also brought the radar altimeter into position so that it could begin monitoring the height above the terrain.

A steady banter of communications filled the quarter million miles between the event and those monitoring the critical systems on Earth. A three-second gap existed between the transmission of a signal and the receipt of the first possible response to that signal.

"Eagle, Houston. You are GO... you are GO for continued powered descent."

"Roger."

"And Eagle, Houston. We've got data dropout. You're still looking good." Houston advised that, although they were not getting a steady stream of telemetry data, what they had showed all factors were nominal.

"PGNCS we got good lock on. Altitude lights out. Delta H is minus 2900." With these cryptic comments, Eagle was indicating that the Primary Guidance and Control System (PGNCS) was getting good data from the radar and confirming that the PDI was within specification.

The computers in Houston were crosschecking the performance data with the remaining fuel and "altitude yet to go" to ensure that the event was within the capabilities of the spacecraft.

Eagle then sent the first of several advisories regarding computer alarms. *"1202, 1202."* The 1202 indicated that the onboard computer was unable to keep up with all of the data being presented. However, the programmer in Houston knew his software code and was quick to assure the flight controller that the computer would be able to continue its primary function by ignoring data that was not of immediate value.

"We're GO on that alarm," was the response issued by Houston.

As Eagle descended through 24,000 feet traveling at just under 1,000 mph, the DPS engine began to throttle down. One minute later, Eagle passed through 15,000 feet and the DPS was down to 55 percent of rated thrust.

With three minutes and thirty seconds until touchdown, Eagle was 7,000 feet above the surface and descending at 90 mph. Armstrong could now see the landing site through his window. A new program, called P64, that would allow manual control of the flight was called up on the computer.

"Eagle you're looking great, coming up on 9 minutes."

One more poll of the flight control stations by Gene Kranz, the primary controller, of Fido, Guidance, Control, G&C, and Surgeon produced the immediate and positive response—*"GO!"*

Then Eagle reported, *"1201 alarm... 1201."*

Astronaut Charlie Duke, assigned as CAPCOM (Capsule Communicator) for the mission, had been relaying the directives from the various mission controllers to the crew. He relayed the over-ride to Eagle, *"We're GO, PGNCS high, we're GO!"*

Armstrong now switched to another computer program, P66, that would allow him to literally hover and move across the lunarscape using his thruster controls while the DPS held altitude. Eagle was slowly descending through 350 feet, and its pitch was almost vertical. From this point P66 would lower the spacecraft the final distance to the surface while moving forward about one quarter mile. As the descent continued, Armstrong could see that the intended landing area had several large boulders and elected to begin manual control. Now the communication was virtually one-way. Both Armstrong and Aldrin described what was happening with their responses: *"Eleven forward, coming down nicely. Five-percent Quantity light. Seventy-five feet, things looking good. Down a half. Six forward."* (The numbers represented feet per second.)

CAPCOM interjected the fuel remaining: *"60 seconds."*

Eagle confirmed the warning. *"Light's on. Down 2 ½...picking up some dust...faint shadow... four forward."*

CAPCOM sent the next scheduled warning of fuel remaining: *"Thirty seconds."* The fuel gage and the telemetry-sending unit only had an accuracy of 2 percent. Thus, at this point the fuel tanks could have run dry. However, the abort sensor was set to detect an interrupt in the supply to the engine, and so not until that absolute point was reached would the APS initiate an abort.

Armstrong knew that he was flying on borrowed time at that point, but his landing approach continued. *"Forward. Drifting right...contact light!"* The aluminum rod that hung beneath one landing pad by six feet had sensed the surface and communicated the event by illuminating a yellow light in front of Armstrong. His response was immediate: *"Okay, engine stop."* Now the 1/6 G of lunar gravity settled Eagle onto the surface as Armstrong continued with the shutdown procedure. *"ACA out of detent. Modes control both auto, descent engine command override off. Engine arm off. Four-one-three is in."*

CAPCOM responded, *"We copy you down, Eagle."*

"Houston, Tranquility Base here, the Eagle has landed."

"Roger, Tranquility, we copy you on the ground. You've got a bunch of guys about to turn blue. We're breathing again, thanks a lot."

The landing occurred as planned on the Sea of Tranquility, the dark Mare that had been referred to as a sea by the ancient astronomers; the first manned scientific station was named for that prominent feature—Tranquility Base.

The actual landing had occurred about 1,000 feet farther west than planned and about 300 feet to the left of the intended track. For the first time in history, a manned spaceship had set down on another body in the solar system.

A series of events now quickly transpired to "safe" the spacecraft and prepare it for immediate take-off if any systems check discovered a critical flaw that might require a quick return to Columbia. Referred to as "stay/no stay" (T-1 and T-2), these contingencies might have to be acted upon if any spacecraft damage was found. The call came from CAPCOM: *"Eagle you are stay for T-1."* T-2 also passed uneventfully a few minutes later, and Tranquility Base was ready for its next phase.

The flight plan called for the astronauts to enter a sleep period before their Moonwalk, but it was obvious that, with their heightened state of awareness, there would be no sleep at this point. Besides, it was now early evening in Europe (where countless millions were following the flight), and, rather than keeping half of that continent awake through the night, it was decided to allow the surface excursion to begin. Following a brief prayer of thanksgiving from Aldrin, the two ate a quick meal and then each attached his Portable Life Support System (PLSS) to his spacesuit. Houston confirmed that they were *"GO for depressurization."*

The hatch was opened at T+109 hours. Armstrong, down on his knees, began backing out though the small 32 inch wide hatch onto the small platform that was referred to as the "front porch." *"Okay Houston, I'm on the Porch."* As there was no video at this point, Armstrong wanted to keep Mission Control advised of his progress. He then pulled a lanyard that allowed one of the equipment bays on the descent module to swing open, and a camera, directed towards the ladder, began to beam images back to Earth. Because the video had to share the communications channel, the resolution was restricted to 320 lines per frame and 10 frames per second. Combined with the stark lighting conditions, the images received were almost ghost-like.

At first the image was upside-down because of the way the camera had been packed, but within seconds the video engineers electronically righted the images, and Armstrong could be seen slowly making his way down the nine steps of the ladder. As he reached its end, there was still a drop of almost three feet to the foot of the LM as the landing gear had not been compressed to any degree by the soft touchdown. In 1/6 gravity the fuzzy image could be seen in a slow motion movement traversing that final distance. He then made a quick hop back up to the last step to verify that there would be no problem and commented, *"Okay, I just checked getting back up to that first step, Buzz; it's not even collapsed too far, but it's adequate to get back up."*

He was aware of the historical importance of his first words as he stepped onto the lunar surface, so he ensured that the world would know when that event happened: *"I'm at the foot of the ladder...the surface appears to be very, very fine grained, as you get close to it. It's almost like a powder."*

Then the momentous words: *"I'm going to step off the LM now. That's one small step for a man; one giant leap for mankind."* The phrase would live in perpetuity.

Armstrong quickly scooped a small quantity of soil into his pocket as a "contingency sample" should an emergency require an immediate return to the LM. Aldrin then followed to the lunar surface. The camera was moved to a tripod so that the viewers could see the complete LM and the activities of the astronauts. The American flag was then planted. A plaque on one of the legs of the descent module that would remain on the surface was read: *"Here men from the planet Earth first set foot upon the Moon, July 1969 AD. We came in peace for all mankind."* President Nixon used the occasion to talk briefly with the astronauts and the world. Scientific instruments were set about. Samples of the lunar soil and rocks were gathered. All too soon, the 2 hour 30 minute moonwalk was over.

Armstrong and Aldrin returned to the LM with their box containing 46 pounds of lunar samples. After re-pressurizing and connecting the space suits to the ascent stage environmental system, the LM was depressurized one last time to allow unneeded equipment, including the heavy PLSS units, to be left on the lunar surface—reducing the weight that would have to be propelled back into lunar orbit. The act was reminiscent of a scene in the science fiction movie *Destination Moon* in which the similar actions were taken—some 19 years earlier. Contemporary science continued to stalk science fiction. The inside of the cramped Ascent stage was now filled with lunar dust that had clung to their boots and spacesuits, and the two began a restless and uncomfortable six hour sleep period.

Luna 15 could no longer beat the Americans back to Earth with the prized cargo of lunar soil, but it could still accomplish essentially the same mission for considerably less dollars and risk. This was what the Soviets were now counting on as the final descent was initiated just two hours prior to the scheduled launch of Apollo 11 from the Moon. The descent was scheduled to take six minutes, but suddenly all communications with Luna 15 ceased at the four minute mark. Luna 15 had hit the side of a mountain. The Soviet news agency TASS reported that it had completed its planned flight and had landed on the Moon.

Although the official Soviet response to *Apollo* 11 was generous, their TV carried no video or film of the launch nor the video images of Armstrong and Aldrin on the Moon. To the average citizen, the *Apollo* 11 lunar landing was just another terse report that was sandwiched between everyday events of the world.

Now, for the first time ever, a rocket would be launched from the Moon and into orbit around it. Lift-off occurred 124 hours and 22 minutes into the mission; a little more than 21 hours had been spent on the Moon. The ascent stage of Eagle, now weighing only 11,000 pounds (down from the initial weight of 33,000 pounds), lifted from the descent stage and ascended into the black lunar sky. In seven minutes and twenty seconds, Armstrong and Aldrin were accelerated to orbital velocity—4,162 mph.

The rendezvous and docking with Columbia went smoothly. To avoid getting Moon dust tracked into the CSM, it was over pressurized slightly with respect to the LM so that the air flowed from the CSM into the LM when the hatch was cracked. In addition, a small vacuum cleaner helped reduce the amount of particulates. This was an important procedure because the dirt could clog sensitive filters in the Environmental Control System. Then the two tired crewmembers of Eagle transferred themselves and their precious cargo of Moon rocks to the CSM and resealed the hatch. On the 29th lunar orbit, Columbia undocked from Eagle. Trans Earth Injection (TEI) occurred at T+135 hours and 34 minutes. The trio was on their way home.

Because there was the possibility that some form of life might have adapted to the extreme temperatures and radiation on the lunar surface, it was decided that the astronauts and their cargo of Moon rocks would be quarantined for a period of three weeks to determine if any such micro-organisms might exist. This meant that on their return they would have to be sequestered in a trailer-like Mobil Quarantine Facility. It also required that the recovery force provide the three travelers with coveralls and breathing apparatus during the transfer from the Command Module to the quarantine facility. It was a less-than-perfect process, but under the conditions it was a reasonable effort to avoid earthly contamination.

Splashdown was uneventful. The decontamination suits were quickly passed to the crew through the momentarily opened hatch as they bobbed around on the ocean, and the *Apollo* 11 spacecraft was lifted to the deck of the aircraft carrier Hornet. The three astronauts, garbed in decontamination suits and breathing apparatus, then walked to the Mobil Quarantine Facility where President Nixon awaited them for a welcoming ceremony. Mission Control, which had been awash with cheers, now started to empty as the controllers and visitors said their last congratulatory remarks and began departing for splashdown parties or just a good night's sleep. One of the large displays at the front of the room had President Kennedy's message prominently displayed: *"I believe this nation should commit itself to achieving the goal before this decade is out of landing a man on the Moon and returning him safely to the Earth."* The *Apollo* 11 emblem was projected next to it with the inscription above it reading: *"Task Accomplished... July, 1969."*

Final Soviet Lunar Efforts

The 13,000 pound unmanned Zond 7 launched on August 8, 1969, two weeks after the return of *Apollo* 11, looped around the Moon, and, after taking color photos of Earth and the Moon, successfully returned to the Earth. Zond 8 launched a year later on October 20, 1970, and was the last of the Zond series. Following a succession of meetings, the Soviets decided to terminate the L1 manned circumlunar program—without a manned flight. The *N1*/L3 lunar landing project would continue, but not for long. Plans also called for continued multiple *Soyuz* earth-orbital flights, but there was nothing now that could truly recoup the lost prestige, and a final effort was made by the Soviets to deflect the true nature of their failure.

A third Ye-8-5 lunar sampler was launched on September 23, 1969. Yet another problem in the fourth stage caused it to burn up in the atmosphere. The launch attempt on February 6, 1970, again resulted in failure. An evaluation of the *UR-500* was then undertaken and some changes made, and success was finally achieved with the launch of *Luna* 16 on September 12, 1970. Following a soft landing, a drill bored into the lunar surface, and the core sample was transferred to the ascent stage. Unfortunately, most of the material was lost during the transfer, and only a minuscule 106 grams (4 ounces) would finally be retuned to the Earth. The path followed by the spacecraft was essentially direct—no atmospheric skip to moderate the temperatures and G forces. The reentry module was reported to have experienced temperatures in excess of 10,000 degrees F and 35 Gs of deceleration. It would be the only lunar return for the Soviets.

A second Re-8 "Crawler," *Luna* 17, launched on September 14, 1971. It was the last, and perhaps the most successful, Soviet lunar mission for several decades. The eight-wheeled vehicle, named *Lunokhod I,* safely landed in a large crater in the Sea of Rains and operated for eight months. It traveled over 6 miles during its lifetime and again presented the Soviet image of automation to explore the Moon.

The third *N1* was launched on June 27, 1971, and failed 48 seconds into the flight. The fourth and last *N1* to be flight-tested rose from the launch pad on November 23, 1972, and all appeared to proceed properly until just 7 seconds before the first stage was to have shut down. At that point, an explosion in the tail section destroyed the rocket. Following a considerable period of investigation and discussion, the decision was made to terminate the *N1* program in June of 1974. There would be no Soviet manned lunar landing.

Why the Soviet Lunar Program Failed

There was much speculation following the successful *Apollo* 11 mission as America (and the world) congratulated itself on the accomplishment—and of beating the Soviets to the Moon. Those in the West who had fought against the manned lunar program were quick to point out that there had been no race. They believed that the Soviets were intent on exploring the Moon with robots and the whole idea of Kennedy's challenge was foolishness and a tremendous waste of money. The Soviets themselves fed that premise continually, and even one of America's most ardent enthusiasts of the space program, CBS reporter Walter Cronkite, jumped on the band-wagon and announced to the nation that *"there had been no space race."* With that kind of endorsement from "the most trusted man in America," many around the world assumed that to be the truth.

The Soviets had hidden their intent and their failures behind a veil of secrecy so well that only by intense scrutiny could anyone have deduced otherwise. For more than 20 years that followed Neal Armstrong's step onto the Moon, it was assumed by many that the competition to land a man on the Moon had all been one sided.

With the fall of the Iron Curtain in 1989, the truth began to trickle out. Many who were a part of the cover-up came forward to reveal the details of the Soviet Moon project and to provide documentation of the failures, although there was still some obfuscation by some to avoid any personal responsibility. It was a bitter pill for a proud nation which had overcome significant adversity to accomplish spectacular feats. The basic reason given by the primary participants in the Soviet space program for the failure to compete successfully was that they were never able to achieve the required reliability of the UR-500 or the *N1*.

However, the intrigue goes beyond that plausible explanation, as there appears to be some "missing links" in the basic story. One aspect deals with the payload being carried by the first two failed *N1* launches in 1969. Because of its inability to carry a full lunar payload, was it carrying a part of a lunar vehicle? Was a *UR-500* scheduled to subsequently launch the manned component that would rendezvous and dock with it in Earth orbit and then perform a TLI burn? Was the launch of *Luna* 15 a part of a conspiracy to claim that the Soviets had made a manned landing and produce the soil samples from *Luna* 15 as proof? Perhaps more time must pass before all of the factors are known.

An equally interesting question is how the Soviets were able to get as close as they did to beating the Americans with the limited technology level and relatively poor industrial and managerial capability that the Soviets possessed. There were many problems in the Soviet style of leadership, in their organizational structure and in the constant requirement to placate the military—who did not support the basic premise of the space program. The vast bureaucracy, coupled with the inflexibility of a planning process, failed to address the Kennedy challenge with an organized, long-term program that had specific goals. It was more than three years after America's commitment that a Soviet manned lunar landing was approved and work begun on structuring hardware for the selected mission modality.

Insufficient funding (estimates ranged between four and thirteen billion American dollars—no more than half of the *Apollo* commitment) led to the decision to flight test complex systems instead of extensively ground test them. This was exacerbated by the lack of quality control processes and an under-educated workforce. The results were devastating.

The inability to develop LH2 upper stages resulted in an inefficient *N1* that had one-third greater take-off thrust but could carry only two-thirds of the *Saturn* V payload. The gross lift-off mass was about the same as the *Saturn* V, but the empty weight of the stages was more than twice as great.

As for the personalities, the lack of visionary leadership following Korolev's death and the constant conflict among the Chief Designers ensured that there would be no consensus on technical direction and a constant squabble for the meager funds.

When the editorials in such prestigious newspapers as the New York Times, following the launch of the first two *Sputnik*s in 1957, are re-read in light of the events that followed over the next 12 years, it is clear that America's own course of action was negatively affected by what some might categorize as a "knee-jerk" reaction. The Soviets had not developed a superior scientific base supported by an advanced educational system as these editorials claimed. That American industry, education and resourcefulness were able to successfully accomplish an incredible task in such a short time period was a testament not only to America's technological and economic superiority but also to a culture that could gather the resources and skills and organize an assault on a problem so effectively.

It is this same mind set that, with its vast technology and economy, America can do anything. Some have called it "The *Apollo* Syndrome." It has led many to believe that the approach to technological problems could be applied to social and behavioral problems. This has not proved to be the case, and these tribulations continue to plague American culture 40 years later.

The failure of the Soviet manned Moon landing simply reflected what had occurred with almost every large technological project in the Soviet Union since World War II. From nuclear power plants to the supersonic transport, the Soviet system failed to produce results—with the primary exception of the *R-7*. It was not until American-style project management and quality assurance techniques were instituted (largely because of the close collaboration during the *Apollo-Soyuz* Test Project) that the Soviet system achieved progress in these areas.

The crew of Apollo 11, Neil Armstrong, Michael Collins and Buzz Aldrin

The Saturn V, America's Moon Rocket

Well before *Apollo* 11 set down on the surface of the Moon, there were those who looked on the Kennedy goal as simply the opening round in establishing a Moon base—not simply a "neat scientific trick." The scientific community, who had taken a back seat to the political urgencies of the space race, now moved forward with determination.

However, with the goal of beating the Soviets to the Moon accomplished, many Americans lost interest in the space program.

As early as 1965, the Lunar Module was seen as the basis for developing living quarters and as a lunar taxi to allow the establishment of an initial base of operations for extended exploration periods. The modest instrumentation and exploration carried out by *Apollo* 11 was expanded for subsequent *Apollo* flights. NASA had anticipated the period beyond the first lunar landing and had established a modest exploration plan for the nine authorized missions that remained.

Not knowing how many attempts it would take to achieve success, planning and funding were approved for ten lunar flights. *Apollo* 11 had targeted a landing area that represented the least risk in terms of unobstructed terrain and power requirements—within 10 degrees of the lunar equator. Emboldened by the success of *Apollo* 11 and the continued performance improvements in the *Saturn* V, two immediate and more demanding objectives were established. The first was to demonstrate the ability of *Apollo* technology to perform a pinpoint landing. Establishment of a lunar base was not possible if provisions and equipment could not be set down on the Moon in a predetermined location. Even before *Apollo* 11, tentative planning pushed for the ability to aim for a specific landing area. The second *Surveyor* III spacecraft that had set down in April 1967 would provide that target.

The second objective was to explore areas that were more mountainous and those in higher latitudes, which were of more interest to the geologist but required greater power reserves and higher risks. In addition, sufficient weight carrying margin in the *Saturn* V and the LM could support up to three days on the lunar surface and return considerably more lunar samples. These capabilities would be integrated incrementally into each successive flight.

Apollo 12: Lightning Strike

Since it was not known whether *Apollo* 11 (AS-506) would successfully complete its mission, AS-507 was assembled in the Vehicle Assembly Building during the summer of 1969 for a targeted launch in September. This would provide insurance for the Kennedy goal of a lunar landing "before this decade is out." The success of Armstrong, Aldrin and Collins in July relieved the pressure, and work schedules at the Cape began a much more relaxed atmosphere—even a 40-hour workweek. As a result, preparations for *Apollo* 12 led to a launch in November.

CSM-108 and LM-6 were virtually identical with the previous mission except for the addition of a nuclear-powered electrical generator for the lunar ALSEP experimental package. That set of scientific instrumentation increased from the 220 pounds carried by Eagle to 365 pounds for the *Apollo* 12 LM named Intrepid.

The actual touchdown point was planned for a quarter mile from *Surveyor* III. Eagle had set down about four miles from its predetermined landing spot because of inaccurate positioning of the LM at PDI. Mission Control in Houston had only 17 minutes from the time that Eagle appeared over the lunar horizon to determine its path and upload the appropriate coordinates to the LM computer for the descent profile. The location of *Surveyor*, farther to the west on the lunar surface, and the timing of the flight would allow 37 minutes for the task.

To achieve its objective, the mission would have to deviate from a "free-return" path early in the flight. A free-return is an initial trajectory that will simply loop around the Moon and return to Earth if the SPS should fail after Translunar Injection. However, NASA was confident of the reliability and redundancy of the systems—coupled with the ability to use the LM engine as an emergency backup during the trip outbound.

The weather was questionable on November 14, 1969, the day of the launch. Rain showers and thunderstorms prevailed as Charles Conrad, Alan Bean and Richard Gordon ascended the elevator to the "white room" 40 stories above the sandy terrain. The low overcast and steady rain persisted as the count-down reached zero. Although the ceiling (base of the clouds) was less than 1,000 feet, the mission rules did not preclude the launch. The giant rocket was quickly enveloped in the low-level overcast, but its fiery tail still licked at the empty launch pad even after it disappeared into the grey mist above. At T+36 seconds into the flight, at an altitude of 10,000 feet, the *Apollo* 12 crew observed a brilliant white light that enveloped the spacecraft, and caution and warning lights illuminated on their flight display. Those on the ground, including President Nixon, clearly saw a lightning bolt travel down the exhaust plume that yet projected from the cloud.

Conrad immediately communicated the status of the ominous red lights. *"Okay, we just lost the platform, gang; don't know what happened here; we had everything in the world drop out!"* CAPCOM responded tersely, *"Roger,"* as those on the ground set about trying to understand what had caused the sudden electrical power failure and its immediate implications

Conrad continued to describe his situation. *"Fuel cell lights and AC bus light, fuel cell disconnect, AC bus overload, 1 and 2 out, main bus A and B out."*

The possibility of an abort was uppermost in everyone's thoughts. The "platform" that Conrad referred to was the *Apollo* guidance unit—the electrical interface to the gyroscopic reference had failed. Fortunately, the *Saturn* V guidance unit in the S-IVB had remained on-line. Had that been interrupted, the rocket would have met a catastrophic end.

As the *Saturn* V continued along its path towards space, the crew reset circuit breakers to bring the electrical systems back. By the time the first stage had completed its burn, Conrad reported, *"We are weeding out our problems here; I don't know what happened; I'm not sure that we didn't get hit by lightning."* He added, *"I think we need to do a little more all-weather testing."* After communicating more status, he noted with some humor, *"That's one of the better sims [simulations], believe me!"* The astronauts were often presented with "what if" scenarios during their training in the simulators, which on occasion they felt bordered on the ridiculous. Conrad's comment was meant to dispel that notion. CAPCOM responded, *"We've had a couple of cardiac arrests down here too, Pete."*

The S-II stage with its five J-2 engines and one million pounds of thrust now continued to power the "stack" towards orbit. The ride was rough for some reason prompting Conrad to comment, *"She's chugging along here, minding her own business."* The reference to the vehicle with a personal female pronoun was often a mannerism of the test pilot.

A short first burn of the S-IVB and the *Yankee Clipper* was in orbit. Now the crew could settle down and take the time, along with Houston, to evaluate what had happened and if the mission in fact should proceed to the Moon or return to Earth. There were voices who expressed concern that there could easily be unknown damage that would not surface until later sequences and that the astronauts' lives should not be risked. Kennedy's goal had been accomplished, why push the issue? The guidance gyros were caged and the platform successfully realigned. All of the electrical systems appeared normal. The decision was made to *"Go for TLI."*

The S-IVB was fired a second time after almost two orbits of the Earth, and *Apollo* 12 accelerated to escape velocity. Following the TLI burn, the Yankee Clipper separated, and extracted the LM from the S-IVB stage. Residual oxygen from the propellant supply then vented through the single J-2 combustion chamber to produce a small thrusting. This would change the path of the S-IVB so that it would not interfere with the assembled spacecraft.

Compared with the lighting strike on launch, the cruise phase to the Moon was uneventful. There was some concern that the crew could not see the exhaust plumes nor hear the low-powered thrusters (25 pounds of thrust) when they fired, but Armstrong, who was in Mission Control, advised that as long as the rate gyros recorded the event there was no problem—his crew had not heard them either. In contrast, the high-powered thrusters (100 pound thrust) were definitely audible as Conrad noted, *"... this thing shakes, rattles and rolls when you fire the thrusters; its like being on a jerking train."*

They also decided to open up Intrepid a day early to make sure there were no surprises waiting within. This was important as the "no-return" burn of the Service Propulsion System motor was scheduled. Several video periods were beamed back to Earth during the 88-hour trip to the Moon.

On arrival, the SPS fired on the backside of the Moon to place the CSM and its ungainly LM into orbit, and three more men were held spellbound by the opportunity to gaze upon the craggy lunar surface. Al Bean appeared to be suffering from a head cold, exacerbated by the weightless conditions and

100 percent oxygen environment.

At the appointed time for vehicle separation, care was taken not to disturb the orbit of Intrepid so the orbital parameters could be more accurately determined. Real-time video of this event and others provided even more vicarious opportunities for the earth-bound public.

The powered descent went as planned, and soon the LM was enveloped in billowing dust as the final few feet of the landing were determined by the radar altimeter. Gordon, who remained in the CSM, reported he had a visual on the LM and that it was resting on the rim of the *Surveyor* crater in the "Ocean of Storms."

Conrad and Bean completed preparation for the first of two surface excursions. As Conrad made the traverse down the LM ladder, making the three-foot drop to the foot pad, his remarks were not as profound as Armstrong's but equally quotable, *"Whoopee! Man, that may have been one small step for Neil but that was a long one for me."* Looking over into the crater he exclaimed, *"Boy, you'll never believe it. Guess what I see setting on the other side of the crater. The old Surveyor."* The *Apollo* program had remarkably demonstrated that man could not only travel to the Moon, but land at any pre-selected point!

As Conrad was repositioning the color camera, he inadvertently pointed it at the sun and damaged the vidicon tube. The large S-Band antenna that had been set-up to allow much improved video signals never got the chance to show the millions watching a much clearer view of the lunar landscape and the activities of the astronauts. It was a major disappointment for NASA.

The two astronauts spent four hours on their first Moonwalk and then returned to the LM for sleep and a replenishment of their oxygen supply. The second Moonwalk took them on a mile and a half journey around several small craters and finally to the *Surveyor*. There they took pictures and removed several items from the now inert spacecraft before returning to the LM and preparing it for launch. The return to Earth was uneventful, and there was an air of routine in the sending of men to the Moon. This would change.

Apollo 13: "Houston, We've Had A Problem"

The crawler-transporter carried AS-508 and its *Apollo* 13 payload out to pad 39A on December 15, 1969. The unhurried schedule called for a launch in April. Few anomalies marred the testing of the rocket during the interim period. One small but aggravating problem was the inability to empty the number 2 oxygen tank of the fuel cell system. Several attempts and techniques were tried before the technicians were satisfied. The problem was attributed to a poorly fitting fill tube. It would take a significant effort to replace the tank, and that would delay the flight until the next lunar launch period. Because it posed no in-flight hazard, it was decided to go with the existing hardware.

The crew of Jim Lovell, Thomas K. Mattingly, and Fred Haise had been assigned the flight the previous August. A few days preceding the launch, it became known that all three had been exposed to rubella, a form of measles, by Charlie Duke from the backup crew. However, only Mattingly showed no immunity and would have to be replaced. A complete crew replacement was not possible at this point, so Jack Swigert from the back-up crew would fly in place of Mattingly. But, replacing a member of a close knit *Apollo* flight team was not a matter to be taken lightly. A series of quick simulator sessions allowed Lovell and Haise to observe Swigert closely—as they would make the final decision. They approved.

The media no longer considered a flight to the Moon of great importance. News coverage of the flight was lackluster, although all three networks managed to interrupt their regularly scheduled programs to show the *Apollo* 13 launch live at 13 minutes past the hour on April 11, 1970. The vibration of the J-2 cluster in the second stage again provided a rough ride—so uneven, in fact, that the center engine shut down more than two minutes early due to low pressure in the LOX line. The guidance system automatically commanded the remaining four engines to burn longer. However, the S-IVB also had to extend its burn nine seconds to help make-up the deficit in velocity. This caused a quick recalculation to verify that enough fuel remained for the TLI burn—there was.

Following a checkout of critical systems while in Earth orbit, the S-IVB again fired to provide the TLI, and the astronauts were on their way. It was Lovell's fourth space flight and his second to the Moon, as he had been a part of *Apollo* 8. Swigert and Haise were rookies. At little more than 27 hours into the flight, the crew passed the halfway point, and a burn of the SPS engine negated the "free return" path. A fifty-minute video show followed, and the crew turned-in for the night.

The next morning, April 13, 1970, the capacity of the number 2 oxygen tank for the fuel cells, the one that had given problems during the Count Down Demonstration Test, showed 100 percent. The evening before it had shown 82 percent—about what was expected. Following a brief conversation with Mission Control, the consensus was that the gauge was bad. Later that day another video session was beamed down, but it was not carried live by any of the networks—space travel was not as exciting as the Dick Van Dyke Show.

Following the closeout of the video, ground sent up a request just short of 56 hours into the flight. *"13, we've got one more item for you when you get a chance. We'd like you to stir up your cryo tanks."*

Swigert replied, *"Okay,"* and drifted over to the console and flipped the four switches that sent electrical current to the small heaters in each tank and the fans used to move the oxygen past the heaters. In the weightless condition of space, there is no heat convection to move the super cold liquid. A few minutes later the crew heard a loud report, and the 44 ton vehicle reacted violently.

Swigert immediately keyed the mic, *"OK, Houston, we've had a problem here."*

CAPCOM responded, *"This is Houston say again please."*

Swigert repeated as he scanned the instruments in front of him, *"Houston, we've had a problem...We've had a Main B undervolt."*

"Roger, Main B undervolt. Okay, standby 13 we're looking at it."

Haise then added, *"Okay, right now Houston the voltage is looking good. And we had a pretty large bang associated with the caution and warning there. And if I recall Main B was the one that had an amp spike on it once before."*

Within minutes, it was obvious from the instrumentation readings that they had lost fuel cells two and three. As Lovell scanned the oxygen supply gauges for the fuel cells, he saw that number one was slowly but steadily dropping as well. He looked out the hatch window and reported even more bad news to Mission Control. *"And it looks to me... that we are venting something. We are venting something out into space."* The fuel cells not only provided their electrical energy but the crew's oxygen as well, and the service module was rapidly loosing their precious supply. The spewing tank generated enough thrusting to cause the spacecraft to perform a continuous roll.

The Command Module's "surge tank" was quickly isolated so that the small oxygen reservoir that it held (about 5 pounds) would not also vent into space. It was their only source of air during the reentry following separation from the LM—several days away.

Flight Director Gene Krantz had just come on duty and asked the question to which he was afraid he already knew the answer: *"Okay, lets everybody think of the kinds of things we'd be venting. G&C [Guidance and Control] you got anything that looks abnormal in your system?"* G&C replied, *"Negative, Flight."*

With two fuel cells out it was obvious that the lunar landing was no longer possible. Krantz immediately realized that they were facing a life-threatening situation. *"Okay, now let's everybody keep cool; we got LM still attached, let's make sure we don't blow the whole mission."*

A short circuit in the heater of number 2 oxygen tank caused a fire resulting in the tank exploding and damaging the other tanks. With the electrical power essentially disabled in the Command Module, whose call sign was Odyssey, the LM, named Aquarius, was now the only source of power and oxygen.

Almost immediately, that option was recognized and actions taken to bring that resource to bear on the problem. The Command Module was powered down and only critical systems in the LM powered up. However, the LM ran off batteries that had been designed only for the duration of the lunar excursion, and its oxygen supply and carbon dioxide scrubbers had been configured for only two men for two days.

With significant damage to the Service Module, it was doubtful that the SPS engine could be used. The two spacecraft were locked together on a path that would take them around the Moon before it would be possible to begin the journey back towards the Earth.

The situation, as it developed over the next few hours, presented Mission Control with an almost impossible task—how to bring the astronauts back alive.

Forty-five minutes after the explosion, oxygen tank 1 was still venting, and its pressure was well below the minimum necessary to generate power. It was obvious that it, too, would soon be depleted, leaving only the LM for power. The decision to move to the LM was communicated. *"It's still going down to zero, and we're starting to think about the LM lifeboat."* The passageway between the two spacecraft was opened.

The guidance unit of Aquarius was aligned before the unit in Odyssey was powered down. As the procedures continued to transfer the life support and other critical functions to the LM, the various contractors from around the United States were called to confirm capabilities and to inquire about added longevity in certain key components. The spacecraft was just over 200,000 miles from Earth, and moving away from it. With the inability to use the big SPS engine, *Apollo 13* was committed to completing its journey around the Moon.

An important task was to use the LM descent engine to change the trajectory back to a "free-return." But, with an almost a full load of fuel remaining in the Service Module, the ability of the Descent Stage to provide significant velocity changes was limited. There was some thought of jettisoning the SM but that would expose the heat shield of Odyssey to the direct and hostile elements of space. It was not known what degradation might take place in the three-day return to the Earth.

A survey of the consumables was one of the first tasks to be completed, for without a clear understanding of what was available, future plans could not be made. The vital supplies consisted of water (for the cooling system), oxygen (for the crew), electrical power to sustain the environmental and communications systems, and the lithium hydroxide units for removing the carbon dioxide from the cabin air.

The LM was designed to provide for two men over a period of 35 hours. It was now being called upon to sustain three for perhaps 100 hours. With the ability to use the oxygen in the two backpacks allocated for the Moonwalk, there was approximately 50 pounds available, which would last well over 150 hours. The 345 pounds of cooling water would also be adequate for 110 hours, assuming that the heat generated by the electrical equipment was significantly reduced. The 2,200 amp-hours of electrical battery supply appeared to be a critical entity. The LM normally used 50 amps, but this was immediately cut to between 10 and 15 amps. Some equipment had to remain powered up such as the gyros to assure that they did not freeze.

From the point the calculation was made of the available consumables, it was going to take 100 hours to return to Earth. That might be too long. It was possible for the LM to provide another shove after they rounded the Moon to shorten the trip by about 10 hours. As it would turn out, this would represent the critical factor.

Within an hour of powering up Aquarius, the astronauts completed the powering down of Odyssey and the power from the failing fuel cell was irrevocably terminated.

Back in Houston and the Cape, astronauts were powering up the simulators as new "pad" sheets (revised sequences of activities) to be communicated to the astronauts were prepared.

With a plan to conserve electrical power, oxygen, and water, only the build up of carbon dioxide became a pressing issue. The available replacement containers for Odyssey were a different size than those of Aquarius. A team at Houston, equipped with the same resources available to the astronauts, devised a means of using duct tape and cardboard to adapt the lithium canisters. Normally these canisters are replaced when the CO_2 level reached 7.5 mm Hg. As this was a conservative figure, it was decided to allow these levels to reach twice that number before replacing them.

At T+61:30, the first burn of the DPS engine restored Odyssey and Aquarius to a free-return path. The next burn, called the Pericynthion+2, would take place two hours after the closest approach to the Moon to accelerate the long journey home. It would shave almost 10 hours off the return time with a reentry at T+142 hours—allowing the most critical consumable, the LM batteries, to perform their life saving activity.

As the spacecraft swung around the Moon, the three astronauts pressed themselves to the windows to witness a sight that only nine other humans had ever witnessed—the Moon passing silently beneath their vantage point. A once in a lifetime opportunity for Lovell and Haise to walk the lunar surface had been lost.

Odyssey was now a dark cold shell, and the three huddled in the tight confines of the Aquarius—built only for two. The temperature quickly dropped to 35 degrees, and the moisture from their breath condensed on the interior surfaces. During their sleep periods, Odyssey provided the "bedroom" so that any activities by the astronaut "on alert" in the LM would not disturb those trying to sleep. But sleep did not come easy due to the damp, chilly environment—not to mention the prospects for survival residing within the thoughts of each of them.

The four-minute 23-second Pericynthion+2 burn at T+79:27 was successful, and now the prospects for a safe return improved considerably. Nevertheless, two and a half days yet remained of the flight, and there was no margin for error or system failure. This may have been the most stressful time for the crew because at this point there was no longer anything that could be done—but wait in the cold emptiness of space.

The astronauts understood the risks they faced and the extraordinary work being performed by thousands of support people back on Earth. Several times during the mission, both the astronauts and Mission Control had offered prayers. Atheist Madalyn Murray O'Hare had been a driving force in removing prayer from the public schools just a few years earlier. Her subsequent effort to ban prayer from being uttered across the federally funded communications that linked the astronauts to their Earthly support people had likewise been rejected by the Supreme Court.

As *Apollo* 13 accelerated back towards the Earth, the tracking data showed that the spacecraft would reenter at an angle of 6.03 degrees, which was well within the tolerances of 5.5 to 7.5 degrees. However, Mission Control was concerned that when the Service Module was jettisoned it might impart a slight velocity vector to Odyssey. So a short burn was scheduled to move the angle more toward the nominal figure. But because the LM was still attached, the burn could not require a positive velocity trim by the LM as the forward facing thrusters would impinge on Odyssey. Therefore, the final 22-second mid-course correction that came at T+137:40 was designed to leave a slightly negative *Delta* V, allowing the aft thrusters of Aquarius to make up the difference. The burn was successful.

The next step was to separate from the Service Module and make the slight velocity correction with Aquarius. The charges were fired by Swigert in Odyssey to sever the connection, while Haise and Lovell used the LM thrusters to make a positive separation. The three astronauts took to the windows to observe and photograph whatever damage might be visible.

As the long dead Service Module drifted into view, Lovell made the first observation, *"And there's one whole side of that spacecraft missing!"* CAPCOM replied incredulously, *"Is that right?"*

Lovell continued his description, *"Right by the high-gain antenna, the whole panel is blown out, almost from the base to the engine."* Inside equipment Bay 4, the astronauts could see the damage that had been done that had included the SPS engine nozzle.

As the trio started to power up Odyssey, the guidance alignment was transferred from Aquarius, and they began preparations for setting their "lifeboat" adrift. The Pericynthion+2 burn had been the life saving factor, for without that push that reduced the flight by ten hours, the water and battery power would have been depleted.

With the life support responsibilities now transferred to the reentry consumables of Odyssey, the hatches between the LM and the Command Module were secured and pressure checked. Aquarius was released at T+141:30 with the comment, *"Farewell Aquarius, we thank you."*

The reentry and recovery were without incident only a few miles from the prime recovery ship. A dramatic chapter in the annals of space flight had come to a close. But as life threatening as it was, it would be years before much of the world would truly understand the magnitude of the problem and the rescue that had been achieved. Flight Director Gene Krantz highlighted the episode in his book aptly entitled *Failure Is Not an Option*. It would be a quarter century before a feature length movie, *Apollo 13*, gave a more complete portrayal of the dramatic events of the flight. Many agree that it was perhaps one of the very few movies that Hollywood ever made that was historically and technically accurate.

As for the cause of the explosion, an in-depth inquiry revealed that the tank heater had been correctly manufactured in 1966 by Beech Aircraft Corporation to the original specifications for 28 volts. However, the voltage was changed a year later to 65 volts, but paperwork and hardware had not been modified. The error might not have caused the explosion had it not been for the problem encountered in attempting to drain the tank following the ground test at the Cape a few weeks before launch. The tank purge procedure had caused the switch to become "welded" in the closed position, and it subsequently could not perform its function as a protective thermostat. An over-pressurization resulted, and the tank exploded. Because of the manner in which all of the various elements of the fuel cell systems were tightly packaged, extensive collateral damage resulted.

Changes were made to the Service Module to make it less susceptible to explosive effects by adding a third oxygen tank in another part of the module. Battery reserves were increased along with several other changes. Fortunately, none of these changes would ever be called upon for the remaining flights.

Apollo 14:   Science and Golf

The flight of *Apollo* 14 was unique for several reasons. It marked the beginning of more intensive science with regard to the expedition; previous missions had concentrated on engineering feedback. It also was the only trip to the Moon by an original *Mercury* astronaut. Alan B. Shepard had been given the privilege of being America's first astronaut in May 1961 when he rode the MR-3 flight on a simple 115-mile ballistic arc. Although celebrated at the time, his accomplishment had already been eclipsed by Yuri Gagarin's flight three weeks earlier, which had been a true orbital journey. Following his flight, Shepard was grounded by an ear problem that kept him from actively participating in either the *Gemini* or *Apollo* flights.

Unlike some of the other astronauts who left NASA when it was obvious that they would no longer fly, Shepard, along with Deke Slayton, remained. Shepard became Chief of the Astronaut Office. However, he never gave up hope of some medical procedure that would return him to flight status. Early in 1969, he became aware of an operation that could repair the ear damage and had it performed in May of that year. The results were more than satisfactory, and Shepard's perseverance paid off. He was assigned to the *Apollo* 14 flight team in August of 1969 along with Stuart Roosa and Edgar Mitchell.

The Command and Service Module CSM-110 and Lunar Module LM-8 became the spacecraft "Kitty Hawk" and "Antares." They were launched on January 31, 1971, following ten months of changes dictated by the almost fatal *Apollo* 13 mission. The early shutdown of the center J-2 on the second stage of *Apollo* 13 had been caused by a return of the "pogo" effect, and that, too, was addressed with modifications to the liquid oxygen plumbing.

The powered segment of the Saturn V (AS-509) flight was without problems. Following the TLI burn, Roosa had difficulty docking with the LM. Three times he tried to nose into the docking collar over a period of more than an hour. As the time required to accomplish the docking wore on, there was concern that the batteries in the S-IVB would be depleted. Should that occur, the S-IVB would not be able to hold a stable attitude, and the mission would be canceled. With the fifth attempt, which bypassed the soft-dock interim stage, a hard-dock finally was achieved. Visual inspection of the mechanism revealed no direct reason for the problem, and the mission continued toward the Moon.

Because of the scientific interest in the Fra Mauro area of the Moon to which *Apollo* 13 had been targeted, Shepard, Roosa, and Mitchell had their destination changed to land there. To accomplish that task, it was necessary to use the big engine of the CSM to lower the LM closer to the Moon before it began the PDI burn. As the astronauts descended to a pericynthion of only 10 miles (50,000 feet), they were in awe. Mitchell commented, *"Looks like we're getting mighty low here. It's a very different sight from the higher altitude."* CAPCOM reassured them that they would be no lower than 40,000 feet above the highest point. Mitch responded, *"Well, I'm glad to hear you say we're that high. It looks like we're quite a bit lower. As a matter of fact we're below some of the peaks on the horizon, but that's only an illusion."*

After separating from Antares, Kitty Hawk fired its big SPS engine to climb back to a higher orbit of 70 miles and wait for the expedition to return from the lunar surface. In preparing for the landing sequence to initiate, Shepard suddenly announced, *"Hey, Houston, our abort program has kicked in."* A faulty abort switch was quickly diagnosed, and they were again cleared for the landing. Antares then began the PDI that was again almost aborted when the radar could not get a lock on the terrain. Shepard quickly reset a breaker, and the descent continued. The daring duo landed an estimated 150 feet from the designated spot on the lunar surface of Fra Mauro.

This region of the Moon was believed to have material from several miles beneath the surface that had been thrown up by a meteoroid impact. The color camera was set up, and Shepard and Roosa began the first of two excursions that lasted four hours. The second EVA involved pulling a small two-wheeled cart called the Modularized Equipment Transporter (MET) on which tools and samples were carried.

On the second EVA, Shepard produced the head of a number six golf club that he affixed to the handle of the contingency sampler. Dropping a small white ball to the surface, he proceeded to make the first golf shot on the Moon. As this first shot was not quite as good as he expected, he allowed himself a "Mulligan" with a second ball.

The return to the CSM was uneventful as the rendezvous was made using the M=1 technique first used with *Gemini* XI. Several scientific experiments were performed during the journey back from the Moon, and, with the splashdown in the Pacific, another successful lunar mission had been accomplished.

Apollo 15: The Moonbuggy

With the completion of *Apollo* 14, America had demonstrated a commanding technological lead over the Soviets, and many believed that NASA should not send any more expeditions to the Moon. The narrow escape of *Apollo* 13, coupled with the tremendous cost of each flight, was seen as justification to suspend the remaining three flights (*Apollo* 18-20 had already been cancelled). However, to the scientific community, *Apollo* was just beginning to return its $20 billion dollar investment.

NASA had recognized that there was more stretch in the hardware and had prepared the J-series missions that would wring the most from the capabilities of the *Saturn* V. Even before *Apollo* 11 had successfully completed the first landing, preliminary contracts had been let to develop a Lunar Roving Vehicle (LRV)—a small Moonbuggy that would allow the astronauts to cover more of the lunar landscape. To accommodate the increased payload weight, the *Saturn* V had to shed weight and increase its performance. Four of the eight retrorockets used to separate the first and second stages were removed along with the four ullage rockets used to settle the propellants (residual thrust of the spent first stage was deemed sufficient). A modified trajectory into a lower Earth orbit allowed yet another 660 pounds to be accelerated to the Moon. A slightly higher propellant flow rate in the F-1 engines completed the tweaking of the *Saturn* V, and the payload for *Apollo* 15 was increased by more than 5,000 pounds.

The LRV "rover" was an extremely lightweight, four-wheeled electrically-powered vehicle. Perhaps more impressive than its innovative engineering was the fact that the ten foot long buggy weighed less than 500 pounds and could be deployed from one of the four LM equipment bays by simply pulling two lanyards. A longer burn of the descent engine, whose propellant tanks were enlarged 6.3 percent, accommodated the 2,600 pound increased weight of the LM. The longer burn also required the silica ablative coating on the engine to be changed to quartz.

To enable the astronauts to range more than ten times as far from the LM than on previous excursions, the life support systems of the space suits were enhanced to provide three EVAs and 20 hours of surface time. The suits were improved to allow greater mobility and increased drinking water. The added capability was also needed should the Moonbuggy breakdown at its furthest point from the LM and require that the astronauts walk back.

Apollo 15 was targeted for a most difficult and dangerous landing area. It would require a higher descent angle to clear the mountains that surrounded the region called Hadley-Apennine. CSM-112 was teamed with LM-10 atop *Saturn* AS-510 for the launch on July 26, 1971. The call sign Endeavor was chosen for the CSM in recognition of English Captain James Cook's 1768 scientific foray to the South Pacific. The LM was aptly named Falcon.

The launch was uneventful, and Dave Scott, Al Worden, and Jim Irwin were propelled into an initial lunar trajectory that, for the first time, did not provide for a "free return." A potentially serious problem was an indication that the SPS propellant valves were in the wrong positions. This was ultimately proved to be a short circuit in a wiper switch and was worked around. A series of other less threatening anomalies were likewise dealt with as the resiliency of the designs of the various systems again showed the strength of the engineering.

As the spacecraft reappeared from the backside of the Moon, Dave Scott reported the successful burn of the SPS engine that now placed them in lunar orbit. *"Hello, Houston, Apollo 15. The Falcon is on its perch."* Preparations for the landing were completed, and the PDI brought Scott and Irwin into the designated landing area. The touchdown was somewhat more pronounced than the previous landings as the exhaust nozzle of the descent stage contacted the surface and was deformed—as it was designed to do. The LM sat backward at an angle of 8 degrees, but this was well within the design tolerance to deploy the rover. An initial indication of an oxygen leak in the LM proved to be an improperly closed urine dump valve.

Scott's egress onto the lunar terrain prompted his comment, *"Okay, Houston, As I stand out here among the wonders of the unknown at Hadley, I realize there's a fundamental truth to our nature—Man must explore. And this is exploration at its greatest."*

The first EVA involved deploying the rover and lunar communications replay unit. This set of antennas would allow the astronauts to transmit voice, data, and video from the rover (and their suits) directly back to Earth when they were out of sight of the LM. With the camera mounted in the rover (and controlled from the Earth), viewers could follow along with the explorers. Scientists at Mission Control provided real-time discussions with the astronauts, who served as their remote sensors in picking up samples and describing their characteristics.

The rover drove well in its first test in the lunar environment although the front wheel steering failed to function. As the rover had both front and rear steering capability, this was not a problem. It was recognized that navigation on the lunar surface, with its undulating slopes and deceptive distance estimates, would be a problem. A directional gyroscope, an odometer, and a computer that integrated the track and time to show the astronauts where they were and the direction to travel to return to the LM was provided.

Driving at a maximum speed of about 5 mph, Scott enjoyed the mobility and driving qualities of the rover. *"Okay Joe [CAPCOM], the rover handles quite well...It's got very low damping compared to the 1G rover [the unit used on Earth for training]... It negotiates the small craters quite well...It feels like we need the seat belts..."* He later added, when queried about the constant movement of the steering, *"It's just there are a lot of craters and it's just sporty driving. I've got to keep my eye on the road every second."*

One of the first objectives on their traverse was to peek down into the Hadley rill, a sinuous narrow canyon that meandered across the lunar landscape. It was a spectacular sight for Irwin and Scott as they stood on the rim, looking across the gorge estimated to be about 700 feet deep and about as wide. The telemetry noted an increase in their heart rate and respiration.

At the end of the first EVA, the ALSEP scientific package was deployed along with a laser retroreflector. An attempt to drill two holes 10 feet into the surface for a heat flow experiment resulted in the drill binding-up after only half that distance. Scott noted, *"I had the impression we were drilling through solid rock."* The EVA ended after 6 hours and 34 minutes.

Following a sleep period, the two were again out on the surface for the second driving excursion. This time the front steering was restored when a circuit breaker on the rover was reset. Returning to the LM after more than seven hours, with another load of lunar samples, they had their second rest period. The physical strain on both astronauts was noted in the biomedical data being reviewed by the medical team in Houston. Both had experienced cardiac arrhythmias on more than one occasion because of the physical exertion.

On the third and final EVA, Scott produced a feather and proceeded to demonstrate a legendary scientific principle. *"In my left hand I have a feather. In my right hand a hammer. I guess one of the reasons we are here today was because of the gentleman named Galileo a long time ago who made a rather significant discovery about falling objects in gravity fields. And we thought what better place to confirm his findings than on the Moon. And so we thought we'd try it here for you. The feather happens to be appropriately a Falcon feather... and I'll drop the two of them here and hopefully, they'll hit the ground at the same time."*

As the viewers looked on, Scott released the hammer and the feather at the same time and they dropped to the surface—landing at the same time. It had taken 500 years for mankind to travel to a location where such a demonstration could be performed in a natural environment.

Following the ceremonial cancellation of some postal stamps (with a special lunar postmark) and the laying of a plaque bearing the names of 14 Soviet and American astronauts who had lost their lives in the pursuit of space flight, the astronauts returned to the LM. Preparations were completed for liftoff, and, at the appointed time, viewers on Earth were treated to the first televised launching of a rocket from the Moon. The rover was left parked in such a manner that its camera could observe the launch.

The rendezvous was completed and Endeavor was soon on its way toward Earth. Just after passing the equigravisphere, Worden perform a 38-minute spacewalk, the first to be performed in deep space, to retrieve film canisters from the Service Module. The descent into the Pacific landing was a bit faster than planned as a result of one chute failing after deployment; the RCS propellant that was dumped during the descent ate through the risers.

The flight would have been remembered for its scientific success had the trio not succumbed to the temptation to make $150,000 dollars by including more postal covers to be illegally marketed to collectors. An additional indiscretion involved a deal to sell replicas of the small "fallen astronaut" figure that had also been left on the lunar surface. None of the three would ever fly again, and they were "retired" or moved to other NASA centers to close out their careers after voluntarily backing out of all the questionable business arrangements.

Apollo 16:   'Wow, Wild Man, Look at That'

John Young, Ken Mattingly, and Charlie Duke were delayed by several months due to a variety of problems that included some changes dictated by the performance of Apollo 15 and a bout with pneumonia experienced by Duke. Changes to the parachute system of CSM-113 and the reinstatement of the solid-fuel retrograde motors in the first stage of the Saturn V were but some of the modifications to AS-511.

The launch on April 16, 1972, was uneventful, but CSM-113 (Casper) and LM-11 (Orion) experienced a series of problems that delayed the descent to the lunar surface and could have aborted the mission. The most troublesome of these was an over-pressurization in the RCS thrusters of the LM due to an apparent leak in the regulator and oscillations in the yaw gimbals of the SPS engine. If there was a problem with the SPS engine, the crew might need the LM to provide the power to return the CSM back to Earth as it had with Apollo 13. The problems were experienced after the two vehicles had separated in preparation for the PDI burn, and the descent was put on hold. The LM thrusters were fired several times to reduce the quantity of propellant allowing more room for the leaking helium as the problem was analyzed. The decision was finally made that neither problem was of a critical nature, and the Descent engine came to life more than five hours late.

The landing was again without problems, and the normally subdued John Young was enthusiastic with the comments: "Wow, wild man, look at that. Old Orion has finally hit it, Houston. Fantastic!" If flights to the Moon had become mundane for those left back on Earth, they certainly were anything but that for those who were fortunate enough to make the trip.

Duke and Young carried out three EVAs with the assistance of the Moonbuggy. In moving among the deployed experiments, Young inadvertently walked through the ribbon cable for the heat-flow experiment, tearing it from the ALSEP unit. There was nothing that could be done to repair the damage, and Young was angry with himself. But the three EVAs were highly successful with the two spending 20 hours, traveling more than 16 miles exploring the region, and returning 170 pounds of lunar samples.

Leaking orange juice in the space suit combined with the lunar dust made the suiting and un-suiting a more difficult chore in the tight confines of the LM. Despite the loss of the heat-flow experiment, the remainder of the mission was successfully concluded.

Apollo 17:   Last Men on the Moon

There was nothing particularly different about the Apollo 17 launch on December 7, 1972, that would set it apart from the preceding six missions that set out to land on the Moon—except that it would be the last manned mission to the Moon for more than half a century. While those who participated in the planning and execution went about their tasks, as they had for the previous lunar excursions, for many it would be the end of their jobs. For the Grumman team, who had engineered the lunar module itself, there was no follow-on program to take advantage of the skills and knowledge of the workers. They placed a sign at the LM level of the giant red gantry that read, *"This may be our last but it will be our best."* Likewise, most of the Boeing employees in Michoud, Mississippi, who had built the giant *Saturn* V, had sought other employment years earlier as their AS-512 creation built in 1967, waited in storage five years for its brief eleven minute flight. Employment in the space program, which reached its peak in 1966 at almost a half million, would dwindle to about 100,000 by early 1973.

The pending Skylab project would see the last *Saturn* V fly five months later, supported by three *Saturn IB*'s and the *Apollo* spacecraft originally allocated for lunar missions 18-20. The last *Apollo* would be flown for the *Apollo-Soyuz* mission in 1975. And then there would be no more manned flights in the American space program for six years until the first Shuttle.

Gene Cernan, a veteran of GT-9 and *Apollo* 10, would command the mission with geologist Harrison Schmitt occupying the other LM position. This was an important crew change in that Schmitt would be the first non-pilot to fly in the American space program. The desire to get an accredited scientist to the Moon caused Joe Engle to be bumped. Ron Evans rounded out the crew.

The target for *Apollo* 17 was the Taurus-Littrow area of the Moon—thought to have cinder cones indicative of volcanic activity. A wide assortment of scientific equipment was carried, including the unfulfilled heat-transfer experiment of *Apollo* 16. The CSM, Christened "America" by Cernan, would also be heavily instrumented to allow Evans to perform a wide variety of scientific observations as he orbited the Moon waiting for his companions to return from the surface.

A hold at T-30 seconds caused by failure of an automatic sequencer to perform a critical function resulted in a manual work-around and a lift-off that was two hours and 40 minutes late but within the 3 hour 38 minute launch window. The area for miles around the launch pad was illuminated brighter than day for this 12:25 A.M. first night launch of a *Saturn* V. The exhaust plume itself could be seen as far away as 400 miles as the rocket climbed into the black sky. The parking orbit was achieved, and the condition of the spacecraft was verified prior to receiving the call from CAPCOM, *"Guys, I've got the word you wanted to hear. You are GO for TLI—you are go for the Moon."*

A growing confidence in the various systems allowed mission planners to select a less forgiving trajectory for *Apollo* 17. The previous two missions had been established in a non-return path that could be powered into a free-return path by the RCS thrusters for the first five hours of the mission. After that point, the LM engine could provide the needed thrust if the SPS failed. *Apollo* 17, however, was propelled into a path that would not allow the RCS or the LM engines to return the craft to Earth should the SPS fail. In fact, the initial non-return trajectory actually would impact the Moon if the SPS burn failed. The first 1.7-second burn 33 hours into the flight increased their speed so that they would now miss the Moon by about 60 miles.

Eighty-Six hours into the mission, the SPS fired to place *Apollo* 17 into orbit around the Moon, followed by a sleep period for the crew. The following day the crew prepared the LM for the lunar landing mission, unfolding its four spindly legs and separating from the CSM America. On the far side of the Moon, the LM, now being referred to by its radio call sign, "Challenger," executed the twelve-minute descent burn; Cernan and Schmitt were on their way to the lunar surface. As Challenger slowly pitched forward, the cratered surface of the Moon appeared to the astronauts as it had many times in the simulations run back at Houston. However, this time there was an air of reality that the "sim" could not reproduce.

Cernan commented, *"OK, there it is Houston, there's Camelot right on target"* (a distinctive crater used as an aide to visually identify the landing area). As with all the other Moon landings, the last 50 feet produced a cloud of dust as the LM backed down until the four-foot feeler probe touched the surface and illuminated the "contact light." *"OK, Houston, the Challenger has landed'.*

There was a moment of uncertainty as the rear leg settled into a shallow crater, and the LM tipped back 5 degrees. Again, it was well within the tolerances of the LM. As they set about verifying the condition of the LM, Cernan was obviously pointing to various gauges and confirming their readings. *"That hasn't changed... it looks good...manifold hasn't changed...the RCS hasn't changed... ascent water hasn't changed...the batteries haven't changed. Oh, by golly, only we have changed!"*

Preparation for the first EVA was soon completed, and Cernan was out the hatch and down on the ladder. *"I'm on the footpad. And, Houston, as I step off at the surface of Taurus Littrow, I'd like to dedicate the first step of Apollo 17 to all those who made it possible... Oh, my golly! Unbelievable!"*

The first of three seven-hour excursions was begun with the rover deployed, the flag erected, and an extensive set of scientific devices powered by a nuclear power supply set about the landing area. Experiments to measure the lunar atmosphere (recording gas molecules near the lunar surface), a lunar surface gravimeter, a lunar ejecta and micro-meteorite facility, and a lunar seismic profiling experiment were the primary ALSEP apparatus. The holes were drilled for the heat flow experiment, and all of the connections were completed.

Schmitt was in his element, describing in detail his professional observations of the rock structures as he spoke of *"fine grained regolith... lighter colored ejectas... scarps... and penetrating bedrock."* Mysterious *"orange soil,"* which he believed might be oxidized material from sub-surface venting, was not the result of a chemical interaction of the lunar soil with oxygen when later examined back on Earth.

The three EVAs allowed the astronauts to spend a record 22 hours exploring the surface, while the rover took them almost 23 miles. They took 2,200 pictures and collected 242 pounds of lunar samples. Cernan and Schmitt spent more than 75 hours on the Moon. During this time, the sun progressed from 15 degrees above the horizon when they had arrived to 40 degrees when it was time to leave. They noted how dramatically the lunar landscape changed with the varying sun angle.

Before they climbed back into the LM, Cernan took a few minutes to note the historic aspect of the event. *"To commemorate not just Apollo 17's visit to the valley of Taurus Littrow, but as everlasting commemoration of what the real meaning of Apollo is to the world, we'd like to uncover a plaque that has been on the leg of the spacecraft..."* It read: *"Here man completed his first exploration of the Moon, December 1972 AD. May the spirit of peace in which we came be reflected in the lives of all mankind."* With the signatures of the three astronauts and that of President Nixon, the plaque portrayed the outline of the Earth's continents.

NASA Director Fletcher said a few brief words from Houston, and then Cernan hesitated before ascending the ladder. He was looking at another smaller plaque placed where the astronauts would see it each time they moved up or down the ladder. He said, *"...while we've got a quiet moment here...I'd just like to say that any part of Apollo that has been a success thus far is probably... due to the thousands of people in the aerospace industry who have given a great deal besides dedication and besides effort and besides professionalism... and I would like to thank them. I guess there might be someone else that has had something to do with it too.* He then read the engraving on the small plaque, *"God speed the crew of Apollo 17... and I'd like to thank Him too."*

Cernan then added, as he stepped from the surface to the LM footpad, *"... as I take these last steps from the surface...for sometime to come but we believe not too long into the future, I'd like to record that America's challenge of today has forged man's destiny of tomorrow. And as we leave the Moon at Taurus Littrow, we leave as we came, and, God willing, as we shall return, with peace and hope for all mankind."*

Following the record stay on the lunar surface and a record number of manned orbits around the Moon, the CSM *America* made the burn on the far side that would return the last mission to Earth—almost four years to the day when man first reached that distant body. Twenty-seven men had traveled to the Moon...twelve had walked its dusty surface... all had returned safely to the Earth.

The Moon is Different

The lunar environment presented some interesting surprises with respect to sight, sound, touch, and smell. Shadows on the Moon, for example, are blacker than those on Earth, making it more difficult to see and work in any area not in direct sunlight. On Earth a shadow is somewhat illuminated by light scattered through the atmosphere. Without omni directional scattering, the lunar shadow allows little discernment of objects that lie within its boundary, although there is some reflected sunlight off the terrain.

Neil Armstrong commented on the phenomena as he attempted to do work in the shadow of the LM during *Apollo 11*: *"It's quite dark here in the shadow and a little hard for me to see that I have good footing."* The astronauts were attempting to access the equipment lockers on the side of the LM. Armstrong continued, *"It is very easy to see in the shadows after you adapt for a while."* Aldrin noted that, *"continually moving back and forth from sunlight to shadow should be avoided because it's going to cost you some time in perception ability."*

Apollo 14 astronaut Alan Shepard was attempting to unload the ALSEP experiments bolted to a pallet. Called "Boyd bolts," each is released by inserting a special tool into a recess and rotating it to release the bolt. However, the recessed sleeves filled with Moondust, and the tool would not go in far enough. Because of the shadow, Shepard could not see into the recess to recognize that it had Moondust keeping the tool from being inserted. When the problem was recognized, there was the frustration of not being able to simply blow into the hole to clear it. They had to turn the pallet upside down and shake out the Moondust. The astronauts were continually thwarted by the smallest shadows working to obscure any recessed feature or gauge, and this cost them valuable time.

The shadow phenomenon also created a problem estimating distances and relationships. *Apollo* 12 astronaut Al Bean related, *"I thought it* [the Surveyor III] *was on a slope of 40 degrees. How are we going to get down there? I remember us talking about it in the cabin, about having to use ropes."* As the sun moved slightly higher in the lunar sky, the perspective changed. When they actually went over to it, they discovered the slope was more like 10 degrees.

Apollo astronauts quickly discovered that Moondust is much finer than beach sand—more like a fine powdery snow. This resulted in the dust easily invading virtually every possible crevice of the suit. It was also very abrasive, cutting through seals and creating air leaks, binding joints and scratching polished surfaces. Because it readily clung to the EVA suits, the astronauts tracked it back into the LM. Eugene Cernan thought its strong odor smelled like gunpowder while John Young thought its taste, *"not half bad."* Moondust was a significant problem.

Jack Schmitt experienced hay-fever-like symptoms that developed very quickly. The severity subsided with his second and third EVA. It was surmised that the dust from the lunar surface reacted with the warm, humid, oxygen-rich content in the LM to produce the smell and the allergic reaction. Unfortunately, the seal on the containers, which should have provided an airtight closure, were damaged by the sharp edges of the dust. As a result, the lunar samples were prematurely exposed to the Earth's atmosphere on their return and could not be properly analyzed with respect to selected chemical reactions back on Earth.

The weight of lunar samples steadily increased from the 46 pounds returned on *Apollo* 11 to the 242 pounds of *Apollo* 17. Likewise, the total duration of *Apollo* 11 on the lunar surface was 21 hours and 36 minutes while *Apollo* 17 remained for 75 hours. The single EVA of Armstrong and Aldrin of 2 hours and 40 minutes was extended to 22 hours over three excursions for Schmitt and Cernan. The *Apollo* 11 astronauts remained within 800 feet of the LM while the Lunar Rover took Schmitt and Cernan more than 22 miles across the lunar surface. The CSM/LM structure in lunar orbit grew from 74,087 pounds to 77,433 pounds. The *Saturn* V increased its Earth orbiting ability from 297,200 pounds of *Apollo* 11 to 310,000 pounds of *Apollo* 17.

The End of the Space race

One of President Kennedy's concerns in establishing the Apollo Moon Program as a national objective was that, if the two countries both achieved that goal within a short period, then being first would not have a significant effect. He was betting the Soviet system would not be able to keep up with the United States over the long run, and that would be the compelling argument demonstrating the fallacy of Communism to the uncommitted countries. He was correct. The USSR could not sustain the level of effort necessary to compete. However, by the time the event happened, most of the world had already accepted as fact that there really was no Moon race, and some of its significance was lost.

Post-*Apollo* planning had called for a manned lunar base, a large orbiting space station, and a manned flight to Mars by 1981. Vice President Spiro Agnew declared, *"It is my individual feeling that we should articulate a simple, ambitious, optimistic goal of a manned flight to Mars by the end of this century."* He was careful not to link his opinion with that of President Nixon, who had very reserved feelings about the next step in space. With the Soviets presenting less technological threat following the Apollo program, there was no political motivation to move these projects forward.

However, the Kennedy challenge had already reshaped the emphasis on space exploration for both the United States and the Soviet Union for the two decades that followed. The political climate in the United States during the decade of the 1970s that followed the lunar landing was one of cynicism. The

Vietnam War, the Watergate Scandal, the perceived oil shortage, and racial unrest continued to take center stage, while the American manned-space-flight effort ground to a halt by 1975. It would be six years before another American astronaut would orbit. For the Soviets, the decade of the 70s was one of frustration as they attempted to move forward with the manned space station.

President Kennedy's own misgivings about manned space-flight expressed in 1962 were echoed by his brother, the Democratic Senator from Massachusetts, in 1969. Like his brother, the late president, Senator Edward Kennedy was not an ardent supporter of a manned space program. Edward Kennedy sought to reduce America's commitment to space when he said, *"I think... the space program ought to fit into our other national priorities."* The cost of these extensive adjunct goals beyond *Apollo* was too much for Congress—who refused to approve an expanded lunar research program.

The Apollo lunar program also proved to be the high point in Wernher von Braun's career. He had moved on to NASA's headquarters in Washington, D.C., to do future mission planning. However, after less than two years of dealing with the ever encroaching bureaucracy of NASA, von Braun could see that the nation was not supportive of his grand visions for flights to Mars and beyond. The *Saturn* V would be the culmination of his career, although he did participate in some of the later changes that would define the Space Shuttle. He retired from government service in 1972 to become the Corporate Vice President of Engineering and Development for Fairchild Industries—a position that moved him far from the creative drafting boards and roaring test stands that had been his domain for the preceding 40 years. The preeminent visionary and early pioneer of rocketry died of cancer in 1977 at the age of 65. His questionable involvement with the Nazi Party during WWII, the use of slave labor to build the V-2, and its use as a "terror weapon" was, for the most part, buried with him.

While the successful completion of the Apollo program essentially marked the end of the space race as far as America and most of the world was concerned, the Soviets continued their effort to compete with the United States. Even though they would still perform some exciting events in space, after the Moon landing, their efforts were anticlimactic. With a few exceptions, the apathy of the world towards space exploration would continue for the next half century.

Launch of the STS-1 Space Shuttle Columbia in April 1981

With the *Apollo* program's objective of landing a man on the Moon before the end of 1969 a reality, NASA turned its attention to what lay beyond. There had been some preliminary planning, but, until the fulfillment of the Kennedy goal, it had been held in the background. The time had come to look aggressively to the future.

The main theme for the NASA was "man in space"—a large space station, a colony on the Moon, and an expedition to Mars were all considered primary objectives. The cost estimates for these adventures were staggering. Whatever the next goal, it would require the ability to lift heavy payloads to low Earth orbit. Using the *Saturn* V, the cost was about $1,000 per pound (in 1969 dollars). To lower this cost, it appeared obvious that some way had to be found to return the various stages of a rocket safely back to Earth so that they could be reused rather than being "expended." Wernher Von Braun, the eminent rocket scientist who had immigrated from Germany to the United States after World War II, had foreseen this and had envisioned that his giant "Ferry Rocket" (featured in a 1953 *Collier's* magazine article) would have some means of recovering all three stages.

As early as 1957, the United States Department of Defense (DoD) had issued study requirements to several contractors to investigate what it would take to convert existing rockets, such as the *Atlas* ICBM and later the *Saturn* I, with a return-to-the-launch-site capability. The basic idea was to add wings and turbojets to the vehicle and have it land horizontally like an airplane.

It was also desirable that, at the very least, the manned spacecraft should be able to "fly back" from space. Several previous programs had provided considerable research regarding winged reentry including the X-15, the X-20, and a novel concept called the "lifting body."

Estimates showed that at least two launches per month would be required to make the effort worthwhile. However, the down side of the concept was that the booster would lose perhaps as much as 25 percent (or more) of its payload to the weight of the structures for the recovery effort.

While most of these programs had the human occupant as the central theme, there was still vocal opposition to continuing to send man into space. The Soviets had demonstrated considerable capability with robotics. With the advent of integrated circuits, the ability to make computers significantly smaller provided extensive on-board programming of events—and even some autonomous decision making. The manned verses unmanned debate would continue for decades to come, but for the 1970s the allure of man exploring the unknown continued to prevail.

Lifting Bodies

During the 1950s, when pioneering research was being performed on reentry shapes for ballistic missile warheads for the ICBMs, it was realized that by contouring these shapes in the form of half-cones, noticeable lift could be generated in the order of 1.5 to 1 (for each foot lost in altitude, the lift provided would move it forward one-and-one-half feet). These special shapes would have to be carefully oriented during the reentry to supply the lifting capability. While this amount of lift was not much compared to a typical airplane that exhibits a glide ratio of 8:1, it would provide the ability to maneuver the reentry vehicle to achieve a landing footprint of approximately 230 miles cross-range and 700 miles down range. These half cones had blunt shaped noses to take advantage of the phenomena of moving the shock wave away from the structure and thereby keeping much of the heat generated during reentry from the spacecraft. They would also have abbreviated aerodynamic fins to provide stability once inside the atmosphere.

By 1958 the "lifting body," as it was called, was considered for the first American manned spacecraft—*Mercury*. However, much more study and development was needed to achieve a stable and controllable vehicle. Because time was of the essence, *Mercury* became a pure ballistic reentry capsule. Nevertheless, work continued on the concept, and the first test flight of a lifting body was aboard an *Atlas* ICBM as part of an Air Force program called ASSET. A subsequent series of research activities led to PRIME (Precision Recovery Including Maneuvering Reentry) that concluded with the successful recovery of a small lifting body after it had satisfied all of the objectives of its acronym.

However, designing a shape that could withstand reentry while maneuvering to a precise recovery area still did not provide the ability to be flown subsonic to a normal "airplane-like" landing. To address this, NASA's Flight Research Center funded several small radio controlled models to develop specific shapes and controls that would allow the spacecraft to perform a relatively normal landing. Once these had shown clear promise, a manned glider that was 20 feet long and 14 feet wide was built of lightweight wood material. Designated M2-F1, the 1,138 pound craft was towed until airborne behind a special high performance Pontiac automobile at Edwards AFB in April 1963.

These "captive" flights were followed by free flight using the venerable Douglas DC-3 to provide an air-tow. On these excursions, the M2-F1 was taken to 10,000 feet and then released to glide to the runway over a period of about four minutes, landing at a speed of 85 mph. To maintain the desired glide speed, the descent angle was in the order of 18 degrees as opposed to the traditional airplane that uses a much flatter 3-5 degrees. Thus, the final "flare"— the transitioning from the descent to the runway touchdown—was a critical maneuver.

The next step was to explore heavier structures and to install a rocket engine to allow flight speeds up to the transonic region. The all aluminum M2-F2, built by Northrop Aircraft, used the reliable XLR-11 rocket engine (as first used in the X-1). The craft was 22 feet long and 10 feet wide and was carried aloft under the wing of NASA's B-52 that was also used for the X-15.

The first gliding free flights were made in July of 1966 from 45,000 feet and at speeds of 450 mph with test pilot Milt Thompson at the controls. The first powered flight was made almost a year later in May. During this flight, the little craft performed several planned maneuvers but then developed a wild rolling motion. The test pilot on this occasion was Bruce Peterson, and he managed to regain control. However, during the final approach to landing, he became distracted by the position of one of the recovery helicopters, and the timing of the critical flare maneuver was off.

The M2-F2 hit hard and bounced back into the air. Peterson lost directional control and the craft proceeded to tumble at over 200 mph down the dry lakebed that was used as a runway. Peterson was severely injured, but, through a series of surgical procedures, he returned to limited flight status. The reconstruction of his body became the central theme of a science fiction book, *Cyborg*, by author Martin Caidin, and a TV series called *The Six Million Dollar Man*. Each week the viewer saw the film of the actual crash used in the opening credits. The M2-F2 was rebuilt as the M2-F3 with more effective airflow control, and by December 1972 it was flying at Mach 1.6 and to altitudes of 71,000 feet.

A similar lifting body designated the HL-10 had been developed in parallel, and, following extensive wind tunnel evaluation, it was first flown as a glider in December 1966. Test pilot Bruce Peterson quickly discovered that the aerodynamic qualities were poor, and the craft underwent 15 months of modifications. The revised HL-10 went supersonic in May 1969 and ultimately flew to Mach 1.86 and 90,000 feet. Further experiments were aimed at determining the feasibility of a small rocket system that would be used to flatten the steep 18-degree approach path to 6 degrees. However, it was determined that the added complexity of the system did not justify its possible function.

A third lifting body constructed during this period was the X-24A that initially looked somewhat similar to the M2-F2. This program, funded by the Air Force, had as its goal a hypersonic cruise aircraft that would be married to the Scramjet (Supersonic Combustion Ramjet) planned for tests on later flights of the X-15. Built by Martin Marietta Corporation, the X-24A was delivered in August 1967 and was again powered by the XLR-11. However, the first glide tests were not made until April 1969, following an extensive series of wind tunnel and ground tests. The first powered flight was in March 1970, and maximum speeds of up to Mach 1.6 and altitudes of 71,000 feet were obtained.

The X-24A was then rebuilt as the X-24B with a much more streamlined nose and a 78-degree double *Delta* wing planform. Following glide tests, the first powered flight occurred in November 1973 and concluded in 1975 after the X-24B had flown to Mach 1.76 and altitudes of 74,000 feet.

It is important to recall that the primary purpose of these lifting body experiments was to determine the aerodynamic qualities and controllability of the craft as it decelerated through supersonic to subsonic flight. They were not intended to compete with the high speed hypersonic X-15. Although a follow-on X-24C had been proposed that would have used either the X-15's XLR-99 engine or the yet-to-be-perfected Scramjet, neither variant was built. Hypersonic cruise flight reached its apex with the X-15, and, although revisited over the next several decades, no viable power plant or structure was deemed acceptable for the mission.

Birth of a Dream

As work on the *Saturn* V moved to completion in 1966, the Marshall Space Flight Center (MSFC) was able to spend more of its resources looking to the follow-on. Working with several aerospace contractors, NASA scientists and engineers studied a wide variety of reusable launch vehicle configurations. However, as in the past, these were all paper proposals since the allocation of money for large projects would have to wait for the successful conclusion of the *Apollo* project. NASA could not attract Congressional funding because of its commitment to the Moon project, and the Air Force continued to lack justification for sending men into space.

The concept of a reusable rocket had been seriously studied for almost a decade. The debate then began in earnest as to the desirability of a fully or partially reusable rocket. Another consideration was a single-stage-to-orbit, or a two-stage-to-orbit configuration. The point of view is determined by which cost was most easily justified and absorbed. Initial development costs were high for the fully reusable vehicle, but these were offset if a use factor of two flights per week were contemplated. The partially reusable configuration had the advantage of lower development costs, but a higher incremental cost of each flight.

A wide variety of proposals was considered. There was a vocal faction who believed that any level of reusability did not bring noticeable savings, but might, in fact, result in higher costs. Until the mission of the vehicle could be defined, there was no realistic way to determine how many flights might take place during a calendar year or during the projected 10-15 year life of the vehicle.

However, there was no doubt that a low-cost and reliable launcher was the key to repetitive applications such as a manned lunar base or a large space station that would have to be built and maintained by the rocket. Nevertheless, without a specific goal, as with Kennedy's Moon landing, there remained a lack of Congressional motivation. In addition, none of the three proposed lunar follow-on projects, including a manned flight to Mars, had the public's attention let alone that of Congress. The Vietnam War and the racial unrest in the cities had virtually paralyzed America's thoughts on the future. Many preferred simply to wait for both of these problems to abate before venturing farther into space.

However, America's space technology infrastructure, built over the ten years that followed *Sputnik*, was already in jeopardy. Tens of thousands of skilled engineers and technicians that had filled the pipeline during the heyday of the early 60s were now being given "pink slips" as there were no large follow-on projects that could use their talents. It was obvious that it was not realistic to put the space program on hold for however many years it might take America to resolve its foreign and domestic problems.

In August 1968, NASA Administrator George Mueller made a presentation to the British Interplanetary Society in which he outlined the future of manned space travel and set forth some basic tenets: *"...there is a real requirement for an efficient Earth-to-orbit transportation system—an economical space shuttle..."* Although the word "shuttle" had been used in reference to a reusable rocket, this was its first formal introduction, and the phrase caught on. From this point forward, reference to a space transportation system would be associated with "Space Shuttle."

In October of 1968, MSFC went out to industry with a "request for proposal" for yet another study on the Shuttle. However, the basic specifications as to the size of the payload bay and the weight to be carried had yet to be solidified. The Air Force was pressured into agreeing to use the shuttle in order to provide additional justification for its being built. The Air Force wanted a 60 foot long by 22-foot wide payload bay that would allow up to 50,000 pounds to be delivered to low Earth orbit (LEO). NASA was content with a 10 by 30 foot payload bay and 25,000 pounds. However, that was not all—the Air Force needed a landing footprint that would permit launches into polar orbit from Vandenberg AFB in California, with the capability of returning to Vandenberg or Edwards AFB after the first orbit. Since the Earth's rotation would have shifted the orbital track almost 1,300 miles to the west, the military needed a large wing that would provide that potential. NASA needed only a modest 300 mile cross range capability.

Until specific aspects of its intended support mission were established, even NASA was unsure of the size of the payload bay or its cross range capability. Virtually all of the proposals evaluated were based on the lifting body, although a few used deployable wings after reentry.

November 1968 saw a change in the political administration of the United States with the election of Richard Nixon to the Presidency. Nixon's focus was extricating the United States from the quagmire of Vietnam, and he was not a strong supporter of the space program. He appointed Charles H. Townes to chair a "Task Force on Space" to help him decide the direction that would be taken following the yet to be concluded *Apollo* program. Their recommendation came quickly in January 1969, and it was simply not to move forward with the Shuttle at that point in time.

February 1969 saw more study contracts awarded and the formation of a new Space Task Group (STG) chaired by Vice President Spiro Agnew. The STG asked both NASA and the DoD to define their requirements, and then a joint group would reconcile the needs of each with a specific program.

NASA decided that the Shuttle would not only support a projected space station but it would be the prime means of launching all future satellites. This caused their payload requirement to increase to the same 50,000 pounds of the Air Force and the 22-foot diameter needed to match the *Saturn* S-IVB stage to which future large satellites were being designed (it was expected that the efficient S-IVB would probably be the upper stage of any follow-on rocket). NASA also envisioned a highly automated computerized checkout to eliminate the majority of people who were required to prepare a large rocket for flight.

NASA had initially focused on a fully reusable two-stage shuttle in August 1969. However, the Space Shuttle Task Group, formed to provide input to the STG, recognized that there were as many as six possible roles for the shuttle and defined several potential configurations that might be pursued in designing it.

The basic issue of "fully reusable" verses "partially reusable" again arose as well as whether it would use a piloted fly-back booster or an expendable first stage, and if existing or new engines would be employed. Further, would sequential staging or parallel-burn provide the best arrangement?

The cross range capability (300 miles or 1500 miles) would dictate whether a winged body or lifting body would be employed—the larger footprint would be the most costly to develop. Finally, the size of the shuttle would be determined by the payload capability of up to 50,000 pounds.

The number of anticipated missions fluctuated between a low of 30 per year to as many as 140 per year depending on the future goals of the American space program. Moreover, since the programs that the Shuttle would support had not been established, there was no definitive means of costing the various options. But it seemed sure that, no matter what configuration or capability was finally determined, the shuttle should drop the cost of going into space by at least a factor of 5 to 15 times—if the cost estimates were based on realistic assumptions.

By September 1969, with the first Moon landing accomplished, the STG submitted its report. It also presented three new national goals to the President. The first was an $8 billion per year expenditure for a manned Mars expedition, a 50-man space station, a lunar orbiting space station, and the reusable Shuttle. It did not recommend any specific Shuttle configuration except to state that it would be a chemically-fueled space transportation system that would provide for virtually all of America's needs in space for the next 20 years.

The second program was somewhat less costly in that it deleted the lunar orbiting space station, while the third option (at $5 billion per year) would consist only of the earth-orbiting space station and the Shuttle. President Nixon did not like any of these options. He was, in fact, at odds with his Vice President, who was strongly advocating the Mars mission. Planning for the Mars mission had progressed to the point of determining that the earliest launch window was November 12, 1981—twelve years into the future.

NASA recognized that the Nixon administration would be a hard sell. It was also obvious that virtually any future goals would be dependent on the Shuttle. Therefore, NASA set its sights on that objective. The new Administrator (appointed in 1971), James Fletcher, also saw that, with the current state of the federal budget which was pulled between the Vietnam War and a greatly expanded social program, every effort had to be made to reduce the cost of the Space Transportation System (STS—as it was now officially designated). It was also critical to get the Air Force to become a full partner in the Shuttle.

The Office of Management and Budget (OMB) would oversee the financial and economic analysis of the Shuttle proposal. However, the more the OMB looked into the whole concept of the shuttle, the less economical it appeared to be. Traditionally, American space projects had cost several times their initial projections, especially when moving the state-of-the-art forward. Thus OMB was very critical of the numbers that were being presented by NASA and the various proposals submitted by industry.

Even when NASA encouraged other countries to become a part of the STS, there were as many negative aspects as positive—especially with the military security requirements. However, an important milestone occurred in 1971 when the Air Force agreed not to develop any new expendable boosters. They would rely on the shuttle for all their future requirements, although they would continue to purchase existing expendables in parallel with the shuttle for the first few years of its operation.

A Major Change

By 1971, it was obvious that there were significant technological problems with the Shuttle as a two-stage-to-orbit vehicle with a fly-back booster. The Boeing 747 was the largest aircraft that had been built, and it weighed 300,000 pounds empty and flew at 600 mph. The proposed Shuttle first stage would weigh 400,000 pounds empty and would have to "fly" at more than ten times that speed. It would carry the manned spacecraft (referred to as the Orbiter) and have to perform separation between the two in a region of speed and space for which there was little data but that extrapolated from the X-15 flights. In addition, although considerable progress had been made with heat protection, the specific method to cover either the first or the second stage had yet to be proven. With so many unknowns, it was easy to see why the OMB had little confidence in the costing and schedules for the project. The OMB had essentially given the Shuttle project a goal of one billion dollars per year. This was less than a third of the initial proposals. Both technologically and financially, the project as defined was beyond what was possible with the state-of-the-art.

New thoughts about how to downsize both the Orbiter and its first stage carrier resulted in the revisiting of expendable tanks. After all, it was the engines and their turbo pumps that were the costly aspect not the relatively inexpensive tankage. This line of thinking also simplified the problem of possible residual propellants in the tanks. If the tanks were removed from the returning craft, the problems of venting, weight, and thermal protection were significantly eased. At first, the Liquid Hydrogen (LH2) was a candidate for an external expendable tank because it comprised about 75 percent of the propellant volume. However, since it only accounted for 18 percent of the weight, the Liquid Oxygen (LOX) was soon included as well. In addition, with the reduced weight, the staging could occur at a lower speed—simplifying the speed problem. It was determined that the optimal staging velocity was about 4,500 mph.

NASA formulated a series of external tank designs that explored a variety of configurations for the Orbiter. This culminated with the most promising (designated MSC-40C) being assigned to several contractors for revising their proposals. The booster itself was the province of the contractor, and it was not decreed that it be reusable. While these efforts reduced the annual expenditures by about 1/3, they were still far above what OMB felt the budget could handle.

The thought of using a phased development in which some systems would be delayed in order to reduce the initial funding impulse was considered. The Orbiter could be flown as the second stage of the *Saturn* V whose development costs had already been absorbed in the *Apollo* program. The reusable fly-back booster could then be developed several years down the road, perhaps after the Vietnam War and its expenses ended, easing the funding burden. It was also at this time that the possibility of using a simplified liquid-fuel engine, or perhaps even solid-fuel "strap-on" boosters as were being used on the *Titan* III-C, was considered, which would be recovered by parachute.

The idea of using solid-fuel boosters with a manned spacecraft had not been actively pursued for several reasons. Solid-fuels did not have the high energy levels (Specific Impulse—Isp) typically needed for efficient booster performance. Once they began their burn, they could not be shutdown—limiting the possible escape system scenarios.

Nevertheless, solids had made significant strides by the early 1970s with the introduction of Nitronium Perchlorate fuels, and Specific Impulse numbers approaching 300 seconds were attainable. Exceptionally large sizes could be achieved by fabricating solid-fuel rockets in segments that could more easily transported and then "stacked" at the launch site. Motor sizes were measured not only in the thrust they produced but also by the diameter and number of segments that comprised the motor. During the late 1960s, significant progress had been made, and motors as large as 260 inches in diameter and five segments in length had been tested. Thrust levels as high as six million pounds had been achieved as well. The ability to provide thrust vectoring to control direction had also made noticeable improvements. In addition, progress had been made with refurbishing motors so that they could be reused after parachute recovery. The simplicity and relatively low cost made the solid-fuel booster an attractive possibility.

The funding issue was so intense that consideration was given to abandoning the Shuttle and using the existing *Apollo* and *Gemini* technology. A new study was instituted to review the current progress and make recommendations. Chaired by Alexander Flax, the committee arrived at the conclusion that there was no economic justification for the Shuttle—the same conclusion that OMB had come to. It was also interesting to note that the estimated Shuttle costs, as defined by NASA, had been almost halved since the first estimates, and yet the capability remained the same. This point would be resurrected decades later when the Shuttle cost was again revisited in retrospect.

Had the Soviets succeeded with their *N1* super booster and a manned Moon landing, even a year or so after *Apollo* 11, there would have been no question that the United States would have moved forward with the Shuttle. However, it seemed obvious that the United States had distanced itself technologically from its arch rival, and the question was "Is the Shuttle really needed?"

However, Shuttle supporters held that if the launch of *Sputnik* in 1957 had taught the United States anything, it should be that the country must continue to be in the forefront of science and technology—cost should not be the overriding issue. In a 1971 meeting between NASA Administrator Fletcher and Deputy Secretary of Defense David Packard, Fletcher was told that NASA was trying to sell the Shuttle on the basis of cost and that was not the correct path. Packard was quoted as saying, *"…the real point has to do with national security and an intangible thing which might be called man's presence in space."* It was the national security aspect of the Shuttle that convinced President Nixon to provide more support than he had previously. Perhaps he was recalling that he had been the Vice President under Eisenhower during the *Sputnik* era and that he had been called on to defend that administration's austere policies during his failed bid for election against Kennedy in 1960. The possible employment factors in key electoral states might also have been an influence on Nixon's opinion of the Shuttle as the 1972 elections moved closer.

The Final Configuration

The move towards solid-fuel boosters was strengthened by the recovery aspect. The liquid-fuel systems would be much more fragile and susceptible to damage, especially from salt-water—the prime recovery area. However, Wernher von Braun in particular was not enamored with solids. The use of the *Saturn* V as a recoverable booster continued to dominate the thinking at MSFC until it was established that the effort to convert it to a reusable system was unrealistic.

The parallel burn of the booster and the recoverable Orbiter had the advantage of being able to assure the ignition and stable burning of the second stage before launch commit. However, the configuration raised the risk of crew escape from a catastrophic failure.

With respect to the large *Delta* wing configuration needed to support the Air Force cross-range requirements, NASA finally decided that launch abort scenarios and some mission profiles could be enhanced with the big wing. The *Delta* wing was by far a more aerodynamic solution that offered additional mission flexibility. NASA decided to use the big wing and parallel staging. Had the planners and decision makers known that the vehicle being configured would be the only path to space for the United States' manned program for almost 40 years, they might have had second thoughts.

In December of 1971, NASA sent a summary of four possible configurations and their cost (which ranged from $4.7 billion to $5.5 billion) to George Shultz, who was then Director of OMB. By that point, Shultz had accepted the arguments and probably believed that NASA had squeezed all that was economically possible from the design efforts. On January 5, 1972, President Nixon announced that he would propose the Shuttle funding to Congress—neither he nor Shultz wanted to define the configuration and only stressed that the costs be held as close as possible to the central estimate of $5 billion.

Nixon also had the option of naming the project but preferred to use the name that it had become known as—the Space Shuttle.

To pay for use of the Shuttle, NASA had established that the using agencies would fund specific launches. They estimated that over the first 12-year period (1979-1990) 580 flights would take place, almost equally divided between scientific research, the Department of Defense, and commercial satellites. Using these estimates and a per Shuttle launch cost of $11 million dollars, the Shuttle would place a pound of payload into low earth orbit for an amazingly low $175 per pound (excluding the R&D costs). This bargain would be too good to be true.

Once approved by Congress, NASA was quick to move forward with defining those aspects of the Shuttle left in limbo. The final configuration would consist of the manned Orbiter, a large external tank (ET) and a parallel-boost first stage that would use two solid-fuel rocket boosters (SRB). The Orbiter would contain the high-energy LH2 primary propulsion unit called the Space Shuttle Maine Engine (SSME)—three engines each producing 375,000 pounds of thrust. The Orbiter would be protected from the thermal affects of reentry by a combination of heat resistant tiles and high temperature metals. The ET contained the cryogenic propellants—LH2 and LOX insulated by a spray-on blanket of foam—and would be the only expendable item.

Four contractors competed for the Orbiter: Lockheed, McDonnell Douglas, North American (NAA), and Grumman. North American's proposal scored the highest points with Grumman a close second—NAA received the initial $2.6 billion contract on July 26, 1972, for the first two Orbiters. The Rocketdyne Division of North American Aviation had also been selected to build the Shuttle's LH2 main engine the preceding year. The awarding of virtually all of the major contracts to a single company resulted in controversy and some legal challenges. The initial specifications called for five Orbiters, each capable of 500 flights over a ten year life. This would later be lowered to 100 flights per Orbiter, and then to 55. Following a critical design review in February 1975, the essential engineering of America's Space Transportation System was complete, and fabrication could move forward.

Solid Rocket Boosters: SRB

The parallel first stage uses two 144-inch diameter, 149 foot long, solid-fuel rocket boosters—SRBs. The SRBs provide 2.65 million pound thrust—71 percent of the thrust at lift off to an altitude of about 150,000 feet. The solid propellant mixture is ammonium perchlorate (oxidizer) and aluminum (fuel) shaped into an 11-point star. This configuration provides higher thrust at ignition that gradually is reduced by one-third 50 seconds after lift-off to reduce stress during maximum dynamic pressure. The SRBs are performance matched by loading each from the same batch of propellant ingredients.

The SRBs support the entire weight of the Shuttle on the mobile launcher platform. Each booster is attached to the launch platform by four 28 inch long, 3.5 inch diameter frangible bolts that are severed by small explosives at lift-off.

A study submitted to NASA by the McDonnell Douglas Company defined their experience with strap-on solid boosters used with the *Thor* and *Delta* rockets, foreshadowing future events. They summarized all of the failure scenarios and indicated only one possible fatal malfunction—the failure of the segment joint that results in a "burn through." United Technologies Corporation was the only SRB bidder who proposed a single segment (monolithic) motor, arguing for that safety aspect. However, Thiokol Corporation was awarded the contract for a seven-segment motor that used a standard tang and clevis joint and a double "O" ring seal.

Only seven tests were conducted between 1977 and 1980 to qualify the SRB for manned flight. None of these tests demonstrated the dynamic conditions of the boost environment, nor established an ambient operating temperature range. Both of these factors would play a role in the pending disaster that would cripple the program in 1986.

"Mechanical thrust vectoring" using an 8-degree movement of the nozzle was incorporated. "Liquid thrust vectoring," had been considered but was not practical due to the large size of the motor. Following burnout of the propellant, the SRBs separate from the ET and descend under three parachutes to a water landing in the Atlantic, some 50 miles down range, to be recovered and refurbished for reuse.

The SRBs were to include a thrust termination capability allowing an abort at any point in the powered phase. This was to be achieved by blowing a hole in the forward facing end of the rocket to allow the combustion gasses to escape and negate the thrust. However, it soon became clear that this technique was fraught with possible catastrophic problems and it was deleted.

The Orbiter

The Orbiter would be the manned component of the Shuttle. It would house the main engines and the guidance and control, and contain the payload compartment for the cargo carried into space. The payload bay was 15 feet by 60 feet with a capacity of 56,000 pounds to low-earth orbit from Cape Canaveral and 38,000 pounds into Polar orbit from Vandenberg. The Orbiter, in these initial specifications, still retained air breathing engines so that it could extend its loiter time prior to landing as well as use these engines to ferry itself across the country.

A set of solid-fuel rockets mounted to the aft side of the Orbiter provided for emergency escape of the Orbiter from the rest of the stack during the first 30 seconds of flight. These were deleted from the specification in 1974 as a cost and weight saving measure. A drag chute to reduce the landing roll was also deleted, as it was felt that the dry lakebeds at Edwards Air Force base (the prime recovery area) were sufficiently long. This feature was added back when the brakes on the Orbiter proved inadequate.

The aerodynamic control surfaces on the Orbiter, used following reentry during its glide to a landing, were digital "fly-by-wire." This meant there were no cables or pushrods between the control stick and the control surfaces. The stick provided input to a computer that determined how much movement to provide to hydraulic actuators based on the speed and altitude of the Orbiter. Because the Orbiter was an inherently unstable aircraft that required the constant input of the flight controls, the importance of a computer was obvious. Four identical computers would work in parallel, and if one produced an answer that was in conflict with the others, its data was ignored.

The computer initially selected was an IBM AP-101—a 32-bit machine with 32K words of ferrite core memory and a floating-point instruction set. It weighed 47 pounds and consumed 370 watts of power. At $87,000 each, it was a bargain for its time.

The crew was provided with a sea level atmosphere of 20 percent oxygen and 80 percent nitrogen at 14.7 pounds per square inch. This was a marked departure from the first three generations of American spacecraft (*Mercury*, *Gemini*, and *Apollo*) that used pure oxygen at 5.5 pounds of pressure.

Providing some means for the crew to escape from the vehicle should a catastrophic failure occur presented a significant challenge. The traditional ejection seat had been developed to a high degree for advanced aircraft such as the *SR-71*. However, it was recognized that the large crew contingent of 7-10 astronauts would need a more sophisticated mechanism resembling the separable crew compartment of the F-111 or the B-1 prototype. However, these would cost the Shuttle not only in dollars but also in weight—and weight is the nemesis of rocketry. It would also require that an advanced warning system be developed to ensure that the crew could be propelled away from the possible effects of an explosion of the booster in time to ensure that they suffered no damage. In addition, with parallel staging, there was a time element that might require several seconds to initiate an abort. Catastrophic failures in complex high performance rocket systems could rarely be predicted with that much advanced timing.

With these factors in mind it was determined that, once perfected, the Shuttle would provide a reliability equal to an airliner and that no escape system would be needed. But this "airliner" reliability became a mind-set in the NASA hierarchy that would pervade its thinking about crew escape issues for the next 30 years. It would not, and could not, be substantiated by statistical analysis. Contingency for shutting down a failed SSME engine and flying the vehicle back to the launch site or to an alternate landing site would provide for a variety of abort scenarios.

An escape system would be needed only during the initial development phases when a higher degree of uncertainty existed about the various systems. For these first launches, a minimum crew of two would be used, and these flights would use modified SR-71 ejection seats that provided for escape below 100,000 feet either during the boost phase or following reentry if a landing could not be made on a runway surface.

In the end, it was felt that the added cost and weight of an escape system should be used instead to ensure redundancy and reliability of critical Shuttle systems. With a projected reliability of one failure in one thousand, the risks were deemed acceptable. These projections were later lowered to one failure in one hundred, and then to one in fifty-five—statistically half of the Orbiters would eventually succumb to a catastrophic failure during the life of the program.

Two primary methods of protecting the Orbiter during reentry were examined. The first was the use of ablation material such as used on the previous manned spacecraft. This presented many problems when the reusability factor was considered.

The second material was a high-temperature reusable-surface insulation (HRSI) in the form of tiles. High heat areas (650 to 2,000 degrees F) to include the bottom of the wing, the vertical stabilizer, and forward fuselage would be protected by this material. The thickness of the tile was dependent on the specific location and the temperatures encountered. The tiles were always considered a high-risk item, were easily damaged, and were labor intensive to install and replace.

The tile material is 89 percent porous (i.e. 90 percent air and 10 percent silica) with a basic density of about 15 pounds per cubic foot—but manufactured with different densities based on the amount of heat protection required. This represented a weight factor of about ten percent that of the traditional ablative heat shield. Because of its brittle properties, it could not be fabricated with large surface areas and had to be applied as a complex set of specially shaped tiles, typically no larger than six inches square. Because of the thermal expansion qualities of the basic aluminum structure beneath, a felt pad was used between the tile and the aluminum.

The leading edges of the wings and the nose cap would experience temperatures of 2,300 degrees F. but, because of their shape, could not effectively be protected with heat tiles. A high temperature "hot-structure" material—using rare and expensive super-alloys—along with a composite of pyrolyzed carbon fibers in a pyrolyzed carbon matrix and a silicon carbide coating (commonly known as "carbon-carbon") were thus employed. This material was originally developed for the DynaSoar project in the early 1960s. Data from other old DoD programs, including ASSET and PRIME, provided invaluable assistance in understanding structural heating. The remainder of the orbiter would be covered with a lower temperature version of the HRSI tiles and nomex felt blankets.

As late as 1979 there were still serious questions as to whether the tiles would provide the needed protection when all of the various factors of heat, stress and vibration were considered. More than 30,709 tiles would cover the Shuttle. Their application on the first Orbiter required three shifts working six days a week for over six months.

The initial program called for five Orbiter's to be constructed. The first, Enterprise (OV-101), did not have all the systems and was used for atmospheric glide tests. The other four were *Columbia* (OV-102), *Challenger* (OV-99), *Discovery* (OV-103), and *Atlantis* (OV-104).

External Tank: ET

The external tank (ET) represented a design challenge, considering its apparently simple role. It was desirable to fabricate it at a low cost (because it was the only expendable item) and at the lowest weight (because it was the largest structure). Accommodating the thrust loads of the two SRBs and the Orbiter, complicated these factors. Martin Marietta was awarded the contract, in part because of its experience with the *Titan* III which involved the attachment of large SRBs.

The ET that resulted is 154 feet long and 27 feet in diameter and consists of the two propellant tanks in tandem with the LOX being in the forward tank. Spray-on polyurethane foam covered the entire structure, and an added ablative material was used on the nose and base sections. The nose is exposed to the hottest aerodynamic heating during the ascent, while the base is subjected to the heat given off by the SSMEs.

With the *Saturn* V, ice was allowed to form on the outside of the tank as a result of condensation from the humid Florida air and the cold content of the tanks. The ice acted as an insulator during the countdown and was shed by the vibration and aerodynamic forces at launch. This ice was not considered a hazard. However, ice was a potential hazard for the Shuttle. Being a parallel staged vehicle, the Orbiter was positioned beside the ET, and ice falling from it could impact the wing or sensitive heat tiles. Thus, the insulation performed two important functions of reducing the boil-off during the countdown and assuring that ice did not form. During the ascent, it also reduced the heat transfer to the propellant from aerodynamic friction.

Space Shuttle Main Engine: SSME

The SSME was one of the first components to be defined and was contracted to Rocketdyne in 1971. Along with the heat tiles, it would be the pacing item that would ultimately delay the Shuttle program. The high-energy cryogenic propellants of LH2 and LOX were the obvious choice because of their high Specific Impulse. It was also decided that the use of multiple thrust chambers would increase the reliability while offering more abort options.

To provide for these abort options, the SSMEs would have the ability to be throttled-up to 109 percent of their rated thrust. In doing so, the Shuttle could shutdown one of the main engines due to a detected malfunction and continue—either to orbit or to a contingency landing site—by using the added power. These engines would also be throttled down to 65 percent to minimize Max-Q— the dynamic air pressure and G forces during the ascent.

As the final Shuttle configuration took shape, three 375,000-pound engines (470,000 pounds in vacuum) were defined for the Orbiter. It had been envisioned that the first-stage booster would also use the same basic engine but with a different nozzle expansion area for its lower altitude burn. However, when the parallel burn configuration was chosen, it meant that these engines would have to perform for the entire launch, from sea level to orbital insertion. This 520-second period required that the SSME be optimized for a wide range of conditions.

It was originally specified that the engines would have a life of 100 launches (27,000 firing seconds), but, with the longer duration of the parallel burn, this was reduced to 55 launches. The engines with the turbo-pump assembly are 13.9 feet long with a diameter at the exhaust nozzle of 7.8 feet.

The ability to change thrust was directly related to the propellant supply provided by the turbo-pump. A mixture of LH2 and LOX in about equal ratios produced a hydrogen rich steam that powered the turbo-pump. After driving the turbo-pump, this steam was routed to the injector where it is mixed with more oxygen (6:1 ratio) for final combustion in the thrust chamber. This feature proved particularly difficult to engineer and, along with the "pre-burn" of propellants, was the key to the very high combustion chamber pressures that made the SSME such an efficient power source.

The first full thrust test of the SSME occurred in July of 1975, but continued catastrophic failures of the engine and the turbo-pump led to a constant series of schedule slippages. A complete flight-qualified engine did not pass the final acceptance test until March of 1980. In order to move forward to the flight test it was decided to forego (for the first few missions) the 109 percent capability which had caused significant reliability problems and schedule delays.

Shuttle Enterprise: OV-101

As the Shuttle began to come together, plans for flight-testing were formulated. Extensive ground, wind tunnel, modeling, and simulator testing were accomplished. Particular emphasis was placed on vibration, resonance such as the pogo affect, and aerodynamic flutter.

With the deletion of the "air breathing engines" from the Shuttle, a new means of transporting it between remote landing locations and the Cape had to be found. Of equal importance were the basic glide tests to determine its aerodynamic flight characteristics. In a manner similar to the X-15, the Shuttle would be carried aloft and air dropped for these glide tests. After some consideration of building a special aircraft for the purpose, the Boeing 747 was selected as the Shuttle Carrier Aircraft over the Lockheed C-5, primarily because of availability and cost.

There had been some concern regarding the effect of the aerodynamic wake generated by such a large combination. However, wind tunnel tests confirmed the concept of mating, and separation would not produce any problems that could not be overcome with appropriate procedures. The 747 was easily modified using the attachment points that mated the Orbiter to the external tank. It was determined that the abrupt termination of the Orbiter's tail structure produced a pronounced drag component that caused a significant degradation in the speed of the 747, so a drag-reducing tail-cone was fabricated for ferry flights.

Construction of the first Shuttle orbiter vehicle (OV-101) began in June of 1975. It was to have been named "Constitution" in honor of the 200th birthday of the United States as it would be "rolled out" in 1976. However, a write-in campaign conducted by aficionados of the television program *Star*

Trek caused the White House (President Gerald Ford) to accede to their request, and the first Shuttle was named "Enterprise."

However, Enterprise was not a space worthy craft. It would be used for verifying the ferry capability with the 747, the glide tests, and other activities that required a full-scale model. It did not contain many of the systems, nor did it have the SSME engines or Orbital Maneuvering System (OMS). Enterprise did represent the weight and balance and was expected to glide like the real thing. It also contained ejection seats whose presence emphasized that, even in the glide tests, there were hazards for the two crew members required to fly the Shuttle.

A series of four tests were planned for the mated aircraft. The first was progressively faster taxi tests of the 375,000-pound combination to determine the effect of the load distribution on the 747 during the initial take-off and landing phases to ensure both steering and braking systems were not adversely affected. The gross weight was well below the carrying capacity of the 747.

Next came five airborne flights that began in February of 1977 without the Enterprise crew aboard; these flights produced no problems. The "captive-active" flights with the crew aboard began in June 1977. The Enterprise crew noted that from their cockpit they could not see any part of the massive 747 that was supporting them.

Finally, the drop tests that allowed the crew to fly the Enterprise through a complete landing scenario were performed. Astronauts Fred Haise and Charles 'Gordon' Fullerton were at the controls on August 12, 1977, when the big 747 with the Enterprise perched on its back went into a shallow dive over Edwards AFB in California. With a negative seven-degree pitch angle, the speed built to 310 mph, and Haise moved the controls of the Enterprise to a position calculated to lift the Shuttle clear of the 747. Then at 24,000 feet, he fired the seven explosive bolts that held the mated combination together, and the Enterprise flew free.

Haise banked slightly to the right to clear the 747 and then transitioned to a negative nine-degree pitch that resulted in a 13-degree descent angle. The drag reducing tail cone had a significant effect on the glide angle. Two 90-degree turns were executed, and then Haise lined up with the runway markings on the lakebed. Coming across the threshold at 213 mph, with the speed brakes out, Enterprise touched down smoothly—just 5 minutes and 21 seconds after release. The 150,000-pound ship coasted to a stop after a roll of 11,000 feet.

Following two additional glide flights with the tail cone, it was time to move to the full drag configuration. The fourth flight was without the tail cone and was equally successful. The drag component represented the flight characteristic of a returning Shuttle, but astronauts Joe Engle and Richard Truly easily mastered the 18-degree final approach path.

The final drop test was a bit more exciting when Haise experienced pilot-induced oscillations during the final approach while exercising the gear and speed brakes. Approaching at a higher airspeed of 334 mph, his timing was off for the flare, and the Enterprise hit hard and bounced back into the air. However, the Shuttle had good slow speed control, and Haise was able to reposition the craft for a smoother touchdown on the second flare.

The Enterprise was subsequently used in a variety of vibration and mating tests at MSFC and at the Kennedy Space Center at Cape Canaveral. It was to have been rebuilt as a full-fledged orbiter, but the costs to make the changes were deemed prohibitive, and it was eventually retired to the Smithsonian Air and Space Museum after giving up some of her parts to other orbiters. Over the years, NASA has occasionally returned to the Enterprise for a variety of tests, inspections, and evaluations.

To assist in preparing the Shuttle pilots for landing out of a rather unconventional approach profile, two Grumman G1159 Gulfstream II business jets were modified as Shuttle Training Aircraft (STA). These provided the Shuttle cockpit display and rotational hand controller for the left side pilot station while the right side maintained the traditional flight displays and standard control yoke for the safety pilot. By extending the main gear and placing the twin engines in reverse thrust, a steep descent profile of up to 14,000 fpm at 300 knots were achieved. As the STA itself could not land at these speeds, the touchdown was simulated and the STA was cleaned-up for a go-around—something the Shuttle could not do. The stress placed on these aircraft to perform these "unnatural" maneuvers took a significant toll on the structural life of these aircraft.

Reusability Becomes a Reality

The first flight of the Shuttle was manned. This was an unprecedented decision that leaned heavily on the extensive testing and simulations that had been used to validate the various systems. The first flight-ready Orbiter, named *Columbia*, arrived at the Kennedy Space Center in March 1979, but mating and testing and rework required another two years before it was ready for the first flight.

The first launch carried an extensive array of recording instruments that would transmit data to ground stations during the flight. These measurements would be correlated with modeling data to determine how closely the actual craft corresponded to the wind tunnel results, computerized simulations—and assumptions. Of particular interest were the conditions during the various critical phases of flight: maximum dynamic pressure (Max-Q), SRB separation, and ET separation.

The Shuttle was assembled (stacked) in the same Vehicle Assembly Building (VAB) used for the *Apollo/Saturn* V missions. The SRBs were first raised into position on the launch platform that had been used by the *Saturn* V—modified to accept the significantly different Shuttle configuration. Then the ET was joined to the SRBs in early November 1980, and finally, the Orbiter was raised into position and mated on November 26, 1980. The completed Space Transportation System Flight Number 1 (STS-1) was transported to launch pad 39A on December 29, 1980.

The Flight Readiness Firing (FRF), a static test of the three SSME LH2 engines, was performed on February 20th: a 20-second firing at 100 percent power. This was the first time that all three engines of a complete flight-ready Shuttle had been fired. A part of the test was to determine the "twang." Because the position of the SSMEs was somewhat off-center to allow its thrust vector to be aligned with the center of gravity of the total assembly, the entire Shuttle rocked forward when the SSMEs were ignited. The movement, which amounted to almost two feet at the nose of the Shuttle, was referred to as the "twang." The timing of the SRBs' ignition was set to allow this movement to return the Shuttle to the vertical position.

During a final launch simulation in March, two technicians were asphyxiated when they inadvertently entered the aft section of the Shuttle which was filled with nitrogen to guard against the accumulation of any explosive gases—principally hydrogen.

The flight was originally schedule for April 10th, but a software glitch caused a two day postponement. The crew of astronauts, John W. Young and Robert L. Crippen, brought two of the quietest and most capable men in the astronaut corps together. Young had flown the first *Gemini* with Gus Grissom on GT-3 in 1965 and flew again with Michael Collins on GT-10. He was aboard *Apollo* 10 that made the final dress rehearsal around the Moon in 1969, and he walked on the Moon with Charles Duke on the *Apollo* 16 mission.

STS-1 would be Crippen's first flight. He had originally been selected for the Air Force's cancelled Manned Orbiting Laboratory but had transferred to NASA in 1969. Now, 12 years after that job change, he would finally fly on April 12, 1981—twenty years to the day after the first manned space flight, *Vostok* I with Yuri Gagarin. Spacecraft had progressed a long way during that time.

The astronauts entered the Shuttle two hours before lift-off, and the main side hatch was closed by T-60 minutes. The white room and the pad were cleared at T-30 minutes. The final updates to the various computers were accomplished, and the crew access arm that allowed an emergency escape while the Shuttle was still on the pad was retracted at T-7 minutes.

With less than three minutes until launch, the aerodynamic surfaces of the shuttle were exercised to affirm the functioning of the hydraulic system, and the three SSMEs were moved through their entire gimbal travel. The "beanie cap" that removed the vented hydrogen and oxygen gases from the top of the ET was retracted as the count moved to within two minutes, and the propellant tanks were pressurized.

At T-3.8 seconds a computer began to sequence the engine start process for the three SSMEs. The ignition of each was staggered by about 1/10th of a second. The Shuttle thrust quickly came up to 1.125 million pounds, and the assembly experienced the "twang." As the huge rocket realigned itself with the vertical axis, the SRBs ignited, and the four 3.5 inch diameter explosive hold-down bolts on each were blown. Now the 2.65 million pounds of thrust from each SRB, added to the SSME, gave a total of almost 6.5 million pounds of thrust, and the vehicle left the pad, balanced on two shafts of white SRB smoke and the virtually transparent plumes of the three LH2 engines.

Now for the first time a new machine took to the air—a vehicle that had more research and devel-opment effort (and cost more) than any flying machine man had ever created. This being the very first flight, public interest was high. (Not quite as many people tuned in to watch the live launch as the first Moon shot, however.)

Because of its greater thrust to weight ratio, the 4,457,111-pound vehicle accelerated more rapidly than the *Saturn* V, and the tower was cleared in 6.5 seconds as opposed to the ten seconds of the *Saturn* V. At T+11 seconds, a roll maneuver aligned the Orbiter's longitudinal axis towards the northeast for the orbital inclination of 40.3 degrees.

Forty-four seconds into the flight, the SSMEs were throttled back to 65 percent power to reduce the aerodynamic pressure at Max-Q. The throttle-up began at T+76 seconds as the vehicle arced over to the northeast.

The white exhaust of the SRBs began to show an orange color as they reached burnout two min-utes into the flight and were released from the ET at an altitude of 28 miles and a speed of just over 3,000 mph. As the SRBs began their slow tumble back to Earth, the Shuttle exhaust was virtually invisible as it continued to power the Orbiter.

At T+260 seconds the point-of-no-return was reached. At this point, if there was a failure of one of the SSME engines, the Shuttle could no longer return to the Cape; it would have to press on to one of the emergency landing sites, the first of which was in Spain. The seven-minute mark was the point where the Shuttle could still reach orbit if one SSME engine should fail. In future flights this would be the point where the 109-percent power would be needed. However, *Columbia* was lightly loaded on her maiden flight and could reach orbit with the 100-percent power available from only two main engines.

At T+460 seconds the main engines were again throttled back as the weight of the Shuttle had been reduced to the point where it would exceed the three-G limitation if it continued at 100-percent.

With the engines at 65 percent power, and a speed just shy of orbital velocity for the 138-mile alti-tude, the main engines were shut down at 8 minutes and 30 seconds into the flight. The ET separated and descended back into the atmosphere to burn up over the Indian Ocean.

John Young positioned the Orbiter's attitude for the first burn of the Orbital Maneuvering System (OMS)—two 6,000 pound thrust units, located on either side of the aft fuselage, accounted for the bulge just beneath the vertical stabilizer. Using monomethyl hydrazine and nitrogen tetroxide as pro-pellants, the OMS unit provided the final push into orbit as well as any subsequent orbital maneuvers for future rendezvous missions. As with all previous manned American spacecraft, the Orbiter had a Reaction Control System (RCS) for attitude control. This consisted of a set of thirty-eight 830-pounds and six twenty-five pound thrusters. These thrusters used the same type propellants as the OMS and had their own supply, but could be cross-fed from the OMS tanks if necessary.

The payload bay doors were then opened. Although there was no payload other than the engineer-ing monitors, the inside of the doors contained the cooling radiators for the Environmental Control System and had to be opened on achieving orbit and remain open during the mission.

On-orbit video of those parts of STS-1 that could be seen from crew positions revealed that some tiles had been lost from the OMS unit faring. While there was some concern, these tiles were not deemed critical as the heat generated during reentry to that part of the Shuttle would be considerably less than to the underbody—for which there was no video available. Young and Crippen had to assume that the primary heat tiles on the underside were all still in place.

The two astronauts spent two days and six hours in orbit testing and verifying the various systems, and then it was time to return. A critical operation was the closing of the payload bay doors. If they could not be closed, the Shuttle would not have the structural integrity to survive the reentry. Young then positioned the Shuttle to traveling backwards to the orbital vector, and the OMS thrusters were fired. This reduced the speed of the Shuttle to where it would descend into the upper reaches of the atmosphere. The Shuttle was then reoriented initially to a 40-degree pitch-up attitude that would grad-ually be reduced as it encountered the more dense layers of the atmosphere. As the Shuttle began to feel the deceleration, the G forces increased, and the heat pulse on the nose and wings of the Orbiter generated the ionized sheath of air that disrupted radio communications. The crew was cut off from Mission Control for a period of over five minutes.

During the blackout period, the computer displayed the required S-turns that dissipated the energy and kept the Shuttle from experiencing the heavier eight G forces that other astronauts had had to endure. *Columbia* and her stable mates eased back into the lower layers of the air, subjecting the occupants to less than three Gs. The path projected by the computer also assured that it arrived over Edwards AFB at 60,000 feet in a position to make a normal landing on the Rogers Dry Lake bed that had, for the preceding 35 years, been a safe haven for America's rocket planes.

As *Columbia* arrived over Edwards, she announced her presence with the distinctive double boom, sonic shock wave. As it made its descending turns, *Columbia* was joined by two T-38 chase planes piloted by other astronauts. The descent profile used the computer to manage the Shuttle's energy and place it on the final approach. As it neared the ground, Young began a two-phased flare that positioned it over the end of the runway with the gear down. One of the crew of the T-38 chase planes counted the estimated feet to the runway over the radio to aid the Shuttle pilot.

Following a smooth touchdown and rollout, there was an extended time when the ground crew had to "safe" several systems and ventilate critical areas that might contain toxic gasses. Almost an hour after touchdown, and following some critical comments from the crew about the time it was taking to get them out of the spacecraft, the hatch was finally opened, and two very elated astronauts skipped down the stairway and bounded excitedly around the outside of the Shuttle for a post flight assessment.

The inaugural flight of the Shuttle had been a resounding success, and great things were expected of the long delayed and controversial project. However, the next flight did not immediately follow. A total of 16 tiles were lost from *Columbia*, most likely during the SRB ignition, and 148 more were damaged. There was more work to be done to assure greater reliability of the tiles as well as several other systems.

The second flight of *Columbia* (STS-2) occurred eight months after the first (November 1981), and was crewed by Joe Engle and Richard Truly. The extended turn-around time was needed to analyze the data returned from STS-1 and to make the required changes to various systems. STS-2 marked the first use of the Canadian built Remote Manipulator System (RMS), the highly dexterous arm that was stowed in the payload bay; the arm would be an important part of the Shuttle operation. The failure of a fuel cell caused this flight to be shortened from the planned five days to three. A new sound suppression system that used water to absorb the intense acoustical pulses emitted by the SRBs at ignition was considered successful as there were only 12 damaged tiles found on *Columbia*'s return.

STS-3 with *Columbia* occurred on March 22, 1982, after a turn-around of four months, and it remained in orbit for eight days. Because the Rogers Dry Lake at Edwards AFB was flooded by spring rains, the Shuttle was diverted to the contingency landing site at White Sands Missile Range in New Mexico. This time the acoustical suppression system did not seem to perform as well; 36 tiles were lost and 19 damaged.

STS-4, in June of 1982, was the first to conduct scientific experiments. Unfortunately, both of the SRBs were lost when the parachute recovery systems failed to function. With the first landing on the hard surface runway at Edwards, the Shuttle was declared operational. Those who truly understood the complexity of the machine realized this was an ominous declaration.

STS-5 was the first spacecraft in which American astronauts flew in a "shirt-sleeve" environment— no spacesuits. In addition to the two Shuttle pilots, the first Mission Specialists flew into space. Because there was no emergency escape system for these Mission Specialists, the ejection seats for the Shuttle Pilot and Mission Commander (which would eventually be removed from *Columbia*) were "pinned" to disable them. There would be no moral dilemma of saving only two of the crew if catastrophic failure occurred. The Mission Specialists were astronauts who were not pilots but who were responsible for conducting on-orbit work. STS-5 carried the first commercial satellites: two communications satellites with solid-fuel Payload Assist Modules (PAM) that boosted them into geosynchronous orbits after they were released from *Columbia*'s payload bay.

STS-6 was the inaugural flight of the second Shuttle, *Challenger*, in April 1983 and included the first Shuttle EVA. Aboard STS-6 was the first of a network of Tracking and Data Relay Satellites (TDRSs) that would ultimately replace the ground stations that had been built around the world since the days of the *Mercury* program. The TDRS satellites, in geosynchronous orbits, were used by NASA for communications with both manned and unmanned satellites to increase the time spacecraft were in communication with the ground and enhance the data transmission rates.

STS-7 carried America's first female astronaut, Sally Ride. STS-8 was the first night launch and included the first black American in space, Guion Bluford. The flight concluded with the first night landing of the Shuttle. STS-9 flew with the first Payload Specialist and the first six-member crew who manned the first European built Spacelab housed in the payload bay. STS-10 conducted the first tetherless EVA, and STS-11 performed the first Shuttle orbital rendezvous with the Solar Max satellite. STS-16 was the first flight of *Discovery* and was the first to experience an on-pad abort after the main engines had ignited. It was rescheduled and launched successfully in August 1984.

With the expected demands placed on the Shuttle, the astronaut corps was expanded considerably to well over 100—the roles determined the training and qualifications. Shuttle pilots typically had military experience and were graduates of one of the test pilots schools. The Payload Specialists and Mission Specialists, were not required to have any pilot qualifications. However, all astronauts were required to fly the sleek, white, twin engine Northrop T-38 to maintain proficiency, enhance coordination and judgment, and to solidify teamwork. Non-rated pilots were not soloed in the two place craft, although several completed their requirements to receive FAA Private Pilot certification.

All astronauts were required to be exposed to, and train in, the zero-G environment before their first space flight. This was accomplished by flying parabolic arcs in a KC-135A. The aircraft is pulled into a steep (45-degree) climb and then the pilot "pushes over" to match the gravitational acceleration of the Earth. Periods of up to 25 seconds of weightlessness can be induced, allowing the participants in the padded passenger compartment to float in zero-G—often causing nausea resulting in the aircraft's moniker—the Vomit Comet. At the end of the arc, the aircraft will be in a 45-degree dive that requires a 2.5 G pull out. A typical flight will perform 30-50 of these maneuvers. (The first two KC-135s were retired and replaced by a C-9, the military version of the Douglas DC-9).

NASA continued to improve the Shuttle turn-around time, but problems still plagued the program, and, at the end of the fourth year of operation, the Shuttle had yet to be launched more than eight times in a calendar year. The time and equipment needed to refurbish each Shuttle and return it to flight status caused costs to soar, and the pressure to launch on time was significant.

The crew of STS-1 John Young and Robert Crippen

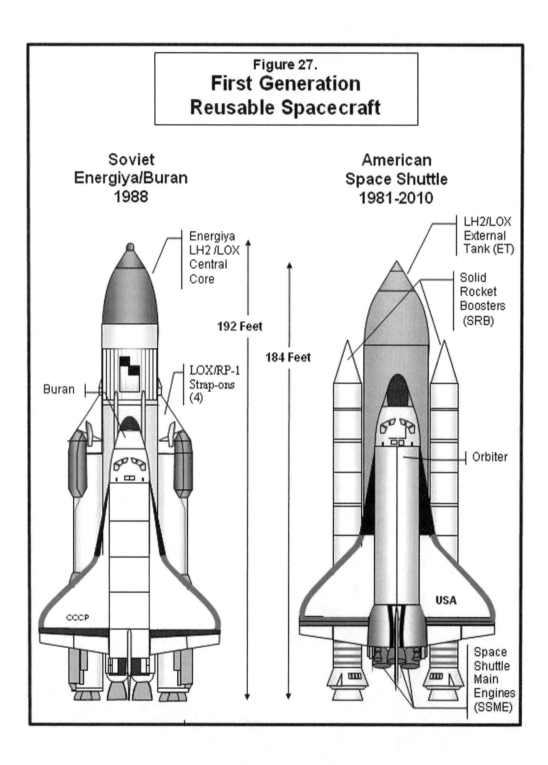

Figure 27.
First Generation
Reusable Spacecraft

Soviet
Energiya/Buran
1988

American
Space Shuttle
1981-2010

Energiya
LH2 /LOX
Central
Core

LH2/LOX
External
Tank (ET)

Solid
Rocket
Boosters
(SRB)

192 Feet

184 Feet

Buran

LOX/RP-1
Strap-ons
(4)

Orbiter

CCCP

USA

Space
Shuttle
Main
Engines
(SSME)

Early Concepts

The concept of a manned platform or station in outer space is not new. Under the genre of science fiction, the *Atlantic Monthly* magazine in 1870 presented the essential idea of a laboratory in space in the form of a serial story. An American minister and writer, Edward Everett Hale (1822-1909), wrote the articles.

The German mathematics teacher, Kurd Lasswitz (1848-1910), in 1897 produced a novel, *Auf zwei Planeten* (On Two Planets), which depicted a Martian space station, supported by antigravity, which provided a stopover for long duration flights.

Konstantin Tsiolkovsky's science fiction work entitled *Reflections on Earth and Heaven and the Effects of Universal Gravitation* (1895) described asteroids and artificial satellites as launching facilities for interplanetary travel. He also provided for artificial gravity on these man-made space stations by rotating them to create centrifugal force—an idea that caught on with subsequent enthusiasts.

Hermann Oberth included a brief description of "observation stations" in his 1923 book and some of their possible uses such as astronomy, Earth observation, military reconnaissance, and use as a refueling station for interplanetary flights. In a subsequent revision of his book published in 1929, *Wege zur Raumschiffahrt* (Ways to Spaceflight), Oberth provided more detail. He presented the concept of the space mirror facility as a part of a space station that could redirect solar energy to a single point on Earth as a military weapon—much as a magnifying glass can be focused to create intense heat. He also thought that, when applied over a wider area, such a mirror could warm northern ports to keep them free of ice in winter or could illuminate large cities at night.

A somewhat mysterious writer whose pen name was Hermann Noordung wrote of a giant space station in his science fiction work published in 1927. He was, in reality, an Austrian Army Officer named Hermann Potocnik, who was born in 1892. Although Noordung had some obvious flaws in his treatment of physics, his detailed design used a rotating wheel to provide artificial gravity and a central "hub" to provide an airlock to gain entry. Electrical power was provided by solar heating of unspecified fluid compounds.

Potocnik foresaw many uses including scientific experiments unimpeded by gravity and astronomy without the interference of the Earth's atmosphere. He recognized that the Earth could be examined for meteorological and military applications. In addition, like Oberth, he wrote of the use of a space mirror to focus the rays of the sun on the Earth. Because his work was widely read (Wernher von Braun had read Noordung as an enthusiastic teenager), he no doubt galvanized the thinking of many space enthusiasts of the period.

Following World War II, both science fiction and tentative scientific planning produced many concepts of a space station. The post-war writings of von Braun in his book *Mars Project* and his *Collier's* magazine articles of 1952, coupled with the prolific writings of Willy Ley, continued to spread the possible capabilities of the space station. Ley prophesied a 250 foot diameter wheel with a 22 foot rim connected to the "hub" by three spokes and manned by a crew of 30 to 60 men (women were conspicuously absent in the male dominated sciences of the 1950s). In von Braun's vision, the space station was the key to pursuing most other aspects of space exploration—following quickly after the initial unmanned satellites. His space station at this point was the culmination of many ideas gathered from the visionaries of the previous decades, and he conceived it as being built in much the same manner as a ship of the sea—in parts welded and riveted together—in space. However, there were others, such as Sergei Korolev, who saw this form of fabrication as unrealistic. They saw the space station as built in prefabricated segments, complete with their interior appointments and joined in a much simpler manner.

With the advent of Sputnik in 1957, the space station concept became a primary interest for the United States and the Soviet Union. Both Korolev and von Braun felt that the space station and trips to Mars would occur within 10-15 years if not sooner. However, by 1961, with the moon established as the goal for the "great space race" by President Kennedy, the focus temporarily shifted away from the space station. With the ability to build giant super rockets such as the *N1* and the Saturn V, and with the Lunar Orbit Rendezvous technique, there was no need to have an intermediary step before reaching for the moon.

The military in both countries, having little interest in the moon race as a national defense priority, continued to pursue the space station. The U.S. Air Force created a project for developing its Manned Orbiting Laboratory (MOL) in December 1963, and the Soviets had Korolev's *Soyuz*-R, a reconnaissance version of the planned *Soyuz* spacecraft. *Soyuz*-R was soon followed in October 1964 by a competitive proposal from Chief Designer Vladimir Chelomey for the Almaz (Russian for "diamond"). Like the MOL, Almaz was a relatively small space station manned by two or three cosmonauts. While the MOL would never fly, the 40,000-pound Almaz was originally scheduled for launch by 1968 and would house cosmonauts for up to one year. Both of these goals were highly optimistic. By the end of 1969, the first launch of Almaz was still not ready despite ten stations in varying degrees of fabrication.

Salyut 1: Haste Makes Waste

As the race to the moon culminated in the spectacular American success, and with the secret Soviet lunar landing effort smoldering in ruins, First Party Secretary Leonid Brezhnev was forced to produce a new strategy to address his country's now second-rate technological position. The triple space flight of *Soyuz* 7, 8 and 9 in 1970 showed the Soviet's ability to put three manned spacecraft into orbit within a few days but demonstrated little else. That the *Soyuz* 9 crew was unable to stand after their record 18 day flight and had to be carried to the recovery helicopters on stretchers indicated that there was yet much to learn about living in space. The effect of weightlessness on the human body was of primary concern. Extensive loss of calcium by as much as 15 percent, coupled with the pooling of blood in the head and upper body and the loss of muscular strength, was high on the list of phenomena to be investigated.

The space station, even in the modest forms being designed in both the Soviet Union and America, would allow relatively long duration manned flights that could emulate a trip to Mars in 30-day increments. By rotating crews to the station, the effects of space flight could be gauged on both the men and the systems required to support them without committing to the hazards of a two or three year mission.

For the Soviets, the space station seemed like an ideal choice to recapture the imagination of the world. It could be touted as a scientific and "peaceful" project and would not depend on the development of any new large boosters. The station, at about 20 tons, could be launched by the *UR-500* (Proton) and crewed and provisioned (once in orbit) by the *Soyuz* atop the continually improved and reliable R-7.

But once again time was of the essence as the Americans had announced that they would be using much of the Apollo hardware to pursue what was referred to as the "Apollo Applications." This included a large space station called Skylab, launched by the Saturn V. As neither the *Soyuz*-R (a reconnaissance version of the *Soyuz*) nor the Almaz (a military space station) was making very good progress, it was decided to move the Almaz (the larger of the two) to Korolev's old design bureau. Vasily Mishin was running this organization—which did not please Vladimir Nikolayevich Chelomey, who headed a competitive design bureau (from whom the project was taken). Almaz had included a two-man spacecraft (almost identical in configuration to the American Gemini) riding into orbit with the space station. But as there was no need to duplicate a capability that *Soyuz* already possessed, that aspect was canceled to accelerate the project.

The space station would not resemble the classic wheel shape as so many of the visionaries of the past had configured it. With the desire to investigate the effect of weightlessness, these first efforts would be simply large cylindrical enclosures that could readily be installed as a payload on existing launch vehicles such as the *UR-500*.

The project had to be completed well before the launch of Skylab for its impact to be significant. As with many Soviet space projects that preceded it, systems would be adapted where possible from other projects with minimal testing. Not only was time a critical factor but the budget had to be kept to a minimum. This meant that some of the more sophisticated scientific equipment would not be installed in order to meet the scheduled launch deadline.

The revised Soviet configuration consisted of a large cylindrical work area, adapted from the

Almaz project, that was 30 feet long and 14 feet in diameter which narrowed to a little over nine feet in diameter after the first nine feet (Figure 28). The space station was named Salyut—the word meaning "salute" in Russian—in honor of the first Soviet cosmonaut, Yuri Gagarin. To accelerate the completion of the Salyut, portions of the *Soyuz* design were fabricated into the space station. The *Soyuz* service module, which contained the on-orbit maneuvering, attitude control and environmental systems, was a seven-foot cylindrical section affixed to the large (aft) end, while a docking and transfer compartment was placed on the opposite (forward) end. This docking mechanism was an improvement over the initial *Soyuz* in that it allowed the crew to transfer directly between spacecraft (through the docking adapter), as had been done in Apollo/LM, without requiring a space walk.

Because a *Soyuz* crew transfer vehicle would be docked to the station, a second hatch on the side of the transfer compartment was provided to allow for a space-walk capability. As had been done with Voskhod I, the *Soyuz* cosmonauts would not wear space suits during their launch—these would be provisioned within the Salyut for later planned space walk activities. Two sets of solar cells (also adapted directly from *Soyuz*) extended from the fore and aft modules to provide 3.6 kilowatts of electrical power.

Although equipped with an Attitude Control System (ACS), the space station employed gravity-gradient positioning, as the larger, heavier aft end would align its longitudinal axis toward the Earth's center. The ACS would be used during some experiments when it was desired to point the station in a particular direction with ion sensors providing the reference.

An "up" and "down" orientation was maintained within the space station to provide a traditional directional reference in the weightless environment for the cosmonauts. As previous experience with *Soyuz* flights had shown a significant degradation in the muscle and bone composition of the cosmonauts for even short, two week excursions, a treadmill was provided that would allow the cosmonauts to get daily physical conditioning with rubber bungees providing the "artificial" gravity for the cosmonaut.

The *UR-500* booster left its pad with Salyut 1 as its payload enclosed in a launch shroud on April 19, 1971—a full two years ahead of the American Skylab. Placed in a 124 by 138 mile orbit with an inclination of 51 degrees, the 42,000-pound space station was the largest object the Soviets had ever placed in orbit. The low orbit was a result of its primary role of military reconnaissance and required the periodic firing of the 1000-pound thrust OMS engine to raise its orbit to avoid premature reentry.

Only one significant problem plagued the unmanned craft as soon as it had reached orbit. The cover for the main telescope failed to jettison, and this extremely large experiment (it occupied most of the aft portion) could not be used. As soon as the Soviets felt confident that Salyut 1 was in a stable orbit and functioning, they made an unprecedented announcement—not only that they had successfully launched the world's first space station but also the date of the launch of its first crew and their names.

However, the scheduled April 22nd launch (three days later) with Lt. Colonel Vladimir Aleksandrovich Shatalov, Alexei Yeliseyev, and Nikolai Rukavishnikov aboard *Soyuz* 10 did not take place. A part of the launch tower failed to retract during the final minutes of the countdown, apparently because of a build-up of ice from heavy rains and the cryogenic temperatures. The launch was canceled. This was a major embarrassment for the Soviets. During the second attempt the following day the segment again failed to retract. Korolev's former First Deputy, Vasiliy Mishin, as the controlling authority, decided to risk the launch and allowed the countdown to continue. Whether it was the vibration generated at ignition or some other quirk, the launch tower segment retracted properly and the cosmonauts were on their way.

Following the uneventful launch, the cosmonauts of *Soyuz* 10 spent the next two days performing the rendezvous with the passive Salyut Space Station. However, failure of the *Soyuz* attitude control gyros and the inability of cosmonaut Shatalov to achieve a "hard dock" after several attempts caused the 22 day mission to be ended after only three days in orbit. The crew returned successfully, and it was determined that the *Soyuz* docking latches had been damaged during the first docking attempt.

Soyuz 11 was scheduled to provide the second crew for Salyut, but problems began before launch when one member of the crew became ill and the entire crew was exchanged for the back-up. The replacement crew of Georgi Dobrovolski, Viktor Patsayev, and Vladislav Volkov complicated the situation as the six spacesuits already aboard the Salyut Space Station were those for the cosmonauts of *Soyuz* 10 and the prime crew for *Soyuz* 11. The unannounced launch of *Soyuz* 11 occurred on June 6,

1971, and its reinforced docking latches appeared to resolve the hard-dock problem.

By now, Salyut had been in orbit two months, and when the *Soyuz* crew entered Salyut, they smelled the distinct odor of what they believed was burned electrical insulation. Not knowing how toxic the atmosphere might be, they quickly replaced six lithium perchlorate cartridges that had failed and turned on the air regeneration system. They elected to return to the *Soyuz* and spent the first night there, isolated from the Salyut environmental system.

The next day the odor had been eliminated, and the crew returned to begin a series of scientific and military experiments. An extensive physical exercise program using the bungee treadmill and other devices for the cosmonauts was closely monitored. As with their American counterparts, the cosmonauts had a difficult time establishing a good work schedule for working and sleeping—trying to stagger the sleep periods was not very successful.

With their initial efforts to get good publicity from the *Soyuz* 10 flight thwarted by the delay in launch and then the inability to rendezvous, the Soviets were more cautious with the information they revealed about *Soyuz* 11. However, after the first week in orbit, a press conference was held in which extensive video tape of the cosmonaut's activities was shown on Soviet television. A second program two days later proved equally popular, and the Soviet leadership was more than pleased with their new-found "openness."

However, the strong smell of an electrical fire caused an alert on the tenth day in orbit, and the cosmonauts took refuge in the *Soyuz* while they discussed their course of action. The only member of the crew who had previous flight experience, Volkov, wanted to return to Earth. But the mission commander, Dobrovolsky, along with Patsayev, returned to Salyut in an attempt to locate the smoldering fire. Because the atmosphere within Salyut was similar to that of the Earth's (75 percent nitrogen and 25 percent oxygen) the fire did not flare as would have been the case with the early American spacecraft. Eventually, the smoldering cable was located and disconnected, but the obvious conflict among the crew had made the situation more difficult.

Of course, none of the contention between the crew members was made public at the time, and the Soviet press was bathing the Soviet people (and the world) in the great success of Salyut. Subsequent TV shows continued to demonstrate various aspects of the mission and its experiments, and all seemed serene on-board. On June 25, the crew broadcast their last TV show and made it clear that they were eager to return. As with all the previous broadcasts, the sessions were not shown live but were taped and edited.

On June 29, 1971, the crew completed the packing of experiments to be returned and secured Salyut in preparation for the next crew. As they prepared to undock, a warning light illuminated indicating that the hatch between the Descent Module and the Orbital Module had not closed and sealed properly. Volkov, who had been on edge for the entire flight, nervously reported, *"The hatch is not pressurized, what should we do, what should we do?"* His voice reflected the seriousness of the situation since the three cosmonauts were not wearing space suits, and if the hatch failed to hold the internal pressure of the Descent Module, they would perish.

Mission Control ordered the crew to re-open and reseal the hatch. However, the warning light still did not go out. Next, they were told to re-open it and clean the gasket as dirt might be keeping it from maintaining a good seal. This they did, and finally the warning light went out. Dobrovolsky separated the *Soyuz* from the Salyut Space Station, and, after another orbit, fired the retrorocket to slow the craft for reentry. However, as they jettisoned the Orbital and Service Modules, the vent in the Descent Module that would allow fresh air into the cabin when they were descending under parachute, opened prematurely. Their precious air was being vented into the vacuum of space—slowly but surely.

The three cosmonauts could hear the hissing sound of the escaping life-sustaining pressure, and all three released their seat harnesses in an attempt to locate and close the valve. Their first inclination was that it was the balky hatch. However, the time spent on it wasted precious air. In an effort to silence other sounds so they could hear the hissing more clearly, they turned off the radio communication that was feeding the sound back through their microphones and into their headsets—interfering with their troubleshooting. Mission Control now had no way of knowing what was going on aboard the *Soyuz* Descent Module.

Within a minute, the pressure had dropped to a value that caused the men to suffer from hypoxia (lack of oxygen), and their efforts were now irrational and uncoordinated. Within another 60 seconds they had passed out—and died a few minutes later. The spacecraft continued on its re-entry and achieved a normal touchdown. Mission Control was concerned that they had received no communications but assumed the mission had been completed successfully. When the recovery crews arrived, the Descent Module was quiet, and on opening the hatch, the heartbreaking discovery was made. Films of the recovery forces show the futile attempts to perform mouth-to-mouth resuscitation.

The families and the world grieved at the loss, and the Soviet news organizations found it difficult to manipulate the information of the tragedy to minimize the negative impact on the perception of the Soviet space program. The *Soyuz* hatch, the separation mechanism and the pressure relief valve were re-engineered. However, it would be more than two years before the next *Soyuz* would fly and the Salyut itself would re-enter and burn up just four months after the tragedy. The promise of the Space Station would have to wait for the next iteration.

Skylab: America's Magnificent Dead End

Following the establishment of NASA in October of 1958, a wide variety of space projects were examined in an effort to ensure that America did not fall further behind the Soviet Union in the hotly contested Space Race. Among the possible projects was a manned Space Station. However, President Kennedy's decision to apply America's technology and funding to the first moon landing delayed any other large-scale projects until that objective was fulfilled. That commitment did not preclude the relatively inexpensive process of preliminary planning.

Throughout the exciting "space race" decade of the 1960s, a variety of configurations for a Space Station were examined with several efforts being made to validate the concept of the rotating wheel as conceived by many of the early space visionaries. As desirable as it appeared, the strength of the structure and the ability to fabricate such a shape in space meant that it would be a very costly project—perhaps more costly than the Apollo lunar landing. A cylindrical structure, more like the Air Force's Manned Orbiting Laboratory (MOL) that was canceled in 1969, seemed to offer the least expensive first path to a manned Space Station. However, MOL was a very small environment being 30 feet long and ten feet in diameter—the physical limits being determined by the *Titan* III launch vehicle. NASA was thinking on a much grander scale.

A cylindrical shape, perhaps matching the 22-foot diameter of the Saturn IB's 60-foot long S-IVB second stage, would allow a large internal volume and the ability to conduct significant experiments. The medical aspects of long duration space flights, as would be necessary for trips to Mars, required a more in-depth understanding of the effects of weightlessness on the human body. It was also clear that the follow-on to the moon program would have to draw heavily on the Apollo hardware to keep development costs and the timetable to reasonable levels.

NASA's Marshall Space Flight Director, Wernher von Braun, had long pushed for the classical rotating wheel, but, as the financial aspect became apparent, he, too, opted for the cylindrical shape. However, he thought in terms of what was then called the "wet tank" concept. A Saturn IB would put its S-IVB second stage in orbit. Then, teams of "construction astronauts" launched aboard subsequent Saturn IBs would purge the remaining fuel in the S-IVB and install the various internal and external systems—including airlocks, power systems, and attitude control. With the Saturn IB's ability to launch 20 tons, it would take several launches to accomplish the task of building a Space Station using the "wet tank," and the hazards of the construction and the need to launch many "construction crew" ships would escalate the costs.

It was clear that this method of constructing the Space Station would be difficult to do in the unforgiving environment of space. Thus the "dry tank" concept was born. A complete Space Station would be built from an S-IVB stage on Earth and then put into orbit by a Saturn V. As only 15 Saturn Vs were built, and, until it was clear that not all of them would be needed to accomplish the first political assault on the moon, there was no way of guaranteeing that any would be available for other missions. Von Braun's new vision would have to wait for a successful lunar landing.

As the moon landing moved closer to reality following the first successful flights of the manned Saturn V in late 1968 and early 1969, the possibility of using a Saturn V for the Space Station, now named Skylab, became more focused. On the return of the Apollo 11 astronauts from the Moon, NASA Administrator Thomas Paine announced the plan and schedule for Skylab—it would use a Saturn V, and it was scheduled for launch in 1972. While this first American Space Station could be used as a springboard for interplanetary travel, a reusable manned spacecraft, for re-supply, was still too grand for the cash-strapped space program that had been giving up its cherished budget to the Vietnam war and the escalating costs of former President Lyndon Johnson's "Great Society" programs. (Even though Johnson was by then out of office, the welfare programs Congress had put in place were already showing dramatic escalation.) Any Space Station would have to use existing flight hardware from the Apollo program. There would be no large follow-on project to Mars or to establish a Moon base.

Already there was talk of curtailing the number of previously scheduled flights to the Moon and mothballing some of the Saturn IBs, Saturn V rockets, and Apollo spacecraft. (In the end, two flight-ready Saturn Vs would become museum pieces for lack of funds to fly them.) However, Administrator Paine was able to get President Richard Nixon to approve the "dry tank" Space Station. While it would not be the grand 250-foot-diameter rotating wheel being featured in the newest science fiction movie *2001—A Space Odyssey*, it would be an impressive structure.

Skylab was fabricated by using the third stage of the Saturn V (the S-IVB) and adding multiple docking adapters that would allow two Apollo spacecraft to be docked simultaneously. It would contain an array of solar telescopes and other scientific apparatus that had previously been too large or heavy for either the Gemini or Apollo spacecraft to carry. Outfitted with the ability to initially support three sets of three-man crews for as long as 90 days in a lavish interior space of 12,700 cubic feet, it would weigh an astounding 165,000 pounds (almost four times that of Salyut). The largest single piece of habitable space equipment ever launched, it would hold that record until well into the next millennium.

To stabilize the huge Space Station, three large gyroscopes (two feet wide and weighing 155-pounds each were oriented to the three axes. With a spin rate of 150 rpm, these would accomplish holding Skylab in a fixed attitude in space by virtue of the gyroscopic property of "rigidity in space." A mechanism allowed the gyroscopes to be tilted so that the space station could achieve a new selected orientation through the use of the gyroscopic property of "precession". An attitude control system using nitrogen thrusters was also installed.

To provide 10 kilowatts of electrical energy, six large solar cell panels were used. Four panels formed a windmill-like structure in the forward part of the station and served as the telescope mount which folded to fit inside a large nose shroud that was jettisoned after achieving orbit. The other two panels folded tightly against the side of the Spacecraft during launch to be extended when in orbit.

The launch of the twelfth and last Saturn V to fly carried the Skylab Space Station into orbit on May 14, 1973—almost a year after its original schedule. The launch was flawless, but once in orbit, the telemetry from the yet unmanned space station indicated a problem with the electrical power. The meteoroid and thermal shielding that wrapped around the cylindrical body of the Space Station had been torn off during the ascent through the atmosphere, and parts of it had apparently jammed the extension of the two solar panels that were folded flat against the sides of Skylab. Although the extent of the problem was not immediately known, the rise in internal temperature of Skylab and the lack of power from the two solar panels were strong indicators of a serious problem aboard the giant craft. The launch of the first crew that consisted of veteran Apollo 12 astronaut Pete Conrad and rookies Joe Kerwin and Paul Weitz, scheduled for the following day, was postponed for 10 days while NASA tried to determine if the project could be saved.

Following an analysis of the data and the preparation of a set of possible fixes, the crew launched on May 25, 1973. The Saturn IB again performed perfectly, and the rendezvous with Skylab occurred without any problems. On arrival, the astronauts immediately confirmed what the ground controllers had presumed. Conrad reported that solar wing two *"... is completely gone off the bird. Solar wing one is in fact partially deployed... there is a bulge of meteor shield underneath it in the middle and it looks to be holding it down."* A single metal strap from the shield had become wedged into the solar wing. This had allowed the solar panel to extend only about 15 degrees of its designed 90-degree movement.

A first space walk with a large 10-foot pry bar failed to dislodge the strap after more than an hour of effort. The crew gave up, and it was decided to dock with the Space Station. However, that maneuver failed five consecutive times. Finally, the astronauts dismantled the soft-docking mechanism from within the Apollo access tunnel and went directly for a hard-dock—which was successful. The crew spent the first night in space in the Apollo without attempting to enter Skylab until the next day. On entering, the crew found the temperatures in the docking adapter and the airlock to be a reasonable 50 degrees. But, the 130 degree temperature in the large workshop area, was stifling.

Using one of the research airlocks on the side of Skylab (that allowed access to the outside), Weitz and Conrad erected a large Mylar "umbrella" sun shade that they had brought with them. The temperature immediately began to drop by about two degrees per hour. By reducing their electrical needs, the astronauts were able to remain on board and perform useful scientific work—primarily medical experiments—-for the next two weeks.

On June 7, during a second space walk by Conrad and Kerwin, a wire cutter on the end of a 25-foot pole was used in an attempt to cut through the metal band that was keeping the solar panel from extending. The problem was getting the right leverage to apply the needed force. After more than an hour of trying, the band was finally cut, and the solar panel abruptly extended another 10 degrees, knocking both Conrad and Kerwin off into space. Fortunately, their 55-foot tethers held, and they were able to pull themselves back. Now they used a cord looped under the panel, and, working together, they pulled until the panel suddenly broke free and extended to its normal deployed position, again propelling the duo into space.

With more electrical power now available, the Space Station crew continued their scientific work for another two weeks. During this time, they observed phenomena that had not been obvious with smaller spacecraft. It was possible for them to drift out of arms reach from the side of the spacecraft—thus they would have no way of propelling themselves. They could actually get "stuck" in the middle of the workspace.

It was also interesting to note that one of the annoyances while trying to sleep was the constant air being blown on them by a number of fans placed around the interior. Without the fans to circulate the air, it was possible (in the weightlessness of space) for the exhaled carbon dioxide to build up in front of their faces—potentially suffocating them.

At the end of a record 28 days, the crew returned to Earth. Because of their dedicated physical conditioning, all three were able to walk from the recovery helicopter. They did, however, take about three weeks to return to normal levels of 1G activity.

The second crew of veteran astronaut Al Bean and rookies Owen Garriott and Jack Lousma launched on July 28, 1973. Almost immediately after docking with Skylab, Lousma experienced space sickness and vomited while Bean and Garriott became dizzy and disoriented. On top of this, thrusters on the Apollo Service Module were leaking, and there was some thought that a rescue mission might have to be launched. However, there were four redundant Attitude Control Systems on Apollo, and, although two eventually had problems, the other two continued to function properly for the reminder of the mission.

The crew's performance, after the first six days of illness, was phenomenal. For 59 days they proceeded to perform a wide variety of scientific experiments and observations while doing some outside repair work to replace the solar heat shield with a larger and more durable structure. As with all previous American missions, starting with the first Apollo, numerous live TV broadcasts were conducted. Inside the cavernous station, a series of tests were performed on several jetpack designs for space walks, and some of these tests were beamed down to the Earthly audience.

In keeping with the tradition of pulling pranks, the voice of Garriott's wife was heard from Skylab about six weeks into the flight. She told CAPCOM that *"the boys hadn't had a good home cooked meal in quite some time so I thought I'd just bring one up."* It seemed to fit into the current schedule and was so natural that Bob Crippen, the astronaut on duty at Mission Control was at a loss for words. Garriott had recorded the session before launch, and the entire crew got a real lift by putting one over on Mission Control. However, during the flight NASA announced that, due to budget cutbacks, a follow-on Skylab II Space Station had been canceled.

After almost two months in space, the crew returned to Earth on September 25, 1973, and, thanks once again to a good physical conditioning program, they were in very good shape—except for the mysterious bone loss that exercise and diet could not control.

The third and last crew—Bill Pogue, Jerry Carr, and Ed Gibson—arrived at Skylab on November 16, 1973. Like the previous crew, they immediately experienced space sickness—Bill Pogue in particular vomited violently. As the station was not in contact with Mission Control at the time, the astronauts held a quick conference among themselves and decided not to reveal the extent of the debilitation to the ground. However, the crew had neglected the fact that all conversations were recorded on tape and downloaded to Houston each day where they were transcribed to print. When Astronaut Chief Alan Shepard read the hardcopy of the discussion the following day, he was livid. *"I just want to tell you that on the matter of your status reports, we think you made a fairly serious error in judgment."* While the crew agreed, the tone for the first month of the mission had been set—one of contention between the crew and the controllers. During this time, the astronauts appeared to be continually behind in the schedule, especially when compared to the exceptional performance of the previous crew.

Following a series of very open discussions between the crew and Mission Control, the relationship was finally restored, and the crew performance achieved very high levels before returning to Earth on February 8, 1974, after 84 days.

Skylab had been inhabited for a total of 171 days and had produced a wealth of information about the long-term effects of space on the human body. It was obvious that a flight to Mars was feasible except for the possible problem with bone loss, which continued to plague the crews. Tentative plans were made to revisit Skylab (which still had enough oxygen and water for another 6 months of habitation) when the Space Shuttle became operational in 1979. However, the Space Shuttle program was two years late, and did not fly until 1981. As a result, Skylab could not be boosted into a higher orbit and re-entered and burned up on July 11, 1979.

Almaz: Analyzing Military Capabilities

The deaths of the three *Soyuz* 10 cosmonauts who had crewed the Salyut 1 Space Station had severely crippled the Soviet space program both technologically and politically. As the Americans prepared to launch the Skylab in early 1973, the Soviets had regrouped and improved not only the *Soyuz* but also the Salyut design—the first of which had been hurriedly assembled simply to beat the Americans in 1971. However, a second Salyut launched on July 29, 1972, failed to orbit when its *UR-500* rocket second stage malfunctioned at T+162 seconds. By December of 1972, Brezhnev himself decided that the Almaz military station (not the civilian Salyut hybrid) would be launched—the goal was still to beat the American Skylab.

The first Almaz attempt on April 3, 1973, (just one month before Skylab) achieved orbit and was named Salyut 2 to disguise its true military nature. However, a problem, most likely an electrical fire, caused its pressure hull to rupture, and the planned launch of its first crew was canceled. A third attempt on May 11, 1973, was referred to as Kosmos 557, to avoid another embarrassment if it proved troublesome. The use of the false name was appropriate as its Attitude Control System malfunctioned and caused all of its ACS fuel to be spent before ground controllers could respond to the problem.

The Kosmos 557 failure, followed three days later by the successful launch of the American Skylab (four times the weight of Salyut), was yet another blow to the deteriorating, Soviet space program. In light of these problems, a major shake-up in the Soviet program occurred, directed from the top echelons of government. Leonid Brezhnev had looked to a new series of space spectaculars to bolster his sagging position, not only in the international arena but at home where he faced stiff competition for control—and four years of successive failures in the manned space program was more than he could bear.

Sergei Korolev's original Design Bureau that had been headed by Vasily Mishin was combined with several others to form a new organization—Energiya Scientific-Production Association. The new director was Valenti Glushko (Korolev's old nemesis), while Mishin was headed for obscurity. It was at this time that the *N1* super booster project was officially terminated, and Glushko moved forward with a proposal to build a Soviet Space Shuttle and a booster for it that would also double as an unmanned heavy launch vehicle. However, work on both Almaz and Salyut continued.

More than a year passed before the next iteration of a Soviet Space Station launched on June 25, 1974. Designated Salyut 3, it was in reality another military Almaz configuration. The *Soyuz*-like service and docking modules were replaced by a single docking port on the large aft end straddled by the orbital maneuvering thrusters. A pair of solar cell panels (larger than those on Salyut 1) was supplemented by two curved panels on the outer surface of the main work area. Total electrical power available was 5 Kilowatts. The surveillance cameras were of very high resolution and capable of distinguishing objects on Earth as small as 12 inches from orbital altitude. As with Skylab, Salyut 3 used a set of large gyroscopes to stabilize the attitude of the space station.

Besides its primary role as a reconnaissance platform, the military aspect was enhanced by the fact that it was the first satellite that actually carried a weapon—a 23mm Nudelmann rapid-fire aircraft cannon that was fixed to the forward segment and aimed by pointing the entire space station.

Although Almaz was originally intended to be launched with a manned spacecraft attached to its forward end, construction of the three-man VA (Merkur) module was still several years from completion. Thus, the station was crewed by linking with *Soyuz* spacecraft launched only after the Salyut (a.k.a. Almaz) had successfully achieved orbit.

The Soviets were not very subtle in disguising the intent of Salyut 3. The amount of information about its experiments and daily routine of the crew was severely restricted compared to Salyut 1. Salyut 3 tested several reconnaissance sensors before the first truly successful visit to a Soviet Space Station was concluded with the completion of the sixteen day *Soyuz* 14 mission. *Soyuz* 15 launched in August for an intended twenty-one day mission, but problems with the Igla rendezvous radar, coupled with the inability to dock with Salyut 3, caused the cosmonauts to return to Earth without completing their mission.

However, the reconnaissance film capsule that had images from the *Soyuz* 14 mission was successfully returned in September 1974. The KSI capsule suffered damage during re-entry, but all the film was recoverable. On January 24, 1975, trials of the on-board 23 mm cannon were conducted under ground control when no cosmonauts were on board the station. The point of the experiment was to provide defensive armament for the craft should the United States try to board or destroy the Space Station—a highly unlikely scenario, but one that fit the paranoia of the Soviet mentality.

The next day the station was commanded to retro-fire to a destructive re-entry over the Pacific Ocean. Although only one of three planned crews managed to board the station, that crew did complete the first successful Soviet space station flight.

With Salyut 4, which launched on December 26, 1974, the Soviets returned to the civilian configuration with the *Soyuz*-like service module on the aft end and the docking port on the bow. The first crew arrived 16 days later aboard *Soyuz* 17, and this mission, which lasted 30 days, was by far the most successful. However, as the Americans had just completed their third and final 84-day Skylab mission, there was no propaganda or prestige gained.

The return of *Soyuz* 17 was not without its moment of concern as the parachute release sequence was late. Georgi Grechko, who had waited almost 10 years for his flight into space, was quoted many years after the harrowing experience as recalling, *"It was absolutely terrifying... sweating, panting like a dog... It was paralyzing, I realized that I had no more than five minutes to live."* His fears were short-lived when the drogue finally deployed followed by the main chute.

The second crew was launched two months later aboard *Soyuz* 18, but it met with a dramatic failure. As the first stage was nearing the end of its burn, the explosive bolts that held it to the second stage fired prematurely, and the sequence ignited the second stage. The entire rocket rotated in an uncontrollable somersault more than 100 miles above the Earth. Because the escape tower, at the top of the rocket, had already been jettisoned, the abort sequence called for the *Soyuz* to be separated from the booster, but the worst was yet to come for the cosmonauts. Following a ballistic arc, the spacecraft experienced more than 15 Gs of deceleration during its sub-orbital re-entry. It landed on a steep snow covered mountainside in western Siberia. Not knowing how long it might take the recovery forces to find them, the crew exited the now frigid reentry module and built a fire. Fortunately, a search party from a nearby village found the crew after only a few hours.

The failure of *Soyuz* 18 again put significant pressure on the Soviet space program managers—by Brezhnev in particular—to send a second crew to Salyut 4. After four years of trying, the Soviets had failed to accomplish what the Americans had performed almost flawlessly with Skylab.

Soyuz 18B (the original *Soyuz* 18 was renamed 18A) launched on May 24, 1975, and was the most ambitious mission yet with a planned stay of 60 days—almost four times that of *Soyuz* 12. As with previous experiments, efforts to grow plants in the zero-G environment continued to elude the cosmonauts. Most seedlings died while the survivors were much smaller than normal, as was the case for a colony of fruit flies.

Because of the degradation of the environmental controls, water was not effectively removed from the space station's atmosphere, and the inside of Salyut 4 was blanketed with a film of moisture. By the 40th day of the mission, a green mold had begun to coat the walls, and the cosmonauts requested that they be brought back early. However, Brezhnev would not allow the termination of the flight until the completion of the Apollo/Soyuz mission that got underway with the launch of *Soyuz* 19 on July 15, 1975—followed 8 hours later by Apollo 18. The political implications of this edict are not clear. The two spacecraft joined on July 17, and the five men from the two counties performed a series of much publicized cooperative experiments. Only after *Soyuz* 19 undocked from Apollo and returned to Earth safely was *Soyuz* 18B allowed to close out its mission and return after 63 days in space.

Although the intent of the Space Station was to provide a long duration facility, all these early Soviet efforts were relatively short lived. Salyut 4 remained unmanned for its last two years, decaying in February 1977. It provided more testing of the station design, its systems and equipment and the conduct of some scientific experiments for the reminder of its lifetime.

Salyut 5 (Almaz 3), launched in June of 1976, was the second successful flight of the Almaz military space station (similar in configuration to Salyut 3). Manned initially by cosmonauts aboard *Soyuz* 21 for a planned 60 days, the crew returned after 50 days when the cosmonauts began having significant disagreements among themselves, and their psychological behavior became a concern to mission control. The relationship between the two cosmonauts, Vitaly Zholobov and Boris Volynov, deteriorated so rapidly toward the end that they were forced to make a night landing on short notice.

Soyuz 22 was not intended to rendezvous with the space station because, like several others in the *Soyuz* sequence, there were numerous individual *Soyuz* missions. As a result, there are gaps in the numerical sequence of the *Soyuz* craft sent to the various space stations.

Soyuz 23, launched 7 weeks after *Soyuz* 21 returned, was to have docked with Salyut 5, but its long-distance rendezvous system failed. The spacecraft landed on Lake Tengiz, which was frozen at the time. The reentry module broke through the ice but floated until recovery forces arrived and attached a flotation collar. However, the two cosmonauts had to wait until daylight to leave the craft.

The third and last flight to visit Salyut 5 arrived four months after the return of the contentious crew of *Soyuz* 21. Because one of that crew had complained of strange odors, the newly arrived cosmonauts of *Soyuz* 24 wore breathing apparatus to guard against possible toxic fumes when they entered. However, examination of the air revealed no traces of toxins, and this tended to support the assertion that the previous crew's experience was a delusion.

Soyuz 24 was forced to land on the night of February 25, 1977, in a region where a snow storm was in progress. Its short, 18-day duration was probably a result of its reconnaissance mission and the need to process the film in a timely manner.

Soyuz 25 was planned, but the mission would have been incomplete due to low orientation fuel on Salyut 5, so it was postponed until Salyut 6. Salyut 5 was the last of the Almaz military configurations and remained aloft for over a year before it was de-orbited to a fiery reentry on August 8, 1977.

In a period of six years, only three of six space stations launched by the Soviets had succeeded in receiving cosmonauts. In addition, of the eight crews sent to these stations, three failed to board. It was a telling commentary on the reliability of Soviet technology of the period. For all the failures and near tragedies suffered by the Soviets, the lessons learned had finally given more focus to their space program. It became evident that there was no military advantage to the manned space station, given the level of Soviet technology. This aspect had been determined by the Americans before they had completed the development of the U.S. Air Force Manned Orbiting Lab (MOL), and it had been canceled before it flew. Almaz continued to fly unmanned missions into the 1980s before the program was concluded.

The Soviet space program had been compelled to include the military in their manned program to assure continued funding. Now that it had been proved that unmanned reconnaissance satellites were much more cost effective and flexible, there was no need to include the military in the manned space flight program if independent funding could be assured. Brezhnev, like Khrushchev before him, realized the propaganda value of competing with the United States in space—although he had also been bitten by its failures. Nevertheless, unlike Khrushchev, Brezhnev, influenced by President Nixon's emphasis on détente, had moved the Soviets away from confrontational policies and towards cooperation as evidenced by the Apollo/Soyuz flight.

Apollo-Soyuz: Detente in Space

Both the Soviets and Americans were eager to improve their image to the world and to ease the tensions between one another. While neither side was actually changing its fundamental distrust of the other, President Nixon had made significant overtures to the Soviet Union (as well as to the Communist Chinese), and the two countries were engaged in negotiations that would ultimately lead to the first Strategic Arms Limitation Treaty (SALT) in 1973. As a part of this initiative towards a more cooperative stance, several suggestions had been put forward to have a joint space project. While the more grand dreams of an international Moon base or a flight to Mars were beyond current consideration, simply having two space craft rendezvous and dock was a first step in a more cooperative effort.

The Nixon Administration, after much soul searching, agreed to a joint space flight. On the Soviet side there was considerable consternation. There was no doubt that Soviet technology was well behind the American's at this point, and a joint flight would open up the Iron Curtain enough so that the Americans could make a reasonable determination of how far behind the Soviets actually were. This aspect made for very hesitant progress in the negotiations. However, the reverse was also true. By working with the Americans, the Soviets could learn more about the technology and the methods the Americans were using to manage and control the quality of their programs.

The basic plan called for the Americans to fabricate a small docking module that would be compatible with the *Soyuz* on one end and the Apollo on the other. This would ultimately lead to a universal adapter that would allow any spacecraft to dock with any other. To this point, there were "male" and "female" receptacles on spacecraft, but it was obvious that there was a need for an androgynous docking mechanism. The schedule was set for the event to take place in 1975.

As the Soviet space scientists had feared, the politically motivated Apollo/Soyuz program coerced them into revealing previously classified data about several of their manned failures including the harrowing experience of *Soyuz* 18.

With the reality generated by a decade long struggle to tame the hostile environment of space came a new perspective of how to pursue that struggle. While the Soviets were forced to open the Iron Curtain to some extent by agreeing to the rendezvous and docking with an Apollo in 1975, there was also much gained by the Soviets themselves. The initial visits to the United States by key Soviet scientists, engineers, and cosmonauts as a part of that program were initially accompanied by a high degree of caution and skepticism. Their first impression was that the Americans had made elaborate schemes to impress them. Their visits to supermarkets, where the shelves were much too well stocked to be real, reinforced this impression as did the lack of long lines at the checkout counter (they obviously were not taken there during a Saturday afternoon). Nevertheless, the sincerity of their American hosts and the almost unlimited access to American space technology soon convinced the Soviets that America was not what their life-long indoctrination had told them.

At the Johnson Space Center in Houston, the Soviets were given a first-hand look at the management structures and quality assurance techniques that had beat them to the Moon. It was as though the Americans were providing a post-graduate course in technology management. In return, the American's learned to appreciate the simplicity of the Soviet systems while the camaraderie brought a new understanding of the Soviet culture. Therefore, while most will admit that Apollo/Soyuz was a public relations stunt for both sides, it turned out to be far more profitable than either side had hoped.

Salyut: The Next Iteration

Salyut 6 went aloft in September 1976 and was a civilian configuration similar to Salyut 4 but with an improved water recycling system. Of more significance, however, was a docking port at each end. This arrangement required the relocation of equipment and OMS fuel tanks to the periphery of the aft module. The obvious advantage was the ability to dock two *Soyuz* spacecraft—or to avoid having to abort a mission if one docking station became unusable. In practice the second port allowed the replenishing of the Salyut by unmanned supply ships that would be called "Progress." These would extend the life of the space station by bringing up supplemental food, water, oxygen, and orbital maneuvering fuel that would permit the station to re-boost periodically to a stable orbit. The perfecting of the ability to transfer fuel was the result of previous experiments in pumping liquids in the zero-G environment.

While Salyut 6 would prove the most successful, the first crew to attempt to board it, in *Soyuz* 25, was unable to achieve a hard dock and had to return after only two days of frustrating attempts. Because the docking mechanism was part of the expendable module jettisoned following the retrograde burn, it disintegrated in the atmosphere, and there was no sure way of determining which part of the latching (*Soyuz* or Salyut) was faulty. The failure of this all-rookie crew resulted in a policy change, dictated by no less an authority than Brezhnev himself—all future crews would have at least one experienced cosmonaut aboard.

Crew assignments were changed for the next attempt with *Soyuz* 26. Georgi Gretchko, who, as a former engineer, knew the construction of the latches, would perform the first Soviet space walk in six years to inspect and possibly repair them. However, because of another edict that each flight have a military commander, he was subordinate to rookie Yuri Romanenko. The personalities of the two were considerably different with the gregarious Gretchko in opposition to the militant and younger disciplinarian, Romanenko. That they had only two short months to prepare for the mission did not improve that relationship.

The flight took place in December 1976, and Romanenko successfully docked to the port on the bow of Salyut. They spent the next eight days activating and checking out the various Salyut systems. Then it was time to perform the space walk to work on the docking port on the aft end of the space station. They helped each other suit up and then entered the docking transfer compartment and depressurized it before opening the hatch to the outside.

Gretchko's inspection revealed no damage to the latches, and both men experienced the exhilaration of floating outside—tethered to the space station by a safety line and the communication link. The suits were of a standard size that allowed any cosmonaut to use them, and the environment was fully self-contained. Previous models for both the Soviets and Americans were custom made for each cosmonaut.

Both men slept at the same time during this first Soviet long-duration flight. As the Americans had also discovered, the activity of the "on-duty" crew member tended to disrupt the sleep patterns of the other. After a month in space, and with the assurance that there was no damage to the docking mechanism of Salyut 6, *Soyuz* 27 arrived in January 1977. The rendezvous and docking were accomplished automatically and without incident. This aspect was critical to the long-range plans for Salyut 6 as the unmanned Progress re-supply ships would have to use the aft docking port that was currently occupied by *Soyuz* 26. Three spacecraft were now joined together for the first time, and some structural dynamics testing was performed to evaluate the rigidity of the structure—the men bounced up and down in unison.

After five days in space, the crew of *Soyuz* 27—Vladimir Dzhanibekov and Oleg Makarov—returned to Earth in *Soyuz* 26, clearing the aft docking port for future unmanned Progress supply ships and providing a fresh *Soyuz* for Gretchko and Romanenko as the *Soyuz* spacecraft, even in its dormant mode attached to the Salyut, only had a two-month life.

Progress spacecraft were similar to the manned *Soyuz* except the descent module contained the fuel and transfer pumps to replenish the Salyut orbital maneuvering system. The first of these, Progress 1, arrived four days after the visiting crew had left. Gretchko and Romanenko then performed the first refueling in space—another milestone envisioned by Tsiolkovsky, Goddard, and Oberth had been achieved. As there was no heat shield, these Progress supply craft burned up when they were de-orbited.

After more than two months in space (exceeding the record of Salyut 4), Romanenko and Gretchko experienced the most threatening encounter with death—fire in the space station. Without warning, smoke began to appear in the main compartment of the Salyut. Romanenko immediately began the evacuation process by grabbing the critical documents and heading to the *Soyuz* to prepare for undocking. Gretchko sought the source of the fire, which was obviously electrical in nature. Because the smoke was so dense, he repeatedly had to return to the *Soyuz* to get a lung full of fresh air like a swimmer diving under water. After several forays, he located the burning cable and disconnected it, undoubtedly saving the mission and perhaps their own lives. The air circulation system was turned up to its maximum setting to filter the smoke particles from the space station atmosphere while the two crew members isolated themselves in the *Soyuz*. As had been demonstrated in Apollo 1, an incident like that in the all oxygen atmosphere of the American spacecraft would have been fatal.

While the personalities of the two cosmonauts occasionally clashed, there was one incident that highlighted their professionalism and selflessness. Gretchko had noticed that Romanenko was taking large amounts of painkiller. Alarmed, he confronted Romanenko, who admitted that he had an extremely painful toothache, and, being the commander, he would not allow it to compromise the mission whose goal was to break the American endurance record of 84 days. However, Gretchko felt they needed some medical advice and so convinced Romanenko that they should communicate the problem to mission control as though it were Gretchko who had the pain. Although the doctors could provide no real relief, the incident solidified the personal relationship between two very different personalities.

The arrival of *Soyuz* 28 at Salyut was noteworthy because of the presence of Vladimir Remek, a Czechoslovakian. While Brezhnev had made positive overtures towards the West during his ten-year tenure, the Soviet hold on its captive Eastern European states was still very oppressive. In an effort to mitigate this situation, Brezhnev used one of the seats on each of the replacement *Soyuz* flights as a propaganda incentive to these caged nations. Remek was the first of many non-Soviets to fly in space with a total of eight cosmonauts visiting Romanenko and Gretchko during their mission.

On March 16, 1977, Gretchko and Romanenko returned from space after setting a new duration record of 96 days in orbit. To ensure that their physical condition was as good as possible before their return, they spent several weeks performing extensive exercises. However, they were still carried from the spacecraft to the waiting helicopters, and it was months before they returned to a normal physical routine. More importantly for the Russians, the Americans, for whom longer periods in space were no longer of scientific or propaganda value, would not challenge the new record. It would be more than a decade before America would join with the Russians and other nations to crew a multi-nation space station. Over the next two years, ten more *Soyuz* and ten Progress supply ships were launched to Salyut 6.

The next to last crew to visit Salyut 6 was delivered by an improved *Soyuz*, designated *Soyuz*-T3, in November 1980. This version of the spacecraft featured several upgrades including an improved emergency escape system. The changes now allowed the *Soyuz* to again carry a crew of three—with enough room for each to wear a pressure suit. Perhaps more importantly, the *Soyuz*-T saw a return to the use of solar cells to recharge its batteries and could remain in space for up to four days (independent of the Salyut). The last crew, launched in March of 1981 aboard *Soyuz*-T4, remained in space for 74 days before returning in May of that year. The use of the new *Soyuz* "T" spacecraft was intermixed with the standard *Soyuz* designations making tracking of the last few flights to Salyut somewhat convoluted.

Soviet cosmonauts were little different from their American counterparts when it came to humor and contriving practical jokes. On one occasion, the cosmonauts inflated an empty spacesuit and had it float through the docking hatch to join the two crew members during a televised session—it created quite a stir among their mission control comrades watching from the ground.

The maturity of the Salyut 6 was evident as it hosted six long-duration crews and 18 manned missions during a five year period although it was originally intended to last only two years. The individual endurance record was extended to five months, and a variety of problems, including a serious fuel leak, was successfully addressed by the crew.

Even towards the end, the role of Salyut 6 was not quite finished. The 15-ton Kosmos 1267, a large transport-support module created by Chelomey's design bureau, was lofted by a *UR-500* in April 1981. The new craft was the forerunner of an expanded means of supporting the Salyut and was named *Kvant* (which translates to *Quantum*) and included the new VA (Merkur) spacecraft which was unmanned. The ship spent four weeks performing orbital maneuvers, and then the VA spacecraft was returned to Earth. The remaining portion of Kosmos 1267 executed more orbital maneuvers before performing an automated docking with the now unmanned Salyut 6 in June of 1981. The combined craft, which weighed over 35 tons, performed several more orbital maneuvers. The space station was de-orbited to its destruction on July 26, 1982, using the propulsion system of Kosmos 1267.

Salyut 7 provided the Soviet space program with the opportunity to demonstrate all that had been learned in over 10 years of operating space stations. Launched in April of 1982, Salyut 7 had many improvements to its systems and to the basic operational philosophy. With more electrical capacity from an improved and expanded solar cell array and a more rugged docking collar, the station was designed to support a series of crews for up to one year in duration.

While the Salyut 7 space station represented the pinnacle of Soviet technology, the crew was again mismatched in their personalities and struggled to work together in the isolated and confining quarters. What was rather amazing was that Anatoli Berezovoi and Valentin Lebedev had exhibited significant interpersonal problems during their training before the flight—yet were allowed to launch in *Soyuz* T-5.

It was during this first manned period of Salyut 7 (May 14 to August 27, 1982) that the first non-Communist block cosmonaut arrived—the Frenchman Jean-Loup Chretien. Brezhnev's desire to influence democratic governments outside the Soviet sphere caused him to make overtures to both India and France—two countries whose strong socialist tendencies were often in sympathy with the Soviets. However, as with the Apollo-Soyuz program, the feedback from the visiting astronauts to their own countries was more negative as they observed first hand the lack of freedom and deteriorating economic infrastructure of the Soviet giant.

The arrival of the *Soyuz*-T6 with French cosmonaut Chretien was another close brush with disaster as the on-board computer went dead during the final automated docking maneuver leaving the *Soyuz*-T6 tumbling as it headed for the Salyut at a closing speed of 15 mph. Soviet cosmonaut Vladimir Dzhanibekov was able to use the small manual thrusters to steer the spacecraft so that it missed Salyut by about 30 feet and eventually used the manual docking technique to join with Salyut 7.

Following the return of Chretien and Dzhanibekov to Earth, the next visiting crew in August 1982 aboard *Soyuz*-T7 included the second Soviet female cosmonaut, Svetlana Savitskaya. Almost 20 years had elapsed since the first female cosmonaut, Valentina Tereshkova, had been launched as a publicity stunt in 1963. Despite the rhetoric, women in the Soviet Union were woefully behind the West in achieving equal rights with men. As the Americans were preparing to launch Sally Ride in the Space Shuttle, the Politburo once again decided it was time to "one-up" the competition. However, when the visiting cosmonauts departed, Berezovoi and Valentin Lebedev resumed their bickering. It seemed that long duration isolation needed the constant invigorating visits of their companions.

However, the Soviets were about to undergo another unplanned change in political leadership as Leonid Brezhnev died in November 1982. Yuri Andropov took control of a country that was in its own death throes. Twenty-five years of trying to compete with the United States in the high-cost world of military armament and space technology had virtually bankrupted the country. While Andropov would continue Brezhnev's domestic and foreign policies—and the space program—the days of the Soviet Union were now numbered.

It was at about this time that cosmonaut Berezovoi began to have severe muscle spasms. However, with only a week to go before setting a new endurance record, neither cosmonaut was eager to return early after all they had endured. Lebedev was instructed by the mission controllers to give Berezovoi an injection of atropine, which he did, and the pain was eased. The two returned to Earth on December 10, 1982, amid an intense snowstorm that required them to wait some ten hours before they could be airlifted to more comfortable surroundings. It has been reported that the two cosmonauts never again spoke to each other.

The next mission to Salyut 7 by *Soyuz*-T8 began poorly when the docking radar antenna was ripped off by the payload shroud separation. A manual docking attempt resulted in a near collision, and the crew of Vladimir Titov, Gennady Strekalov, and Alexander Serebrov were forced to return to Earth when their maneuvering fuel ran low.

Soyuz-T9, with Alexander Alexandrov and Vladimir Lyakhov, was able to dock successfully in June of 1983 and, along with Kosmos 1443 that had been joined the previous March, once again constituted a large three-module station. However, because the station had been boosted into a higher orbit by the Kosmos, only two astronauts could occupy the *Soyuz*-T9 to reduce its weight for the boost phase into orbit. After unloading Kosmos 1443, it was undocked and de-orbited to destruction. Its unmanned VA (Merkur) was returned safely to Earth with film and other experimental results, and the two cosmonauts moved their *Soyuz* from the aft to the bow port so the next Progress could dock.

Then a fuel leak in the Salyut 7 maneuver-thruster supply in September left it drifting helplessly. Although it could be periodically reoriented by the *Soyuz* or Progress thrusters, moving the great mass of Salyut 7 required excessive fuel from their reserves. With the inability to orient the station so that the solar cells could provide the appropriate level of electrical power, the internal temperature soon dropped to 55 degrees, and the humidity rose to virtually 100 percent. This resulted in moisture condensing on the internal walls of the Space Station. Salyut 7 was in a critical situation. There was no hope of repairing the fuel leak at that time, and Salyut 7 would need a trained crew to perform a space walk to attach the new solar panel that had been previously brought up to restore the higher levels of electrical power.

Vladimir Titov and Gennady Strekalov would get another chance to fly much sooner than either had expected since they had been trained to attach the solar array. The two quickly prepared for a launch scheduled for September 27, 1983. However, at T-90 seconds a fire developed in the booster that quickly burned through the automated abort wiring, rendering it useless. The two cosmonauts tried to activate the manual abort system, but it, too, had been disabled by the fire. They lay in the spacecraft helpless as they could see the orange glow of the flames grow more intense. Now only quick action by the controllers who had to simultaneously press two buttons located on opposite sides of the control room (to assure that no one person could abort a mission) could save the cosmonauts. It was perhaps the longest ten seconds in their lives, but the solid-fuel escape rockets finally ignited and pulled the *Soyuz* spacecraft more than 3,000 feet into the air just as the booster exploded beneath them.

Now the Soviets faced a real challenge and a tough set of decisions. There was no way that another booster could be readied in time to rotate the *Soyuz*-T9 that was docked to the space station. If the cosmonauts remained aloft for the full duration, it would mean that a *Soyuz* spacecraft that had been given a life of two months would not be called upon to bring the men back until it had been in space for five months. However, to allow them to stay the full duration, they would also have to install the new solar panel extensions without any training. It was decided to make the critical decisions one at a time and to let them perform the space walk and see if the panels could be set up.

While they were preparing to suit-up, they discovered a tear in the outer layer of one of the suits and had to improvise a repair using duct tape! Finally, they were out in space, and the lure of the fantastic scene was mesmerizing. The two spent more than three hours performing their space walk and successfully completed the solar panel installation. Alexander Alexandrov and Vladimir Lyakhov were able to complete their record-breaking stay in space and returned to Earth three weeks later.

The next crew, launched in February 1984, consisted of three cosmonauts— Leonid Kizim, Vladimir Soloviev, and Oleg Atkov (a cardiologist). With the experience they had thus far gained, the Soviets decided that it might be helpful if the crew was expanded to three and used the new slot to provide more real-time evaluation of the health of the crew with the presence of a medical doctor. It was also determined that perhaps the leaking oxidizer tank could be fixed by a series of space walks.

After they had been in space for more than two months, Kizim and Soloviev began a series of extravehicular activities to the aft end of Salyut where the leaking tank was located but where there were few handholds. A variety of specially designed tools brought up on the Progress supply ships allowed the two cosmonauts to cut holes in the side of the service module and to disconnect a series of fittings and pipes. It took several space walks over periods of hours to prepare the work area. Included in the preparation was a ladder like structure and a TV camera to allow the ground to view the work environment.

Nevertheless, the leak was elusive. Every time they managed to isolate and bypass one leak, another appeared. Finally, they felt they had all but the last leak fixed, and this one would take yet another special tool that had to be fabricated on Earth, tested in the neutral buoyancy water tank, and flown up to Salyut on *Soyuz* T-12.

For the final fix, the special tool was accompanied by female astronaut Svetlana Savitskaya. The decision to send her was obvious. The Americans had planned that astronaut Kathryn Sullivan would perform a space walk as a part of an upcoming Shuttle mission, and the Soviets saw this as another opportunity to scoop the Americans. Savitskaya launched with Dzhanibekov, who had done the testing of the tool on Earth and would instruct Kizim and Soloviev on it use. Savitskaya and Dzhanibekov performed a four-hour space walk to do some other maintenance, and, after ensuring that Kizim and Soloviev were thoroughly trained on the use of the new tool, returned to Earth.

Kizim and Soloviev made their last space walk a week later and sealed the final leak. The space station had been returned to full functionality, and the Soviets had performed an impressive accomplishment. Unfortunately, it would be more than a decade before the magnitude of that space repair was fully appreciated, due to the lack of information released by the Soviets at the time. Their secrecy tended to work against them almost as much as it aided them, but significant changes in the openness of the Soviet space program were soon to come.

Kizim, Soloviev, and Atkov returned from space in October 1983 after a record 237 days in orbit. Again, bone loss ranged from 3 to 15 percent in density in various bones in the lower body but had remained mostly constant in the upper body. All three were back to normal after a month of recuperation and exercise.

The three cosmonauts had left Salyut 7 in good condition, and it was expected that the next crew would extend the space record to 10 months and then the yearlong mission. Before the next crew could be launched, radio communications with the telemetry aboard Salyut 7 suddenly ceased on February 12, 1985. All efforts to regain contact failed, and the Soviets had one large but dead space station in orbit.

A rescue mission was immediately planned, and *Soyuz* T-13 launched on June 6, 1985, with cosmonauts Vladimir Dzhanibekov and S. Savinykh. The rendezvous was flawless, and, as the automated docking system had been removed to save weight and space for repair equipment, Dzhanibekov had to perform as he had in previous missions—with skill and professionalism.

Following the docking, the crew assured that they had a good seal in the docking adapter and began to carefully pressurize it. The pressure held, and the cosmonauts entered the dark and very cold space station (ice coated the inside walls) wearing breathing apparatus to guard against any possible toxic gasses that might be present. However, test showed none, and they soon removed the masks and began to troubleshoot the electrical problem, which was quickly found. A sensor that determined when the batteries needed to be recharged had failed and had simply kept the solar cells from doing their job, and the batteries quickly were depleted. The crew replaced the faulty sensor and the batteries and began to bring the electrical systems on-line. The temperature was 14 degrees F, and it took several days to warm the station back to a livable environment and to remove the water that had accumulated.

After more than three months of work, the space station was again ready to support another long duration crew, and *Soyuz* T-14 was launched on September 18, 1985, with Gretchko, Alexander Volkov, and Vladimir Vasyutin. After a brief eight day visit, Dzhanibekov returned in the *Soyuz* T-13 spacecraft with Gretchko, while Savinykh and Volk remained aboard with Vasyutin.

A large transport-support module, Kosmos 1686, brought three tons of fuel for the attitude control system and more than five tons of cargo, and the trio began what was hoped to be a new space endurance record. However, two months later Vasyutin developed a very high temperature (104 degrees) and had trouble urinating—the pain was often intense. Despite all attempts to resolve the problem aboard Salyut 7, the crew was forced to return on November 21, 1985, due to Vasyutin's inflamed prostate.

As the cosmonauts had reported that Salyut 7 was still suffering from the affects of its long cold deep sleep, the remaining agenda for it was revisited. Plans had called for a large *Kvant* addition and an all female crew. However, there was little to be gained at this point since an entirely new space station, Mir, was in the advanced stages of construction. One more visit to Salyut 7 retrieved the remaining experiments and recoverable equipment.

After almost nine years in orbit, Salyut 7, with Kosmos 1686 still attached, had depleted its attitude and maneuvering fuel. Although ground controllers attempted to de-orbit to a reentry on February 7, 1991, over the Atlantic Ocean, the timing was slightly off. Following an impressive nighttime re-entry that lit up the sky, many pieces of the 35-ton space station survived the fiery plunge and fell on Argentina although there was no reported damage or injuries.

Soviet First Secretary Andropov had passed away after his brief tenure, and his interim replacement, Konstantin Chernenko, had likewise lasted only 13 months before he, too, died. A much younger leader had emerged—Mikhail Gorbachev, who recognized the death throes of the "Evil Empire" (as it had been labeled by the American president, Ronald Reagan). Gorbachev was aware of the critical economic and political situation into which the Soviet Union had fallen. While he wanted to keep the space program as a vital national asset, he understood that the country could not afford the many initiatives that had been undertaken. Unlike his predecessors, he was not enthralled with manned space exploration and personally felt it was a waste of precious resources. There were big changes in store for the Soviet Union and its space program.

The Salyut 7 Space station with Soyuz-T docked at right

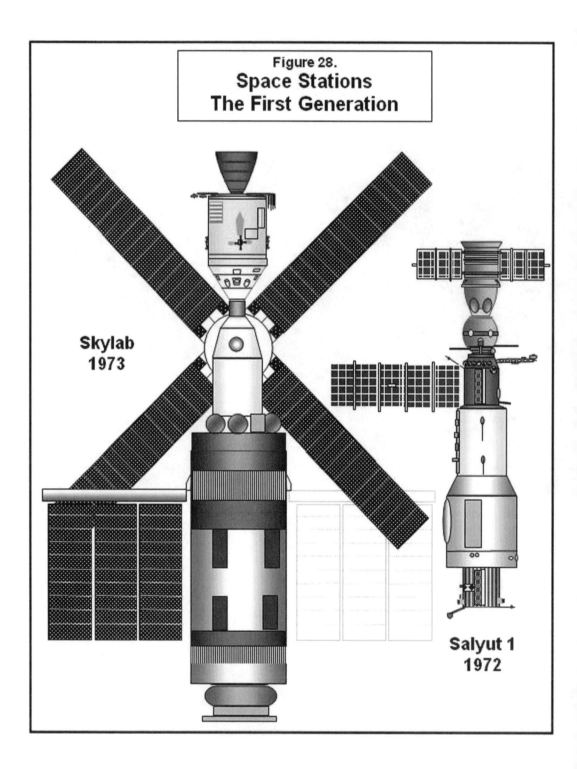

Figure 28.
Space Stations
The First Generation

Skylab
1973

Salyut 1
1972

Chapter 31 — Unfulfilled Promises and Dreams

As the year 1986 began, the number of expendable boosters used for both commercial and military satellite launches continued to decline with the anticipation that the Shuttle would be America's primary means of access to space. Following each flight, the Orbiter returned to the Orbiter Processing Facility (OPF) adjacent to the Vehicle Assembly Building at the Kennedy Space Center where the process of refurbishing and retesting the various systems took place. The next payload manifest would then be loaded and the Orbiter rolled over to the Vehicle Assembly Building for stacking in preparation for the next flight. The time for this rework varied from one to three months—the average being 6 weeks. The stacking in the Vehicle Assembly Building and rollout required another few weeks.

The shortest turn-around between flights during the first five years was two months. At this rate, only 24 launches could be accommodated each calendar year with the four Orbiters. However, the "short," two-month turn-around could not be sustained. The inability to cycle the Orbiter through the OPF in a time scale to permit more rapid launches precluded the previously announced number of 50 Shuttle launches per year, and only eight flights per year would ever be achieved.

The effort expended on the turn-around was measured in thousands of hours, and the replacement costs of various components that failed tests or were upgraded continued to swell the Shuttle "per launch" budget. The cost estimates made ten years earlier had not adequately envisioned the amount of work and the number of specialized people that would be required to keep the Shuttle flying.

Challenger: "A Major Malfunction"

The STS-33 launch had been postponed 4 times, and the morning of the fifth attempt, January 28, 1986, saw a cold but clear blue sky. The designation of STS-33 was no longer in synchronization with the number of Shuttle flights as some had been cancelled. It was in fact the 25th flight of the Space Transportation System and *Challenger*'s tenth foray into space. Since its first mission on April 4, 1983, *Challenger* had spent 62 days in space, completed 995 orbits, and had flown almost 26 million miles.

The temperatures had slid into the upper 20's—an unusually cold snap for central Florida but not a record for the date. Water had been sprayed on various parts of the pad to protect some components from the temperatures, and this had resulted in a significant accumulation of ice. NASA was concerned about the ice that had formed on some of the structure of Pad 39B—the first time this launch facility was used for the Shuttle. An engineering assessment had been made to consider the effect of the temperature, and an "ice team" had surveyed the launch pad and the *Challenger*. Based on the team's recommendation, the launch had been delayed several hours to allow the temperatures to climb to the non-freezing point to melt some of the ice.

While the concerns being explored around the pad related to the interference of the ice with the launch equipment, there was also concern expressed by engineers of the Thiokol Corporation (who built the SRBs) as to the effect the cold would have on the rubber O-rings that sealed the seven segments of the solid-fuel boosters. The engineers recognized that the temperatures experienced would harden the rubber and not allow it to seal as effectively as it should. They strongly recommended that NASA "launch within the established experience factors," meaning that a launch temperature colder than any that had been previously experienced would place the launch in a high risk, "unexplored" region. Half of the previous flights had experienced O-ring erosion, and a few had actually exhibited some "blow-by" of the hot exhaust gases. "Erosion" is the effect of the high temperatures of the solid-fuel combustion, while "blow-by" was actual intrusion of the combustion gasses past the O-rings. Even the president of Rockwell, the builder of the Orbiter, had expressed his lack of support for a launch.

However, this Shuttle carried the first passenger into space—schoolteacher Sharon Christa McAuliffe. A U.S. Senator, Jake Garn, and a Florida Congressman, Bill Nelson, had been aboard previous Shuttle flights, but Christa was to be the first "commoner" to experience space flight. President Ronald Reagan was scheduled to present his State of the Union message that evening, and NASA very much wanted him to be able to refer to her flight. Despite several attempts by lower level management and engineering personnel to bring the added risk factor to the attention of NASA's upper management, the launch took place at 11:38 EST.

To all observers there were no indications of any obvious problems as *Challenger* lifted from the pad and began a flight that had become almost routine. At T + 19 seconds, the engines began to throttle down from their 104 percent thrust to achieve the 65 percent level. The communication, *"Challenger, you are GO for throttle-up,"* at T+35 seconds was met with the last response from Commander Dick Scobee, *"Roger, GO for throttle-up."* Max-Q was encountered at T+59 seconds, just three seconds after the SSMEs began to work hard in their gimbals to compensate for the affect of high altitude wind shear. Then, a large, violent explosion at T+72 seconds marked the point at which *Challenger* ceased to exist. The "voice of Mission Control" was in the process of reading out the speed and altitude parameters and did not notice the event unfolding before the viewers eyes. When he again glanced at the live picture monitor and saw the huge vaporizing fireball, there was a long pause before he could continue with his job of reporting the status. He simply said what was apparent to all, *"...obviously a major malfunction."*

Although many who were watching knew almost instinctively that the end had come for the seven astronauts, there was still a feeling of incredulity that swept persons in attendance as well as those watching at home on their TV sets. Many assumed that the crew had somehow been propelled to safety by an escape system. However, there was no escape from the Shuttle.

The nation mourned the loss of the *Challenger* crew, and the President immediately appointed a commission, headed by William P. Rogers (a former Secretary of State), one week later to determine the cause. The commission consisted of astronauts Neil Armstrong and Sally Ride, Brigadier General Charles "Chuck" Yeager, and several eminent scientists and engineers including physicist Dr. Richard Feynmen. NASA was deep in the process of evaluating the telemetry, film, and video tape when the commission convened. The cause was almost immediately suspected to be the SRB O-rings.

Special cameras, which followed each flight to record the intimate details of events, readily showed the initial puffs of black smoke at ignition from one SRB joint in particular. This indicated that the O-rings had not seated from the internal pressure, but instead had allowed the packing material, a form of putty that protects the rings from the heat of the combustion, to be blown out and expose the rubber rings to the combustion. In the later frames, the flame could be seen protruding from the segmented joint, enlarging, and impinging on the External Tank (ET).

The flame eventually ate through the lower part of the ET and the strut that attached the SRB to the ET. Liquid hydrogen began escaping from the hole burned in the ET, and this was followed almost immediately by the failure of the strut. This caused the right SRB to swing into the top of the ET destroying it and creating lateral movement that caused the Orbiter, traveling at Mach 1.92 at an altitude of 46,000 feet, to experience side loads for which it was not stressed. It was literally torn apart by the aerodynamic forces.

NASA contended that it was doubtful that the crew had any impending knowledge of what was to befall them. As the downlink of communications was severed with the explosion and the disintegration of *Challenger*, only one word was captured that indicated an awareness that something was about to go wrong. Commander Scobee's last words were, *"uh oh."* When the remains of the crew module were recovered several weeks later, there was evidence that at least two of the crew had the presence of mind to activate the emergency oxygen system provided for the individual crewmembers. The pressure vessel of the crew compartment had likely been breached by the destructive forces, so it is probable that rapid decompression immediately followed as the crew compartment continued to soar to over 70,000 feet before beginning its plunge toward the ocean. However, there was no indication that any of the crew were conscious during the long two-minute plunge into the Atlantic. Death of the crew occurred by "blunt force trauma" on impact.

The concerns of key Thiokol employees had proven correct; the failure was the direct result of the inability of the rubber O-rings to properly seat under the temperature conditions experienced. The joint was not well designed, and the bending loads experienced by the SRBs as they rode through strong wind shear aloft accelerated the failure.

However, the problem was deeper in that management had been made aware years earlier of the inability of the O-rings to properly seat during cold ambient temperatures. Almost half of the SRBs had experienced some erosion and burn-through. Yet the potentially fatal anomaly was allowed to continue into a pattern that was called "a normalization of deviancy."

Following the recommendations of the commission, the SRB joint was redesigned and re-certified, and the management structure and reporting paths were changed. The recommendations were reportedly "indistinct" in order to allow the Shuttle to return to flight at the earliest possible date. The reason was obvious—critical "national security payloads" of the Department of Defense now rested solely on the availability of the Shuttle. If the Shuttle had to await an effective crew escape system it would probably have meant a 3-5 year delay. These reconnaissance satellites had been produced to the size of the Shuttle payload and its relatively gentle 3-G acceleration into orbit. The suitability of the limited number of available expendable boosters was severely limited.

Possible emergency escape modes were defined in a two phased approach to the problem. The first, a "bail-out" solution for a stable and sub-sonic gliding Shuttle, was implemented; however, it would not directly have helped the *Challenger* crew. The second phase, the use of a survivable escape module, might have. Yet the Rogers Commission reported *"It is highly unlikely that any of the systems discussed below, or any combination of those systems, would have saved the flight 51-L crew."*

Knowledgeable people point out that the crew module emerged from the fireball intact. Some of the crew had the physical and mental capacity to activate the emergency oxygen supply. They contend that it would be possible to rework the cabin and employ a means of separation and a parachute recovery system. The implementation of the second phase appeared as a viable option. However, those who could have moved it to reality kept anticipating that a Shuttle replacement program would be a better investment of the funds. Sixteen years after *Challenger*, NASA's Aerospace Safety Panel attempted to force the issue, and its membership was told that the phase two recommendations would not be implemented.

Nevertheless, the impact of the *Challenger* explosion went far beyond the Shuttle. The decision to use the Shuttle as a replacement for expendable launch vehicles was revisited and reversed. It was decreed, first by President Reagan's administration and then by federal law, that the Shuttle would be used only for purposes that *"require the presence of man... or requires (sic) the unique capabilities of the Shuttle... or compelling circumstances."*

The entire premise of the Shuttle program had been overturned by the *Challenger*'s accident—inexpensive access to space. However, as reviews of the Shuttle's performance and costs made clear, the Shuttle had not provided inexpensive access. In fact, because of the extensive testing and preparation required for each flight, expenses had constantly exceeded estimates, and new factors continued to add to the cost of each flight. The initial STS development ultimately cost $6.6 billion, 1.6 billion more than the 1973 estimate given at the time the program was approved. Considering inflation and the hi-tech complexity of the program, it was not really out of line. However, when the Shuttle returned to service in 1988, the per-flight cost of $300 million was a hard figure to reconcile. The cost to orbit one pound of payload had risen to $5,000—not the $175 predicted sixteen years earlier. (As there had been significant inflation over that period the $175 figure may be adjusted by a factor of five).

Another victim of the *Challenger* disaster was the "deep space" program, which had planned to use the *Centaur* upper stage launched from the payload bay of the Shuttle. This had been considered a high-risk situation, and many in the scientific community had been calling for a re-thinking of the wisdom of placing the volatile hydrogen fueled *Centaur* in the payload bay. With safety issues dominating the post-*Challenger* period, the use of LH2 was considered too hazardous, and probes such as the unmanned interplanetary missions, Galileo and Ulysses, had to be reduced in weight and reprogrammed for different flight profiles in order to use solid-fuel boost that replaced the LH2 Centaur.

Even the commercial aspects of space flight were changed by the *Challenger* accident. NASA had been competing with the Arianespace Corporation, a company partially funded by the European Space Agency (ESA), for contracts to launch commercial satellites. ESA had been using the French Ariane expendable booster since 1984. In addition, a private U.S. company, Transpace Carriers, had attempted to move into the market using the McDonnell Douglas *Delta* rocket. There was some legal maneuvering by Transpace to stop Arianespace from receiving contracts from U.S. companies because it was financed in part by the French Space Agency (a government entity) and represented unfair competition. This action had been dismissed because it was determined that the Shuttle presented the same subsidy aspect. However, with the Shuttle no longer in the business of launching commercial satellites, the suit was again reopened, and the Office of Commercial Space Transportation began regulating the launching of satellites by issuing licenses.

Nevertheless, the Space Transportation System pressed on. It was, after all, a national asset and represented America at its technological best. A fifth Shuttle was constructed using portions of a test article and named *Endeavor* (OV-105).

For the remainder of the millennium, the Shuttle continued to launch research payloads such as the SpaceLab, SpaceHab, and the Hubble Space Telescope. Military launchings included reconnaissance, signals intelligence, communications, and early warning applications. Nine missions were flown to the Soviet Space Station Mir between 1995 and 1998. Assembly and support of the International Space Station began in December of 1998 with STS-88.

Seventeen years after *Challenger*, the Shuttle continued to be the only manned vehicle for America. By January 2003, 112 Shuttle flights had been accomplished.

Columbia: Foam Insulation

Shuttle *Columbia* had made 27 previous flights into space, and the launch on January 16, 2003, into a 150-mile orbit was apparently flawless. The crew consisted of seven astronauts commanded by Rick Husbands. The primary payload was the SpaceHab-DM Research Mission, along with several "hitch-hiker" experiments.

The 16-day (255 orbit) journey was the third longest (STS-80 held the record with 17 days 9 hours—280 orbits). Because *Columbia* was the first Shuttle constructed, it was also the heaviest and was not able to achieve orbital inclinations greater than 40 degrees with a significant payload. It was, therefore, used for missions other than to provide service to the International Space Station, which is at a 51-degree inclination.

Within days of the launch, engineers examining the video of the ascent noticed that a piece of the foam insulation from the ET had broken off about 80 seconds after liftoff and impacted either the leading edge of the left wing or the Thermal Protection System tiles on the Shuttle's underside. This began an inquiry as to the possible effects. It was estimated that the piece of debris was about 20 inches long by 16 inches wide by 6 inches thick and weighed about 2.5 pounds. It struck the wing at less than a 90-degree angle. An initial analysis was performed by a simulation modeling program called "Crater." Although the engineers did not have much experience with it, the result indicated a possible breach in the Reinforced Carbon Carbon (RCC) leading edge—if that was where the impact occurred. Because of the conservative nature of the software and the belief that the debris was "just foam," management discounted the effect to some possible localized deformation but decided that it did not represent a "safety of flight" issue. However, the 2.5-pound piece had apparently punched a hole, later estimated to be about two inches in diameter, in the leading edge of the left wing. From the point of impact (just 80 seconds into the mission), the crew had almost no hope of returning safely, although they had no knowledge of their impending fate.

It was possible that, as this Shuttle had sat on the launch pad for several weeks and endured some significant rainstorms, perhaps water had collected in a break in the insulation, and, with the super-cold temperatures of the cryogenics, the foam became encrusted in rock-hard ice on the day of the launch. Because the video showed a piece of light-colored material striking the wing and exploding into a cloud of white dust or vapor, some experts suspected the material was ice (or ice-coated) as the foam insulation was orange. Moreover, because the piece broke off at the time the shuttle was near max-Q while moving about Mach 2, the impact might have been more destructive than first thought.

Some NASA officials did not agree with the benign impact report and felt it was potentially a more serious problem. Someone in NASA had made a request for the military to inspect the shuttle with its powerful secret cameras for damage from the debris. But this request was withdrawn by management.

Management again dismissed the impact of the debris on the Shuttle as "inconsequential" in a report issued 11 days into the mission. With respect to the dissenting opinions that predicted greater risk, the Shuttle program manager stated (after the fatal reentry), *"They weren't brought to my attention."*

In the days before *Columbia*'s reentry, concern centered on possible damage to the left wheel-well rather than the leading edge. A loss of tiles in this area might allow the temperatures around the left main gear to impede the gear's operation. Several engineers predicted various warning signs that might be exhibited on re-entry and took steps to understand the degree of danger—like simulating a landing with blown tires. That simulation took place just hours before *Columbia* began its descent, and some officials were alarmed by this last-minute action when they had been "assured" only a few days earlier that there were no significant risks. Depending on the telemetry indications of the conditions in the wheel well, there was consideration of advising the crew to extend the gear early so that, if it failed to lock into place, they could be prepared to use the "bail-out" capability that had been instituted after the *Challenger* disaster, as a gear-up "belly landing" would probably not be survivable.

The landing was planned for February 1, with the de-orbit burn occurring at 8:15 EST over the Pacific Ocean. The Entry Interface (EI), an arbitrary point at 400,000 feet where the first detectable signs of decelerating G- force are encountered, occurred at 8:49 AM. At 8:52 a.m., *Columbia* crossed the coast of California as it encountered the denser layers of the upper atmosphere and entered Roll Reversal #1 while traveling Mach 20.9 at 224,390 ft. Seconds later, the left wheel well showed an "off nominal" temperature rise followed by a loss of sensor data from the Left Inboard Elevon and several other temperature dropouts. Those engineers who had been analyzing the possible affects of the debris strike now knew that this re-entry would not end normally and mentally prepared to advise the crew following the communications blackout period.

At 8:54 a.m. EST, over Eastern California, sensors indicated a continued increase in temperature in the left wheel well and mid-Fuselage. Two minutes later, while over New Mexico, sensors indicated an increase in drag on the left side, and the flight control systems were automatically compensating. A fuzzy photograph, taken by the Air Force from Kirtland AFB as the Shuttle passed over Albuquerque, shows some form of disruption at the leading edge of the wing and a flow trailing off the back, which could have been evidence of plasma vortices from the ever-widening hole.

At 8:59 a.m., the tire pressure sensor caused an onboard alert that was acknowledged by Commander Husbands. It is probable at this point that the Shuttle pilots understood that something was going wrong with the reentry. The final data from telemetry showed that hydraulic pressure had been lost and that the Shuttle's nose yawed left at more than 20 degrees per second (the maximum rate the sensor could determine). The firing of the RCS thrusters did not stop the movement, and communication with the crew and loss of telemetry data occurred seconds later.

The vehicle broke up while traveling at 12,500 mph (Mach 18.3) at an altitude of 207,135 feet over East Central Texas, where it was witnessed by thousands of persons on the ground and videotaped. Debris from the spacecraft began to rain down on a wide swath across the state. As the visual sighting reports began to filter in to Mission Control, and with the inability to regain communications following the anticipated black-out period, it became obvious that the Shuttle had met with fatal circumstances.

During the reentry, the hot ionized gasses were allowed to enter the left wing through the hole punched by the foam debris and proceeded to weaken the internal structure until the aerodynamic loads of reentry caused the wing to fail and the Shuttle to disintegrate. It was not survivable, and (like the *Challenger*), while the end came quickly, there may have been a period where the life support of the space suits delayed the inevitable. Contrary to some reports, the astronauts, while burned, were not totally incinerated by the residual reentry processes—their remains were remarkably intact. It is, of course, conjecture as to whether a "hardened" module, as proposed following the *Challenger* accident, would have allowed the crew to descend to a lower altitude where bailout was possible.

While it is speculative to imagine what might have taken place had the imaging request been completed and the results revealed, it is worthy of consideration as there was another Shuttle in the preparation stages at the Cape. Under normal circumstances, it was in a 21-day launch cycle. Could *Columbia* have remained in orbit several days longer and the launch cycle accelerated to produce a rescue mission?

NASA: "A Broken Safety Culture"

The analysis of the *Columbia* disaster revealed that the findings of the *Challenger* accident, seventeen years earlier, had not been institutionalized. The *Columbia* accident report stated that *"Space Shuttle Program managers rationalized the danger from repeated strikes on the Orbiter's Thermal Protection System…"* and *"the intense pressure the program was under to stay on schedule, driven largely by the self-imposed requirement to complete the International Space Station."*

The same failures in management awareness of potentially fatal problems—in this case falling debris—were still prevalent. The debris problem was likely to occur when the ET, filled with its cryogenic fluids, experienced shrinkage. The cocoon of brittle insulating foam could fracture and portions come lose—possibly striking the fragile underside of the Shuttle. The loss of even a single tile on the bottom of the Shuttle could result in a "zipper effect" that might strip away other tiles and result in a burn-through in the aluminum skin of the orbiter during re-entry. There had been many warnings of impending foam damage.

A report in August 1996 by MSFC stated that the leading edge of a wing of the Space Shuttle *Atlantis* suffered major impact damage in a flight in 1992. In 1997 engineers warned that hardened foam had broken off the ET and had damaged tiles. In October 2002, a piece of foam from the ET of *Atlantis* had broken loose and hit the skirt of a solid-fuel booster rocket. Debris from the ET and SRBs—including ice and hardened foam—had damaged orbiter tiles during the shuttle's first 33 flights. The foam that hit the *Columbia*'s wing came from the "bipod," the attachment between the external fuel tank and the shuttle's nose.

In the final analysis, it was believed that the damage had been done to the leading edge, based on the evidence. Most of the 22 panels that make up the leading edge of the left wing had been found, except for the one closest to where the wing was attached to the fuselage. The pattern of the debris that was scattered over more than 400 miles of central Texas and radar returns from the FAA Air Traffic Control facilities had been used to help determine the sequence of the break-up. The accident report stated that *"the five analytical paths – aerodynamic, thermodynamic, sensor data timeline, debris reconstruction* [84,000 pieces—34 percent of the Shuttle], *and imaging evidence…"* all independently arrived at the same conclusion.

In the beginning of the Shuttle program, NASA safety rules said no debris (to include ice and foam) should be allowed to hit the Orbiter and possibly damage the fragile heat-resistant tiles. Following the *Columbia* accident, NASA identified more than 170 possible sources of liftoff debris. Engineers recognized that they could not eliminate all risk from debris, but they could do a much better job of reducing it. The question must be asked, "Is foam really that complicated?

As with *Challenger*, the *Columbia* accident board issued its report on the causes of the *Columbia* disintegration and was especially critical of a "broken safety culture" at NASA that had grown complacent of many risks. It stated that:

"The organizational causes of this accident are rooted in the Space Shuttle Program's history and culture, including the original compromises that were required to gain approval for the Shuttle, subsequent years of resource constraints, fluctuating priorities, schedule pressures, mischaracterization of the Shuttle as operational rather than developmental, and lack of an agreed national vision for human space flight."

The report went on to say:

"Cultural traits and organizational practices detrimental to safety were allowed to develop, including: reliance on past success as a substitute for sound engineering practices (such as testing to understand why systems were not performing in accordance with requirements); organizational barriers that prevented effective communication of critical safety information and stifled professional differences of opinion; lack of integrated management across program elements; and the evolution of an informal chain of command and decision-making processes that operated outside the organization's rules."

The report concluded that *"[the Shuttle] has never met any of its original requirements for reliability, cost, ease of turnaround, maintainability, or, regrettably, safety."* But, most damningly, it says: *"Based on NASA's history of ignoring external recommendations, or making improvements that atrophy with time, the Board has no confidence that the Space Shuttle can be safely operated for more than a few years based solely on renewed post-accident vigilance."*

Following a stand-down of almost two and one-half years to address the problem, NASA's contention that it had produced the safest External Tank in shuttle history was dispelled two minutes into the launch of the Shuttle *Discovery*'s "return to flight" mission in August, 2005—a 0.9-pound piece of foam was dislodged and struck the Orbiter. It could have led to another catastrophe if it had ripped away a minute sooner, and it forced the immediate suspension of Shuttle flights until the problem could again be addressed. Eleven months would elapse before another flight.

With Shuttle flights now costing a half-billion dollars per launch, and with the Shuttle fleet again diminished to three vehicles, the call went out to end the program and replace the aging "super-star" with something more cost effective and safe. NASA had spent 2.5 years and $14 billion after the *Columbia*'s destruction trying to fix the Shuttle's problems.

It had long been asserted that the Shuttle was never a cost effective means of placing cargo or astronauts into space. Much of the work carried out by the Shuttle could have been accomplished by unmanned spacecraft for far less money. This is not to say that man should stop flying into space—but that it should be done by something less complex and costly.

Few would deny that the Shuttle has been a modern marvel. With more than two million parts, it has advanced the state-of-the-art. However, its basic design leaves no room for survival *when* something goes wrong. The assumption that the Shuttle could be made as safe as an airliner was highly optimistic—in fact unrealistic. It is interesting to note that the two catastrophic failures in the Shuttle's 25-year career were not the direct result of any of its high-tech features such as the LH2 engines, the exotic thermal protection, or the exacting computer control of its fiery reentry. The failures came from very basic components—rubber and foam—that NASA management had been made aware of but had failed to acknowledge.

Following the *Columbia* tragedy, President George W. Bush acknowledged that the time had come to retire the aging relict, and there would be no funding for the Space Shuttle after September 30, 2010—almost 30 years after its first flight. However, in the years that lead to the 2010 suspension, NASA is planning to spend yet another $25 billion dollars to resume the Shuttle flights. Many are now asking if this expenditure is justified in light of the evidence of the Shuttle's shortfalls and the possible investment of that money in a more capable and safer follow-on?

President Bush commented, *"We cannot find any justification to continue the deficit funding of a program that has no application other than proving that with enough money America can do anything."* It was an interesting statement that was followed by: *"The whole world knows that already, so why keep spending money on it?"* In continuing the brutally frank account, Bush related, *"I don't want to see another NASA administrator—appointed on my watch—left to justify a program to Congress based on lies, disinformation, half-truths and sexed up reports."*

These comments were obviously prompted by those who have cynically asserted that the Shuttle came into existence and continued to be funded long after it had been proven incapable of performing the job for which it had been designed. Its over-due departure was the result of a NASA bureaucracy which leaned heavily on its effective lobbying power with congress and successful public relations with the American people. There is perhaps an element of truth in those assertions. However, the Shuttle was but another of man's attempts to provide an outlet for his intellect, curiosity, and creative engineering.

In a rare admission of error, NASA Administrator Michael Griffin, in September 2005, stated that the Space Shuttle and International Space Station were costly strategic mistakes and raised doubts about America's commitment to getting the Shuttle back in flight. When asked if the decision to build the Shuttle was a mistake, Griffin said: *"It is now commonly accepted that was not the right path. We are now trying to change the path while doing as little damage as we can. My opinion is that it was a mistake."* He further added, *"It was a design which was extremely aggressive and just barely possible."*

Nevertheless, the Shuttle is also inextricably tied to the International Space Station, whose construction is dependant on the Shuttle's ability to launch large payloads. Is it possible to complete the ISS, which NASA's pre-*Columbia* schedule indicated at least another 28 flights? The best NASA estimate shows that there may be only 19 possible Shuttle flights until the deadline of 2010. Of these, two are listed as "contingencies," and one is allocated to the yet-to-be-approved Hubble maintenance mission.

While it may be possible to shift some of the ISS assembly to the Soviet Progress and the yet-to-be-built Shuttle successor, it is unlikely that these vehicles and their launchers could be available to carry the mass allocated to the nine canceled Shuttle missions. The "final" configuration of the ISS will undoubtedly lack some laboratory modules, and the crew compliment will probably have to be less than the standard 6-person group.

Is the ISS itself a viable project? As for support for it, in the September 2005 interview, Administrator Griffin said, *"Had the decision been mine, we would not have built the space station we're building in the orbit we're building it in."* These admissions of fault (coming from someone who had no involvement with the initial decisions) call into question not only continued support for the ISS but funding for the return to the moon as requested by President Bush. Bush himself stated—perhaps with tongue-in-cheek—*"We plan to either hold an auction on Ebay or give it away to our international partners."*

Louder voices are now being heard that the exploration of space should be turned over to private corporations and that manned exploration should be limited to those flights that absolutely demand the presence of a human. These will be hard choices for the people of the United States, who were historically risk-takers, but who have become averse to risk. Will the ballot box determine the future of America in space?

Soviet Space Shuttle: Energiya-Buran

Following the cancellation of the *N1* super booster in the early 1970s, the Soviets concentrated their efforts on Earth orbiting projects—the Salyut space station in particular. The *UR-500* launch vehicle and improved versions of the venerable R-7 allowed the Soviets to pursue a variety of scientific and military programs. However, the need for a heavy lift booster in the Saturn V class continued to be studied.

As America moved forward with the development of its Space Shuttle in the mid-1970s, Soviet paranoia continued to see it as a threat that had to be addressed in much the same way as America had perceived the Sputniks of the late 1950's. Perhaps the need was driven more by those in the rocket development business and politicians than by military planners. Several proposals to develop a "reusable space plane" and its attendant launcher were rejected, apparently because Soviet leadership did not perceive it as being as capable as the American Shuttle, which the Soviets had been closely watching since it had been announced.

They carefully studied not only the Shuttle design and capabilities but the economics as well. They arrived at the same conclusion that Shuttle opponents had: the cost numbers used by NASA to justify the program did not add up! The Soviets believed that expendable boosters were still more economical (by far). However, Soviet First Secretary Leonid Brezhnev felt strongly that if the United States thought it would be economical, then it must be. The cross-range capabilities of the American Shuttle were of particular interest because of its military implications. Despite their belief that it was not economically justifiable, and without a clear mission for such a heavy-lift capability, the Soviets embarked on the development of a large reusable rocket and a space plane similar in concept to the American Space Shuttle. It would be the most expensive Soviet space project.

By February 1976 a reusable space system was approved that would provide for a payload of up to 65,000 pounds—about equal to that planned for the American Shuttle. While the Orbiter, which would be called Buran ("snowstorm" or "blizzard"), looked strikingly like its American counterpart, and the launch vehicle would employ similar parallel staging, there were important fundamental differences.

At first glance, Buran looked virtually identical to the Shuttle with a bulbous nose and double *Delta* wing configuration. It used a similar underbody composed of over 38,000 heat resistant tiles. The primary difference between the two craft was that Buran would not contain the main engines—they would be a part of a "universal launch system" called Energiya (pronounced with a hard "g"). Buran would house only the orbital maneuvering engines which provided the final push into orbit—similar to the OMS unit of the American Shuttle.

The Energiya launch vehicle for Buran was comprised of two parts. The central core was powered by four LH2 and LOX thrust chambers that provided almost 1.5 million pounds of thrust. Four LOX-kerosene-powered, "strap-on" parallel booster units, of 1.6 million pounds of thrust each, completed the assembly. Both the core and the booster units ignited at launch in the same manner as the SRBs of the American Shuttle, providing a total of 7.8 million pounds of thrust—more than the Saturn V.

The four booster units were actually a derivative of the first stage of another Soviet launch vehicle called Zenit. These units each contained four RD-170 thrust chambers, which burned for about two minutes before being separated for their descent back to Earth under parachutes for recovery. The central core then continued to provide power until just short of orbital speed. It then separated and was destroyed on its plunge back into the atmosphere. The payload, the Buran or a large unmanned satel-

The Buran stands awaiting launch at Baykonur Cosmodrome

lite, then provided the small added velocity to achieve orbit.

Unlike the American test program, a series of sub-orbital flights were first made with small-scale models of Buran between 1983 and 1986. A full-scale jet powered version was used for 24 glide tests starting in 1984 and continuing into 1988.

As envisioned, the Energiya launcher was to have been used with several unmanned payload vehicles besides Buran. This would allow for most of the payload weight to be allocated to the satellite, which then needed only a small thrusting unit to achieve orbit. The first launch of the Energiya was employed in this manner with an attempt to orbit a satellite called Polyus in May 1987.

As a part of the Soviet "Star Wars" program, the huge 120-foot long, 200,000 pound "Polyus Skif" was constructed in the early 1980s. Most authorities believe that it contained an anti-satellite laser system. Launched by the first Energiya vehicle, it may have been the heaviest object ever placed in orbit—almost. Although the Energiya performed successfully, the attitude control of Polyus failed, and the space tug that accompanied the satellite was unable to provide the final thrusting to achieve orbit. The huge satellite reentered and burned up.

With the Buran as a payload, the second Energiya was launched in November 1988. Buran was unmanned because many of the systems needed to support a crew had yet to be completed. The vehicle was flown in a totally automated manner and flew two orbits (at 156 miles and an inclination of 51 degrees) before reentering and landing perfectly. It was an impressive first flight—but it would be Buran's only flight.

Plans called for the number of booster modules attached to the sides of the core to be variable between two (Energiya M) and eight ("Vulcan") depending on the weight and mission of the payload. Energiya II was to be completely reusable with the core employing wings to glide it to a landing. However, this was not to be. The development of the promising vehicle was cut short by the disintegration of the Soviet Union and was officially terminated in 1993. Two additional Buran shuttles under construction were never finished. In an unfortunate ending to the remains of the program, a hanger in which Buran was stored, along with a full-size model of Energiya, collapsed in 2002, destroying the priceless artifact and killing eight workers.

However, components of the Energiya live on. The strap-on liquid-fuel RD-170 powered boosters are the basis for the Zenit rocket. A less powerful derivative, the RD-180, powers America's *Atlas* V—a Russian engine powering an American rocket. The cost of developing the Energiya and the Buran along with attempts by the Soviets to maintain military parity with the United States are often cited as contributing factors for the Soviet Union's economic collapse.

Orbital Space Plane: OSP

The Orbital Space Plane (OSP) program was designed to support the ISS for crew rescue and transport, and contingency supplies—esentially to suppliment and perhaps replace the Russian *Soyuz* and Progress spacecraft. But it was primarily to be a more economical replacement for the Shuttle and was to be produced in several versions to increase its mission flexibility. The first version was expected to enter service by 2010. However, the OSP was put on hold after the *Columbia* breakup and eventually canceled in favor of the Crew Excursion Vehicle (CVE) program which had been one of the initial OSP proposals. After the *Columbia* accident investigation, it was decided that the spacecraft had to provide a separate escape system which the OSP did not provide.

A part of the OSP was the Demonstration for Autonomous Rendezvous Technology or *DART*. This was a vehicle designed to locate and rendezvous with other spacecraft using only onboard guidance sensors. The Russians had mastered this basic technology more than 30 years before, but NASA has always employed the human element in its rendezvous and docking.

Like the Shuttle and Buran, OSP was yet another technological promise that could not be brought to fruition within the limits of a realistic price tag.

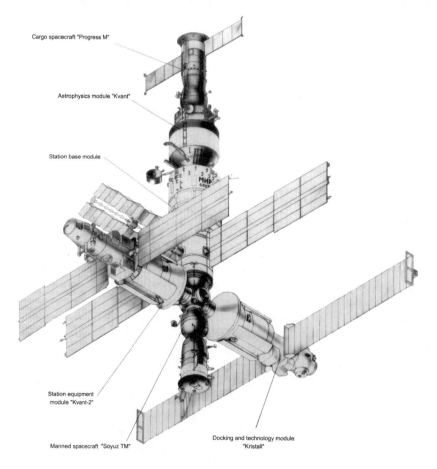

Cargo spacecraft "Progress M"

Astrophysics module "Kvant"

Station base module

Station equipment module "Kvant-2"

Manned spacecraft "Soyuz TM"

Docking and technology module "Kristall"

Official Energia diagram of the completed Mir Space Station

Chapter 32 — A Permanent Presence in Space

Mir: Third-Generation Space Station

The Salyut program had been a difficult but rewarding experience for the Soviets. Over the years, a wide variety of scientific, biological, military, and fabrication process applications was tried. However, there had been a fundamental problem in performing many experiments as specialists were needed who could effectively interpret and modify the experiment in a real-time environment. The American Space Shuttle was able to take advantage of its ability to fly up to eight astronauts, most of whom were payload or mission specialists, along with the experiment. However, the Shuttle was limited to periods of no more than two weeks. To address the Soviet's problem a more sophisticated follow-on was approved in 1976—Mir. With the new Mir station, crew selection was expanded to include such specialists.

The name "Mir" can take on several meanings in the Russian language but was generally interpreted as the word "peace." The Soviets had struggled since the inception of their country in 1917 to overcome the poor image that their militant and often brutal policies conveyed to the rest of the world. The military focus of much of their space program required that they make a concerted effort to change that image, and the naming of Mir (as well as several other programs) was yet another such attempt.

Mir was designed to be the first continuously manned space station supporting a primary crew of two for nine years (except for two brief periods). The duration of each crew was gradually lengthened from an initial four months to one year. During that period, supplies were flown up using improved versions of the R-7 to orbit an unmanned version of *Soyuz* called Progress, and periodic visits were made by visiting crews.

The station was designed to be incrementally built by a series of five modules (arranged in a "T" shape). It would grow to more than 130 tons and measure 107-feet long with docked Progress-M and *Soyuz*-TM spacecraft, and stretch 90-feet wide across its array of solar cells.

The core contained the principal operations area where the crew monitored and controlled the station's systems, science equipment, and facilities. It was also the primary living quarters with areas allocated to a galley (to prepare and heat food) and personal hygiene—with a toilet, sink and shower. Of course, in a zero-G environment these normal daily functions were performed in a much different manner to prevent water or waste from floating throughout the station.

For the first time crewmembers would have a small but private living quarters, though not much larger than a closet. This allowed the cosmonauts their own space, something that had been recognized after several crews experienced the interpersonal problems of living in close proximity. These small areas provided a foldout desk, sleeping bag, and porthole. All of the crew areas of Mir had a distinct floor (designated by its carpeting), walls (which were colored) and a ceiling (which was white) as it was believed that humans still required the traditional orientation despite the fact that there is no up or down in space.

The environmental system allowed the water vapor in the air to be removed and recycled into potable (drinking quality) water. Urine was processed and eventually used with an electrolysis process to separate the oxygen to supplement the primary supplies brought up by each Progress ship. The atmosphere was maintained at a normal 14.7 lbs. per square inch (sea-level equivalent) pressure with the normal 80/20 relationship of nitrogen to oxygen. The temperature inside the space station could be adjusted from 64 degrees to 82 degrees Fahrenheit for a shirtsleeve work environment.

As in the United States, tight budgets continued to plague the Soviet space program, but the first module of Mir was launched on February 19, 1986, (three weeks after the *Challenger* disaster) while Salyut 7 remained unmanned in orbit. In keeping with Soviet President Mikhail Gorbachev's earlier edict of Glasnost, the launch date was released prior to the flight, and the world media was invited to telecast the event. Once again, space exploration was providing an opportunity for national policy to be visibly displayed—in this case a dramatic change.

The timing of the launch was critical in that the Soviets wanted to have both Mir and Salyut 7 in the same orbital plane, as the first crew to visit Mir would also visit Salyut. It required a launch window that was only five seconds in length. The first 44,000-pound, 43-foot-long core module, or "base block" of the new space station, was placed into an elliptical orbit with the now Soviet "standard" inclination of 51 degrees. The orbit was circularized over the next few weeks to 250 miles with a period of 92 minutes. The core looked almost identical to the Salyut in appearance with two cylindrical sections but with a five-port docking adapters on the forward end. There was also a single aft docking port.

As a part of the "service module," two OMS engines and 32 smaller attitude-control thrusters were incorporated within the basic structure. These were primarily used to circularize the initial orbit and stabilize the station. They were not intended to be refueled since the *Soyuz* and Progress ships would provide for periodically restoring Mir to its primary orbit. The size of each module of Mir was determined by the lifting ability of the *UR-500* and its structural dynamics. Although their heavy-lift Energiya launcher was under development, its first flight was still several years off.

Soyuz T-15 with cosmonauts Leonid Kizim and Vladimir Solovyov launched on March 13, 1986, and the mission was designated EO-1. The term EO represented "expedition" numbers to distinguish between the various crews which typically arrived in one *Soyuz* and departed in another—making the tracking somewhat ambiguous at times. To add to the possible confusion, some reports used the designation Mir-"x"—thus EO-26 and Mir-26 represented the same crew. Radio call signs also differentiated the various crews.

After six weeks on Mir, the cosmonauts went back aboard the *Soyuz* T-15, undocked, and rendezvoused with Salyut 7. (At one point in its orbit, Mir passed within five miles of Salyut 7, but there was never any intent for the two space stations to rendezvous.) They made some repairs and removed experiments left by the last Salyut crew who had departed in haste because of the illness of cosmonaut Vasyutin. The T-15 crew went back to Mir on June 26th and finally returned to Earth on July 26, 1986, after 125 days in space. Salyut 7 was never re-visited.

A new automated docking system, called "Kurs," allowed more flexibility during rendezvous in that the space station no longer had to be oriented directly towards the approaching space ship. Computers were playing an expanded role in Mir operations, and seven "Strela" systems were installed that could be remotely programmed from the ground. One computer, called "EVM," provided the ability to hold a specific attitude without the crew's intervention. As had been done in the American Shuttle program, geosynchronous relay satellites called "Luch" replaced most of the fleet of tracking ships to maintain continuous communications.

Kvant-1, a 19-foot-long astronomy observatory module with a diameter of 14 feet was launched in March 1987 as the first addition to the core unit. An EVA (spacewalk) had to be performed before docking could be achieved because of a "garbage bag" problem remaining from a previous Progress cargo ship. The bag had become wedged in the collar area and had kept the docking apparatus from achieving a hard dock to the aft port.

Over the next nine years, Mir continued to grow as modules were added to the five port docking adapter. Kvant-2, a 19-ton, 40-foot-long, 14-foot diameter expansion module (as large as the Mir core itself) was attached in November 1989. It was the first module to permanently be attached to one of the radial ports and essentially doubled the living area. This module had photographic and biotechnology equipment, and its airlock provided EVA capability.

EVA was a common activity, and the Soviets had long surpassed the United States in the number of hours outside of the spacecraft. However, numerous incidents came perilously close to taking the lives of several cosmonauts. In one situation, an outward opening hatch literally blew open before all of the pressure had been evacuated, and the hatch hinge was damaged. The cosmonauts completed their EVA assignment but then had difficulty sealing the hatch and came within minutes of running out of oxygen. In the end, it was not the bent hinge that kept it from sealing but a hatch rim gasket cover that had not been removed.

When the airlock was re-pressurized following an EVA, lithium perchlorate "candles" were ignited to quickly replenish the oxygen. These Solid Fuel Oxygen Generators, or SFOGs (pronounced "ess fogs"), burn for less than a minute with an intense flame and give off 1.74 pounds of oxygen. (The average cosmonaut consumes 1.5 pounds of oxygen on a typical day.) These too, would prove to be a hazard.

A small crane was attached to the outside of Mir. In addition to aiding in EVA activities, it shifted incoming modules from the bow port to one of the radial ports. The aft and bow ports were the only ones to which an initial docking could be made, so incoming modules had to be relocated manually to clear the bow port.

Launched in May 1990—almost two years later than originally planned—the Kristall module was another 19-ton, 40-foot-long structure for biological and materials processing technology. It could produce semiconductor materials that benefit from the low-gravity environment of space. The Kristall also had a greenhouse to grow plants and had a spherical universal docking port with two androgynous docking units for the planned Buran Shuttle. Kristall was relocated to the bottom docking port opposite Kvant 2 in the top port.

Lack of funds caused the remaining two modules to wait in storage almost five years before they could be launched. The Spektr module, added in May 1995, contained equipment for atmospheric research and surface studies. The Priroda remote sensing module, launched in the spring of 1996, contained a variety of infra-red radiometers, and radar and spectrometers for examining the Earth's atmosphere with specific interest in measuring ozone and aerosol concentrations.

Every two months an unmanned Progress freighter would deliver 5,000 pounds of supplies to the station. This included food, compressed air and nitrogen, hydrazine fuel and oxidizer for the attitude control system, supplemental water, and newspapers and mail from home. New scientific experiments, replacement parts and special tools were also provided. The oxygen/nitrogen atmosphere had to be replenished as some was lost when airlocks were opened for space walks. Refueling could be performed without a crew on board the space station. A radar homing transmitter and TV cameras mounted on the outside of Progress helped guide and monitor each module to the space station.

As a part of many Progress freighters, a small recoverable capsule named "Raduga" (rainbow) allowed the safe return of up to 330 pounds of material—relieving the *Soyuz* of some of that task. A ton of trash was generated each month and was loaded into the Progress freighter. Following undocking, the Progress was de-orbited to burn up over the South Pacific Ocean.

EO-2 with Yuri V. Romanenko (launched in February 1987) recorded the first long duration mission of 326 days. His arrival back on Earth was amid a snowstorm that had winds of up to 70 mph, which dragged the spaceship several hundred feet.

In 1988, as the Soviets struggled to regain influence in the Moslem world following their foray into Afghanistan, Abdul Ahad Mohmand, an Afghan, became the first inhabitant of a Soviet spacecraft to recognize the existence of God when his reading from the Koran was broadcast to the public on Soviet television. He was almost stranded in space when the retrorocket of the *Soyuz* returning him failed to provide the required thrust to de-orbit on its first two attempts, but a third attempt on the following day was successful.

The Winds of Political Change

An attempted coup to topple Mikhail Gorbachev in August of 1991 failed as cosmonauts Sergey Krikalyov and Anatoli Artsebarski (who performed a Coca-Cola commercial from space) followed the intrigue that was taking place 200 miles beneath them. The All-Union Treaty that was being hammered out as a result of the dissolution of the USSR would divide the former Soviet Union into fifteen separate nations. This treaty was about to present problems for the Russian space program. The primary launch site at Baykonur just happened to be located in one of the countries scheduled to get its independence—Kazakhstan. Although it was expected that most of these newly liberated countries would participate peacefully in a new Commonwealth of Independent States (CIS) (four eventually refused), the leaders in Kazakhstan knew they had a money-making situation that they could use to negotiate with the Russian republic. Ultimately, an initial 20-year agreement provided for the Russians to continue to use the Baykonur launch site for an annual fee of $155 million.

On December 24, 1991, the Soviet flag was lowered for the last time at the United Nations, and the flag of Russia, as the largest surviving member of the USSR, was given its place. Membership in the Security Council also went to Russia. Gorbachev resigned the following day, with Boris Yeltsin eventually assuming the leadership role. It took some time for the details of the structure of the space program to be reconciled, but eventually "Glavkosmos" (the Soviet space organization) and the Soviet Ministry of General Machine Building were incorporated into the Russian Space Agency (RSA), which was modeled after NASA. The design bureau that had originally been Chelomey's and had been melded into Energiya became a semi-private agency to continue production of the *UR-500* (Proton) launcher. Energiya was also semi-privatized and given the responsibility for the Russian manned space program under the direction of RSA. However, this infrastructure for the space program was not in much better shape than the military or civilian sectors; there was no money to pay even salaries. Funding would eventually come from a source that had been the prime antagonist of the USSR— the United States.

In an effort to assure that its former enemy did not degenerate into a massive civil war that might return a totalitarian government to power, the United States embarked on a program to provide significant financial support. The basic goal was to assure that the thousands of nuclear weapons that were scattered across the now independent 15 states could be controlled and that engineers and scientists in critical positions did not flee to renegade countries such as Iran and Iraq in search of employment. There were several problems with America's financial investment in Russia—not the least of which was the gross misappropriation of the incoming dollars into the pockets of high officials.

In another attempt to secure hard currency, the Russian Space Agency brought more than 200 items to Sotheby's Auction House in New York in December 1992. These items, which ranged from space suits to watches, brought $6 million dollars. One of Gretchko's space suit gloves went for $10,925.

The lack of money caused a delay in Mir replacement crews into the spring of 1992. When cosmonaut Krikalev finally returned to Earth in March 1992, after 313 days in orbit, he landed in the prescribed recovery area—but it was now in a different country—Russia, not the USSR.

In 1992 Presidents Bush and Yeltsin agreed to a further cooperative endeavor in space with the American Space Shuttle rendezvousing with Mir and performing a series of crew exchanges. This would eventually pave the way for the higher level of inter-operation required for the International Space Station. This agreement was further expanded by the Clinton administration to provide $400 million to the Russians for using Mir to gain knowledge and techniques in using the Space Shuttle to service a space station. Mir was ultimately visited by nine Shuttle flights beginning with STS-63. All but the first docked with the space station.

Russian cosmonaut Valery Polyakov recorded the longest stay aboard Mir—438 days in 1994-95. Initially launched on January 8, 1994, aboard *Soyuz* TM-18, he was also among the first to celebrate publicly the Russian Orthodox Christmas from Mir on January 8, 1995. The Godless Communist regime, despite 70 years of trying, had not destroyed the faith of the Russian people.

Elena Kondakova became only the third female to fly in the Russian program (ten years after Svetlana Savitskaya's flight in 1984) when she went aloft in *Soyuz* TM-20 on October 3, 1994, and returned with *Soyuz* TM-20 March 22, 1995 . She made a second flight on STS-84 in May 1997 as a part of the crew of the Space Shuttle *Atlantis* for a two week stay.

Shuttle *Discovery* (STS-63) was the first to rendezvous with Mir in February 1995. No attempt was made to dock although a close "station keeping" was performed to see how the Shuttle's thrusters impinged on Mir. Aboard the Shuttle was cosmonaut Vladimir Titov, Michael Foale (who would eventually occupy Mir on a subsequent journey), and Eileen Collins, who would become the first female to command a Shuttle. The crew aboard Mir commented on how precise Shuttle Commander James Weatherbee was in piloting the 100-ton craft along side the 100-ton Mir. A subliminal Coca-Cola commercial was performed with the dispensing of Diet Coke and regular Coke for a taste test aboard *Discovery*.

The first docking of the Shuttle with Mir occurred with the flight of *Atlantis* in June of 1995. The combined weight of the station, the Shuttle, and a docked *Soyuz* was 250 tons with ten crewmembers.

Americans on Mir

Expedition EO-18 saw the first American, Norman Thagard, fly aboard a Soyuz (March 14, 1995) en route to Mir with cosmonauts Vladimir N. Dezhurov and Gennadi M. Strekalov. Both Strekalov and Thagard were experiencing their fifth flight into space. Thagard's flight as a part of a Russian crew highlighted the considerable differences between the two cultures. Although he had trained with the Russians, it was not until he came aboard that he discovered that, when "on the job," they became very bellicose and condescending—especially the mission commander. These observations would be borne out by subsequent Americans who encountered the problems that multi-cultural crews would experience in the future.

Thagard's flight was also another foreshadowing of the encroaching bureaucracy that was befalling NASA. Perhaps because these early flights to Mir by American astronauts were a "political deal," Thagard received very poor support from NASA's own Mission Control. He returned after 181 days by way of STS-71, which delivered the first Russians to Mir by Shuttle. He reported Mission Control's lack of responsiveness to his requests to none other than the NASA Administrator, Dan Goldin, who acknowledged the problem but failed to make any changes. The *Challenger* disaster of almost ten years earlier had not improved the management ability of NASA.

Shannon Lucid, launched March 22, 1996, aboard STS-76 and returned September 26, 1996, by way of STS-79, found the same belligerence with the Russian crew and neglect from NASA but handled the problem better than her male counterparts. Her personality and her ability to receive a one-hour CNN weekly news update and to periodically email her family eased the isolation. It was during her flight that the cosmonauts performed their Pepsi commercial. She had to avoid any involvement because of the strict limitations NASA had on possible personal-business gain from space flight activities.

American astronaut John Blaha attended the Defense Language Institute in Monterey, California, to learn the Russian language and completed the Russian training program at the Cosmonaut Training Center in January 1995. In September 1996 by way of STS-79, he arrived on Mir, where he spent four months with the EO-22 cosmonauts conducting material science, fluid science, and life science research. Blaha returned to earth on STS-81 in January 1997. When he left Mir, he remarked to his replacement, Jerry Linenger, *"Don't expect any help from the ground. You're up here on your own."*

Linenger was the fourth NASA astronaut to crew Mir, launching aboard STS-81 in January 1997 and returning with STS-84 in May 1997. His statements were even stronger than his predecessors on his return: *"The* [Shuttle-Mir] *program was not primarily concerned with doing good science or advancing our expertise in space operations, but rather was conceived and thrust down NASA's throat by the Clinton administration as a form of foreign aid to Russia."* As a result, NASA's indifference to their obligation to support the program was a way of NASA's "getting even" with the politically appointed Clinton administrators.

However, Linenger had more than just a crusty Russian crew and ignorant bureaucrats at NASA to worry about during his stay. On February 23, a few weeks after Linenger arrived, cosmonaut Aleksandr Lazutkin was in the Kvant module, where he went to light a lithium perchlorate candle to increase the oxygen content of the atmosphere. With six people aboard, the oxygen regeneration system had to be periodically augmented. Instead of producing the intended gas, it burst into flames, as one had in 1994. Unlike the previous fire, which was immediately smothered with a cloth, this one burned furiously. The fire alarm was triggered, and a call was made for a fire extinguisher as thick smoke billowed out.

Linenger was in the Spektr module and was initially unaware of the emergency. He heard the audio alarm, but it had previously triggered when there was a problem with the electrical power, and he proceeded to save the work he was doing on a laptop computer. When he passed into the central docking adapter to see what was happening, he was met by one of the crew who was yelling "fire." As he moved to view the interior of the Kvant module, he saw the flames and smoke which now blocked any access to the second *Soyuz* that was moored to the aft docking port. Only the one *Soyuz* attached to the central docking adapter would be available for abandoning Mir, and it would only hold three of the six men.

The crew donned respirators, but the first one Linenger tried failed to work as did the first fire extinguisher brought out. There were several other extinguishers immediately within reach, but they had been bolted to the walls to secure them during the launch phase several years earlier, and no one had thought to remove the bolts. Linenger positioned himself in the docking tunnel and handed another extinguisher to Korzum. But, as the cosmonaut pulled the trigger, the discharge of its content exercised Newton's third law of motion and he was propelled backward away from the flames. Linenger had to grab Korzum's legs and steady himself against the bulkhead so Korzum could direct the stream of precious extinguishing agent against the base of the fire.

After what was estimated as 10 to 15 minutes, the fire went out. Other than Korzum, who had several fingers burned, there were no casualties. It took a full day before the environmental system could clear the smoke and the respirators could come off. However, it took a while before they had all the soot cleaned up. Another tragedy had narrowly been averted. The Russian crew and Roskosmos (the Russian space agency) suppressed the intensity and seriousness of the fire, and NASA was content to accept the Russian assessment despite Linenger's report to the contrary.

Less than two weeks later, failure in the automated docking system caused yet another near collision with a Progress supply ship. Again, the Russians refused to acknowledge the potentially fatal accident, and NASA was unconcerned. Linenger became so upset with Mission Control that he finally decided not to communicate with them over the voice channels and simply used email.

Astronaut Michael Foale replaced Linenger in May 1997 and found the same conditions that his American predecessors had been complaining about. Foale's personality was, like Shannon Lucid's, more forgiving and easy going, and he found it easier to brush off the cultural affronts of the current Russian contingent, Vasili Tsibliev and Alexander Lazutkin, and get on with his experiments.

A docking test was scheduled toward the end of June to perform a manual procedure. The Progress supply ship was undocked from the aft port, and the maneuvers began. However, the commander lost sight of Progress at a critical moment, and, before he could reestablish its location, cosmonaut Lazutkin observed it closing at a high rate of speed. He yelled, *"My God, here it is already!"* Within seconds the Progress hit Mir between the Kvant module and the docking port. It then somersaulted into the Spektr module, and the crew could feel the pressure inside the space station change as their ears began popping—the pressure hull had been breached.

Inside the Spektr module, the crew could hear the sound of air escaping into the void of space. It should have been a simple task to isolate that module. However, the hatchway was blocked open with 18 air ducts and electrical cables that had been laid between the module and the rest of the space station to provide a variety of connections to other equipment and power. Lazutkin grabbed a knife and tried to cut through the cables but this proved futile. He then entered the Spektr module and began to disconnect them. With the path finally cleared, the hatch was closed, and the module was finally isolated from the rest of the space station.

Some of these disconnected cables, that now lay floating inside Kvant, provided power from Spektr's solar cells to the rest of the space station, and availability of electrical power dropped significantly. In addition, the impact had caused the station to begin a slow rotation that was too forceful for the gyros to stabilize, and the attitude control system had shut down. With the station no longer positioned for its solar cells to receive maximum sunlight from the remaining solar cells, even more electrical power was lost. For more than a full day, Mir drifted with no lights or fans. During the 45 minutes of each orbit when they were in total blackness, they could do nothing, and Foale noted that the three of them would gather *"in front of the big window, looking at incredibly complex, swirling auroras with the galaxy showering down on them, with nothing else to do."*

While the Russians pondered their next course of action, Foale convinced them that they could regain control using his observations of the proximity of the Earth from the window relayed to Tsibliev, who would work the attitude controls in the attached *Soyuz*. It took a while, but eventually the crew got the solar panels pointing towards the sun, and enough power was restored to begin recharging the batteries. While this was obviously the key to saving Mir, the availability of electrical power once again allowed the toilets to work after almost two days, and the crew was much "relieved." However, the stress was too much for Vasili Tsibliev and Alexander Lazutkin, who returned on August 15 and were replaced by a new crew before their scheduled time had been completed. Back on Earth, a new hatch was quickly fabricated that had electrical plugs on both sides so that the power cables could be plugged in to restore the connection to Spektr's solar cells. This was flown up on the next Progress, and an "internal" space walk allowed its installation.

It was somewhat paradoxical that the Russians became more efficient and improved their quality control in their space endeavors using many techniques observed from NASA. Their American mentor (NASA), who had accomplished virtual miracles in space with Apollo and the creation of the Shuttle, had degenerated into a squabbling bureaucracy that was indifferent to the needs of its astronauts. This arrogance and complacency would eventually cripple the American space program and bring it to a virtual standstill in the first few years of the new millennium.

Russian Capitalism

When the head of Energiya, Glushko, passed away in January 1989, Yuri Semenov took the helm with his first objective being to improve funding to enable Mir and the rest of the Soviet space program to continue. At this point only one module had been added to the core in almost three years (Kvant 1—a 19-foot-long astronomy observatory). Even the cost of the bi-monthly Progress ships was a drain, and the station remained unmanned for a significant period of time. This occurred with the departure of Volkov, Krikalev, and Polyakov in April of 1989. A Progress freighter then boosted Mir into a higher orbit to extend its life.

As the Soviet Union disintegrated in 1989, the space program, crippled by the lack of funding, was given permission to engage in a series of financial deals with other countries and western corporations. The decision was an outgrowth of one of Gorbachev's initiatives (called the Enterprise Law) to transform state-run industries into private companies. This included commercials for Coca Cola and Pepsi to be broadcast from Mir. Mission Control in Moscow suddenly sprouted advertisements hanging from the walls.

Cosmonauts returned to Mir in February 1990 aboard *Soyuz* TM-9, partially funded by the producers of the television documentary program NOVA. The launch vehicle now sported advertising logos, and the launch complex had a 150-foot-high sign for an Italian insurance company.

Another endeavor used the faculties aboard Mir to manufacture protein crystals for pharmaceutical companies. An estimated 25-million-ruble profit could result from the sale of 220 lb. of these products. The cosmonauts also produced other organic and inorganic substances whose structures were improved when made in the weightlessness condition aboard Mir. This included glass, metal alloys and semiconductor materials. Unfortunately, most of these proceeds were from other Russian "businesses," so it was simply moving money from one pocket to another and not generating hard cash from outside the country.

The Russians also planned to earn money by charging businesses for testing materials in the *Kristall* "laboratory" and by sending "tourists" to *Mir* for about $10 million per person for a two-week stay. The tourist plan continued in the 21st century, after the end of the *Mir* era, with several space-tourist flights aboard *Soyuz* transports to the International Space Station.

While Energiya and Khrunichev were marketing spacecraft capabilities (the French paid $36 million for three-crew positions while the European Space Agency paid $60 million for two Germans), the launch vehicles were being contracted to foreign agencies to launch satellites by Glavkosmos. These activities provided a significant improvement in needed capital to rebuild and sustain the Russian space program, and the foreign income reached more than $500 million by 1996.

Soyuz Evolution

While emphasis of the Soviet program has shifted to the space station over the past 35 years, the enabling technology has been the *Soyuz* spacecraft. The basic *Soyuz* ("union") spacecraft design has had a long and useful life. Conceived originally by the Korolev team in 1963, the basic structure has undergone many changes, and it has been used in both a manned an unmanned role in a variety of programs. The first manned version of Soyuz flew eight missions as an independent spacecraft (*Soyuz* 1, 3-9). There were 28 flights in support of both the *Salyut* (10-13, 17, 18, 20, 25-40) and *Almaz* (14, 15, 21, 23, 24) space stations. Finally, three flights (16, 19, 22) were accomplished as a part of the Apollo Soyuz Test Program (ASTP)—only the last of these was the mission that made the rendezvous with *Apollo*.

The *Soyuz-T* was modified for use as a manned transport for the Salyut 6 and 7 space stations. The final *Soyuz* T-15 was a *Mir* mission. The *Soyuz-TM* craft allowed 500 pounds more payload, as a result of systems' weight reductions, and could carry either two or three cosmonauts. Twenty-nine were launched to *Mir* (TM-2 to TM-30). The *Soyuz-TMS* included changes which were required by NASA for use as a "lifeboat" for the International Space Station and which provided new cockpit displays, an improved parachute system, and the ability to accommodate a wider height spectrum of crewmembers.

An older variation of *Soyuz*, the 7K-L1 (1967-1969), was used in a direct launch for circumlunar flight and was called *Zond* but was never manned.

Soyuz craft modified as unmanned cargo vessels for delivering supplies to the space stations are called "Progress" (with later variations designated M and M1). Most have no thermal protection system and are destroyed upon reentry.

The *Soyuz* type has flown more times into space than any other type of manned spacecraft. It is currently (2007) the primary support vehicle for the International Space Station.

The End of Mir

Over its 13-year life span of habitation, almost 16,000 experiments were conducted, with the emphasis being the adaptation of humans to long-duration space flight. Mir also experienced more than 1,600 system failures—some were very serious such as the onboard fire in 1997. By 1989 almost half the scientific equipment on Mir was no longer functioning, and by 1992 the Kvant 1 "astrophysics laboratory" module was simply used to store equipment and garbage. There were several near-fatal collisions with Progress cargo ships including the one which left the Spektr module unusable. The solar panels degraded at about five-percent per year and by the end of the century were producing only about 15 KW of electrical power. The windows had become almost opaque from the constant impacts of micro-meteoroids.

The geosynchronous communications satellites used to maintain an almost constant communication with Mir were also in disrepair as were the few tracking ships that remained.

However, even the near disasters and numerous problems due to the age of Mir were not enough to deter the Russians. The Soviet/Russian space program had proven to be resilient and resourceful. Ultimately, the cost of upkeep was more than the Russian economy could handle, in light of their commitments to the upcoming International Space Station.

NASA finally determined that not only was Mir a hazard but it was draining its own budget of funds that were needed for the ISS. When cosmonaut Valeri Ryumin returned to space in 1998, more than 20 years after his last flight to Salyut 6, his impressions of Mir were very down to earth. *"I don't know how they live up here,"* he was quoted as saying. *"This is worse than I imagined. This is unbelievable. This is unsafe."*

EO-27 with cosmonauts Viktor M. Afanasyev, Jean-Pierre Haignere of France, and Ivan Bella of Slovakia launched in February 1999. They returned to Earth in August 1999 completing the final planned Mir mission of 188 days in orbit. When they left Mir aboard Soyuz TM-29 on August 28, 1999, the space station was unmanned for the first time in ten years.

The station was to be de-orbited in early 2000 but MirCorp, an American-Russian international group based in the Netherlands, wanted to use Mir for commercial and tourist purposes. An automated Progress mission in February 2000 boosted Mir's orbit 40 miles higher, while the Russians sought to determine its fate.

EO-28 with cosmonauts Sergei Zalyotin and Aleksandr Y. Kaleri launched April 4, 2000 in *Soyuz-TM* 30. They spent 72 days in space in a flight funded by MirCorp—the only privately funded mission.

On March 23, 2001, the 120-ton Mir was deorbited to a fiery end over the Pacific where an estimated 1,500 pieces impacted without incident. One hundred and four men and women had inhabited it representing twelve countries—four of the men had spent more than one year in space at various times. Mir had traveled more than 2.2 billion miles in 86,000 orbits of the Earth. Sixty-six Progress supply ships had delivered more than 80 tons of cargo as opposed to the 13 tons that had supplied Salyut 6 and the 15 tons sent to Salyut 7.

Cosmonaut Anatoli Solovyov completed 16 space walks, spending 78 hours outside during his five expeditions to Mir. More than 780 hours of EVA were accomplished. Mir had proved to be a significant step in the conquest of space.

International Space Station: Alpha

The oil embargo and resulting inflation of the late 1970s caused a recession that plagued the Carter Presidency. President Reagan faced significant budget deficits during his first administration. With the political reality of tight funding and lukewarm public support in the early 1980's, NASA had accepted that the Shuttle had to precede the space station. For America, whose Skylab effort in 1975 had been cut short by a lack of enthusiasm and dollars, a large continuously manned space station as envisioned by Tsiolkovsky, Oberth, Goddard, and von Braun could finally proceed when the Shuttle was declared "operational" in 1982.

In an effort to galvanize the American people behind such a project, President Ronald Reagan announced his vision during the annual State of the Union message to Congress in January 1984. Leaning on President Kennedy's *"before this decade is out"* theme, Reagan proposed that NASA *"develop a permanently manned space station and do it within a decade."* Unlike Kennedy's desire to compete with the Russians during the Cold War, Reagan saw this as a step in the direction of stimulating commerce in space and allocated $8 billion dollars to accomplish the task. However, some in Reagan's own administration were opposed to the project—there was yet no proven need for man's "permanent presence" in space.

The President was not satisfied with how America was proceeding into space as the bloated and bureaucratic NASA was the antitheses of the private enterprises that he would prefer develop space. Nevertheless, in his view, now was time for the space station.

Reagan's announcement was actually the culmination of many years of planning by NASA. However, it did not provide a specific goal or functionality for such a space station—which was not an end unto itself. Thus began more years of design and re-design of the station. A bureaucratic power struggle occurred between the various NASA centers that fought each other with the vigor that they had previously applied to the technical challenges of the Apollo program 20 years earlier. Each had specific design goals, and, coupled with Congressional oversight and micromanagement, progress was slow. No less than three distinct configurations were formulated over the next six years as the country attempted to define just what functions the space station should have and how it should be constructed and manned.

The first design that emerged in mid-1984 was a 450-foot-long truss to which five habitable modules were attached. Large solar panels would supply 75 KW of electrical power and a crew of six would operate the gravity-gradient oriented station. However, this layout was criticized by scientists as not allowing for effective micro-gravity experiments and as having too little electrical power.

A second arrangement situated the modules in the center of a 508-foot truss and used mirrors to provide a steam powered electrical generator—recalling the concepts of the visionaries of 50 years earlier. By this time the Japanese and European Space Agencies were included as partners, and two of the five modules would be built by them—reducing the cost to the American taxpayer. It was at this point also that the term "International Space Station" was applied—the ISS. However, this design almost doubled the initial estimates, and another redesign occurred. A reshuffling of the structure and the deletion of the expensive steam turbines left the station with only 45 KW of power and an estimated $12 billion dollar price tag.

The name "Freedom" was applied to the space station, and the suggestion was made that, to save money, it need not be continuously manned. Many of the experiments could be conducted without constant tending by a crew. The main truss was shortened to 353 feet, and some of the solar-power panels were eliminated which reduced the power capacity to only 30 KW. The attitude control fuel was changed from hydrogen-peroxide to hydrazine, and the closed-loop environmental system was eliminated—requiring the Shuttle to supply water periodically. The design effort alone had cost the nation almost $4 billion by 1991, and yet nothing had actually been fabricated, let alone launched into space.

Although a spectacular machine, the Shuttle could carry only about 50,000 pounds of payload in its cargo bay—about one-fourth that of the Saturn V which had been abandoned a decade earlier. It would take four or five Shuttle flights just to put a space station the size of Skylab into orbit.

With little progress made by the time Reagan's successor, President George H.W. Bush, left office in January of 1993, the Clinton administration had to pick-up the ISS political hot potato. The Soviet Union was in disarray following its economic collapse, and it was obvious that, despite their accomplishments in long-duration human space flight, they were no longer a serious military threat in space. Their Mir space station, although relatively successful, would not be followed by Mir-2 for lack of funding unless the Russians became a partner in the ISS.

NASA was on its seventh design iteration of the ISS, which continued to grow smaller as the US budget deficit grew larger. Although President Clinton endorsed the ISS, it was nearly canceled by Congress on a vote of 216 to 215 in June of 1993. With its capabilities being severely reduced over the years, it was under fire from the scientific community as being inadequate, and its political meaning had been virtually eliminated by the end of the Cold War. Clinton urged NASA to finalize its negotiations with the Russians to become a full partner—the ISS at that point provided for a crew of only four while the Russian Mir-2 design provided for a crew of six. Clinton agreed to go with the larger basic design of the Mir-2 in September, and by October 1993 the Russians (there was no longer a Soviet Union) agreed to an amalgamation of their Mir-2 with the ISS. The name "Freedom" was dropped as Clinton did not want the ISS to be associated with the Reagan administration, and the somewhat less inspiring name "Alpha" was substituted. On December 16, 1993, NASA and Roskosmos formally announced the Russian inclusion into the joint manned space station project.

However, with the inclusion of the Russians, the orbital inclination of the ISS had to be increased to 51 degrees to be easily accessible by rockets launched from Russia's Baykonur spaceport. This higher orbital inclination effectively reduced the Shuttle's net payload. To improve the Shuttle's performance, NASA developed a new super-lightweight aluminum-lithium propellant tank (ET).

As finally defined, the ISS consisted of 31 modules built by five countries. Each of the modules would be constructed on Earth and launched into orbit and assembled together by the astronauts with EVA activities. A 354-foot long truss provided stability to the structure and acted as a wiring conduit for communications and power.

As originally designed, the ISS would weigh almost one million pounds and have 110 KW of electrical power furnished by an extensive array of solar cell "wings." To accomplish the launch task, more than 50 flights were originally envisioned with the Space Shuttle performing 39 and the Russian UR-500 Proton and ESA's Ariane providing the remainder. As with Mir, the Russians would provide regular unmanned Progress supply vessels. The interior would provide almost 13,000 cubic feet of pressurized space for equipment and living area for six crew members— only a marginal improvement over the Skylab's 12,700.

The first section of the ISS placed in space was Russia's 42,500-pound Zarya (sunrise) Control Module, also known as the Functional Cargo Block (referred to with the Russian acronym FGB). It was launched into a 250-mile high orbit atop a UR-500 Proton rocket on November 20, 1998. The $200 million module was funded by NASA and built by Khrunichev State Research and Production Space Center in Moscow. A competitive bid by America's Lockheed Corporation came in at over $400 million and was rejected.

The Zarya design was adapted from the Mir core module (14-foot diameter and 41 foot length). It included a multiple docking adapter, a pressurized cabin section, and a propulsion/instrument section adjacent to the aft docking port. Although completed on schedule and within budget, its launch was delayed 17 months because the Russians were behind on their Zvezda service module development.

The second segment of the ISS launched was America's 25,500-pound Unity Node, 12 feet in diameter and 34 feet long. Delivered by space shuttle *Endeavor* (STS-88) in December of 1998, Unity had two pressurized mating adapters to provide a six sided docking hub and passageway where major sections of the space station join. Astronauts performed an EVA to lock Zarya and Unity together.

Discovery (STS-96) carried internal logistics and re-supply cargo for the ISS and a 12,000-pound external cargo crane that was mounted on the outside of the Russian station segment for use in spacewalking maintenance activities.

Atlantis (STS-101) launched May 19, 2000, prepared the station for the arrival of the Russian Zvezda (Star) Service Module, the third major component, launched from Baykonur on July 12, 2000. It is the primary Russian contribution to the ISS and provides the life support systems and living quarters (sleeping facilities, galley and toilet). It also has the primary docking port for Progress supply ships and provides attitude control and orbital maneuvering capability for the early station elements.

The first Progress M-1 supply ship (ISS-1P) was launched on August 6, 2000. It remained docked to the Service Module for 73 days. During this time, *Atlantis* (STS-106) was launched on September 8, 2000, and ferried supplies to the station in preparation for the first resident crew. The mission included two spacewalks to connect power and communications cables between the Zvezda and Zarya modules.

Discovery (STS-92) launched on Oct. 11, 2000, to deliver the Integrated Truss Structure ITS-Z1 and Pressurized Mating Adapter-3 (PMA-3). ITS-Z1, an exterior frame on which the first solar arrays were installed, holds the Ku-band and S-band communications systems for transmission of scientific data, audio and video, and station telemetry. It also holds four of the Control Moment Gyros that aid in attitude control.

Expedition Crew 1 was launched to Earth orbit aboard a *Soyuz* 1 on Oct. 31, 2000, and docked to the space station two days later. The three-man crew consisted of Expedition commander, American astronaut William M. "Bill" Shepherd (Capt., U.S. Navy), *Soyuz* commander Russian cosmonaut Yuri Gidzenko (Col., Russian Air Force), and Flight Engineer Sergei Krikalev.

During Expedition One's tenure, Shuttle STS-97 delivered the first of eight solar arrays, while STS-98 provided and installed (using the Shuttle's robot arm) the 28-foot-long Destiny laboratory which expanded the space station's power, life support and attitude control capabilities.

Discovery (STS-102) in March 2001 delivered the 9,000-pound Leonardo Multipurpose Logistics Module (MPLM) built by the Italian Space Agency (ASI). The MPLM is an un-piloted, pressurized module that allows equipment, experiments and supplies to move between the ISS and the Shuttle. *Discovery* also delivered Expedition 2 and returned Expedition 1 to Earth on March 21, 2001. With the second Expedition, the command switched to Russian cosmonaut Yury V. Usachev with American's James S. Voss and Susan J. Helms rounding out the crew.

Endeavour (STS-100), launched on April 19, 2001, ferried the Canadian Space Station Remote Manipulator System (SSRMS) and a second MPLM and the Ultra High Frequency (UHF) antenna. *Soyuz* 2 flight (April 28, 2001) carried Dennis Tito, an American businessman. Tito paid $20 million for the flight as the first ISS space tourist and created a political flap between the American and Russian. NASA was determined that Tito not be allowed to fly, but the Russians boycotted their American training until Tito was allowed to participate. The flight delivered a new *Soyuz* spacecraft to the station for use as an emergency crew-return vehicle. The taxi crew returned to Earth in the old *Soyuz* vehicle. Fare paying visitors were finally given official status with the title of "Space Flight Participant" as construction moved into high gear.

Atlantis (STS-104), on July 12, 2001, carried the Joint Airlock and the High Pressure Gas Assembly for the ISS. The 13,000-pound airlock provides for EVAs from the ISS. *Discovery* (STS-105) delivered the third expedition to the space station (Aug. 10, 2001) and returned the second crew to Earth.

A Progress M (Sept. 14, 2001) transported the Russian Docking Compartment 1 (DC-1) for additional EVA capability and increased space for ships docking under the Zvezda module. *Soyuz* 3 (Oct. 21, 2001) delivered a new *Soyuz* spacecraft to the station for use as an emergency crew-return vehicle.

Endeavor (STS-108), launched Dec. 5, 2001, and delivered the Expedition 4 crew for a five-month mission. *Endeavor* returned the Expedition 3 crew to Earth after almost four months on station.

Atlantis (STS-110), the 13th shuttle mission to visit the ISS, was launched April 8, 2002, with astronaut Jerry L. Ross who made a world record seventh flight to orbit. The shuttle carried aloft the Integrated Truss Structure S0—the center segment of the 300-foot exterior framework. It is 44 feet long, 15 feet wide, and weighs 27,000 lb. STS-110 also delivered the 1,950-lb. Mobile Transporter (MT), a movable base that allows the station's Canadian mechanical arm to travel along the station truss.

Soyuz 4 (April 25, 2002) transported Space Flight Participant Mark Shuttleworth and a new *Soyuz* spacecraft to the station. Shuttleworth, of South Africa, paid $20 million for the flight as the second space tourist. After eight days, the taxi crew (with Shuttleworth) returned to Earth in the old *Soyuz* vehicle.

Endeavor (STS-111) delivered the Expedition 5 Crew on June 5, 2002, (who took three spacewalks), and returned Expedition 4 crewmembers. *Atlantis* (STS-112) followed on Oct. 7, 2002, and delivered the ITS S1 Truss and attached it to the S0 Truss' starboard side during three spacewalks. The S1 was the third component of the 11-piece ITS and allowed for the outward expansion of the station.

Endeavour (STS-113) launched November 23, 2002, to ferry Expedition 6 to the ISS along with the Port 1 (P1) Integrated Truss Segment (ITS) and attached it to the port side of the S0 Truss. The Expedition 5 crewmembers were returned to Earth.

The fiery plunge of Space Shuttle *Columbia* on February 1, 2003, ended the routine visits of the Shuttle to the ISS. NASA recognized that it would be some time—two and one-half years—before another Shuttle crew would visit the ISS. While the ISS remained well behind its initial schedule, significant progress had been made with its assembly. This all changed with the loss of *Columbia*.

Soyuz TMA-2 launched on a taxi flight on April 26, 2003. Crew Commander Yuri Malenchenko and Science Officer Ed Lu delivered themselves to the station in a new *Soyuz*, TMA-2, to become the

Expedition 7 crew. They were the first two-person ISS crew and the first primary crew to travel to the space station on a Russian *Soyuz* spacecraft. By reducing the crew from three to two, ISS supplies would last longer. The reduced crew maintained systems aboard the station, made repairs as needed, and carried on the science experiments. Further construction of the station was postponed. *Soyuz* TMA-2 remained docked with the station as their lifeboat during their six-month ISS tour that lasted until October 2003. The Expedition 6 crew retuned to Earth in *Soyuz* TMA-1, which had been docked with the ISS as their lifeboat.

Medical and industrial research is the central theme for ISS activities. The Microgravity Science Glovebox allows astronauts to perform a wide variety of materials, combustion, fluids and biotechnology experiments in weightlessness. The Pulmonary Function System, delivered in July 2005, provides information about the cardiovascular system in the weightless environment. The Kubik Incubator performs biological experiments, while human physiology is explored with the Cardiocog, Neurocog, and the Eye Tracking Device. The PK-3 device is used for complex plasma experimentation.

In its first five years of habitation, the ISS has received nearly 100 astronauts— seven of which were European. Twenty-nine were members of Expedition crews with long-term stays ranging from 129 and 193 days—totaling almost 16 years of crew-days aboard the ISS. There were 11 *Soyuz* spacecraft and 12 Shuttle flight visits to the Station since the Expedition 1 Crew arrived.

Repercussions of Shuttle Columbia

When *Columbia* was lost, NASA again grounded all Shuttle operations (as it had with *Challenger* 17 years earlier) while it investigated the problem. *Columbia*, being the first and heaviest Shuttle, could not achieve the high-orbital inclination required to reach the ISS with the same payload capability as the others and had been used for non-ISS work. However, the problem that resulted in the loss of *Columbia* was just the start of a new series of problems for NASA.

The ISS was far more expensive than originally projected by NASA, and the General Accounting Office (GAO) was highly critical of NASA's accounting practices that hid many costs. At the time of the *Columbia* accident, ISS was expected to be finished in 2010 - 16 years behind schedule, and to cost about $100 billion over its lifetime for construction and support. However, the scientific community, for whom the station was built, continued to express its discontent with the ISS, estimating that, when completed, it would have considerably less capability than first planned. That NASA expects to abandon the ISS after only seven years of full operation (by 2017) has caused many to question again its viability.

While the assembly has been without any major incidents, and the station completed its first five years of manned operation in 2005, there is still some concern about the safety of the ISS itself—particularly the prospects of either a meteor or space debris puncturing one of the modules. Russian engineers estimated that there was a 23 percent chance over its lifetime that the latter might occur. If it should, there is the possibility that the fracture of the pressure hull could cause the module to explode like a punctured balloon, resulting in the break-up of the station.

With completion of the ISS closely tied to the availability of the Shuttle, there is a growing sentiment in Congress and the space industry that the ISS may never be completed—at least by the United States. The declaration by President Bush that the Shuttle program will end by 2010 and the admission by the current NASA administrator, Michael Griffin, that the decision to build the Shuttle and the ISS, in their present configurations, was fundamentally flawed has led to speculation that the United States may simply back out of the ISS program. Under the current conditions, it is highly unlikely that the Shuttle will fly any more than fifteen flights before it is retired. Current estimates indicate that 28 Shuttle flights are needed to complete the ISS. Griffin has said it will be impossible for the remaining three Shuttles to complete the assembly and has called for a major review of the ISS. Other partners in the ISS have become very concerned that their investment is in jeopardy. Griffin said that the US will keep its commitment to launch all of the non-US modules.

The Russians have indicated that, if America abandons the ISS, they would continue to operate their part, which has a certain amount of autonomy. The ESA commitment to the ISS is represented by only five of its 16 member states because of concerns about the expense or simply lack of interest. ESA contributions include the 40,000-pound Columbus science laboratory module that will be orbited by

2007 after NASA's Space Shuttle returned into service. ESA has committed to pay 8 billion euros in support of the ISS, coming primarily from Germany (41 per cent), France (28 per cent) and Italy (20 per cent).

The ESA has started construction of a space freighter for the ISS, the Automated Transfer vehicle (ATV), with a cargo capacity of 8 metric tons. It is scheduled to begin flights in 2007. After the Shuttle's retirement in 2010 and until its replacement is available (the CEV), the ATV along with Progress and *Soyuz* will be the only link between Earth and the ISS.

When the Shuttle was grounded in 2003, crew rotation and supply of the ISS was carried out by the Russians using the venerable derivatives of Sergei Korolev's R-7 and the *Soyuz* and Progress space-craft. But the science that is being conducted has been severely curtailed as the reduced supply flights allow only three persons to man the station instead of the planned six. When Shuttle flights resumed, the final ISS configuration was scaled back to accommodate the reduced capability, and many of the planned added modules might never leave their storage warehouses on Earth. Cosmonaut Pavel Vinogradov, the Soviet commander of Expedition 13 to the ISS lamented, "I always thought we have to fly in the interests of science, to produce results needed by many people, and all we're doing is keep-ing the station in working order. 62% of our time goes to servicing on-board systems, 15% to person-al needs and only 23% to science."

However, there are other options. The growing disenchantment with the NASA bureaucracy has left the door ajar for private corporations to step in, not only to provide supply flights but also to cre-ate a new manned space vehicle that would be more efficient than the Shuttle.

Another GAO report stated that NASA's logistical support of the ISS with the Shuttle was not prac-tical and that it should consider making use of alternative commercial services. An Alternate Access to Station (AAS) study, a part of NASA's Space Launch Initiative program, was begun in 2000 to deter-mine if other methods of re-supply to the ISS could be more efficient than either the Shuttle or Progress. One consideration was that no new launcher capability would be developed—it would have to use an existing expendable launcher, only the spacecraft itself would be reusable. The European Space Agency (ESA) and the Japanese space agency (NASDA) have been moving forward to develop such a craft.

NASA had reported that contractors being considered for the AAS indicated that it would take 3 to 5 years to develop an alternate launch capability. But the AAS program was canceled before any of the initiative was approved. Now Administrator Michael Griffin has testified to Congress that the AAS was important and should be saved. Constellation Services International, Inc. (CSI) indicated that it could provide supply services in two years or less using existing launch vehicles (such as the *Atlas* or *Delta*) and existing technology to rendezvous and dock with the ISS.

There are few in Congress or industry who doubt that the future in space is important and that the ISS can and should play a role in expanding mankind's horizons. Nevertheless, for the past 30 years, NASA, like virtually all large bureaucratic institutions, has fallen far short of providing the services for which the American taxpayer is being billed. There is a strong call for removing NASA from the pas-senger and payload business and returning its focus to exploration and science.

Because NASA continued to falter in its efforts to resume manned space flight, it was forced to pur-chase services from Roskosmos for $44 million dollars for several *Soyuz* flights to the ISS that may well stretch into 2012—when the Shuttle replacement is, hopefully, available. One of the problems in the arrangement was that Congress had restricted commercial trade with the Russians because of the Iran Nonproliferation Act which it passed when it was determined that the Russians were assisting Iran in developing a nuclear program.

It is humiliating for America, who prides itself on its technological superiority, to admit that it no longer has a viable manned space flight capability.

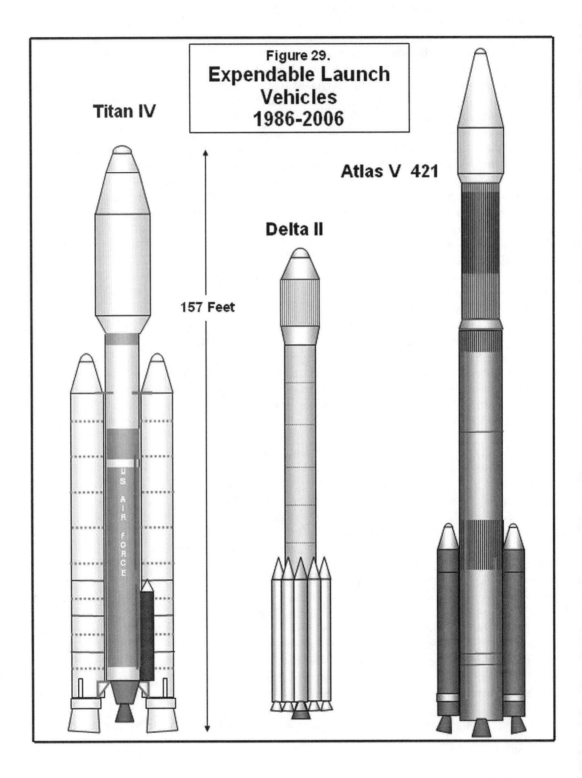

Figure 29.
Expendable Launch Vehicles 1986-2006

Titan IV

Atlas V 421

Delta II

157 Feet

US AIR FORCE

Chapter 33 — Return of the Expendables

As the space age moved into the early 1980s, America's expendable boosters that had been based on the weapons of the 1950s were nearing the end of their life. NASA had decreed that all future commercial satellites and space probes would be launched on the Shuttle (and the military was just completing its own Shuttle launch facility at Vandenberg AFB on the coast of California). The existing stockpile of mothballed military hardware that had been carefully stored away as it was retired from active service in the 1960s (*Atlas*, *Titan*, and *Thor*) had just about been depleted. Had the Shuttle operated with its predicted economies, it would have claimed the market for launching most foreign and domestic satellites and doomed most, if not all, existing and proposed foreign and domestic launch vehicles.

Then came the *Challenger* disaster and the realization of its true costs and the new edict that the Shuttle is to be used only for purposes that *"require the presence of man... or requires the unique capabilities of the Shuttle... or compelling circumstances."* Thus, virtually all commercial and military satellites and space probes represented opportunities for a revived aerospace industry. As a result, the "Evolved Expendable Launch Vehicle" program (EELV) was born—a rocket whose core unit could be augmented to match larger payload requirements.

This dramatic change in strategy would have a profound affect on the survivors of the cold war aerospace industry in America. The complexion of that industry had changed considerably from the hectic years of the dawn of the space age in the late 1950s. Many of the giant corporations had merged or simply closed their doors. Douglas Aircraft (*Thor* and *Delta*) had merged with McDonnell Aircraft (*Mercury* and *Gemini*), and that entity was swallowed up by Boeing (*Satur*n V).

North American Aviation (*Apollo*) merged with Rockwell Standard Corporation in 1968 to become North American Rockwell (*Space Shuttle and SSME*). A reorganization in 1984 merged the Rocketdyne division (SSME) into North American Space Division and the Satellite Division of Rockwell. The space systems, aircraft division, Rocketdyne, Autonetics, missile systems, aircraft modification and other units joined The Boeing Company in 1996. In August 2005, United Technologies Corporation (UTC) acquired Rocketdyne Propulsion & Power from Boeing. They combined Rocketdyne with Pratt & Whitney's Space Propulsion division which they had previously purchased.

In 1962 American-Marietta Company merged with Martin Aircraft (*Titan*) to become Martin Marietta. Convair (*Atlas* and *Centaur*) became General Dynamics and eventually sold their interest to Martin Marietta. Lockheed (*Agena*) combined their operations in a "merger of equals" to become one of the largest aerospace, defense, and technology companies in the world—Lockheed Martin.

A new company, United Space Alliance, initially formed and equally owned by Lockheed Martin and Rockwell, was selected by NASA as the single contractor for launch operations for the Space Shuttle fleet. It is now a Boeing/Lockheed Martin Limited Liability Company (LLC) which performs Space Shuttle launch and landing activities. The name was carefully chosen so that the designation "USA" on the side of various facilities and vehicles now has a dual meaning.

Grumman (*Lunar Module*) and Northrop had likewise merged. By the 1990s there were only three big aerospace corporations left in the United States—Boeing, Lockheed Martin, and Grumman Northrop.

The Thiokol Corporation, known more perhaps for the infamous "O-ring incident" on the Space Shuttle Solid-fuel Rocket Boosters (SRB) of the *Challenger*, was purchased by Alliant Techsystems (ATK) Inc. in 2001. ATK had earlier bought the Hercules Aerospace Co., also a major provider of solid-fuel rockets, and, as a result, ATK now controls the largest share of the U.S. solid-fuel rocket market.

A new consortium in the launch-vehicle market is International Launch Services (ILS). Established in 1995 by Lockheed Martin and the Russian Khrunichev State Research and Production Space Center, these two corporations of the Cold War superpowers are now working together. ILS has become an important player in providing commercial access to space with two vehicles —*Atlas* and *Proton*. It has contracted for more than 100 commercial and U.S. government launches, worth a total of more than $8 billion.

Three major mission profiles provide for the delivery of payloads to low earth orbit (LEO), geosynchronous transfer orbit (GTO), and to escape velocities for lunar and planetary exploration. Within each of these, the weight lifting requirements range from several hundred pounds to tens of thousands of pounds. The latter are referred to as "heavy lift" boosters. The weight capability of the term "heavy lift" has been increased over the years as new and larger boosters became available. The sophistication of the emerging new line of EELVs allows a single launch to accommodate one or more payloads.

Intense competition now exists between the two primary American companies—Boeing (*Delta*) and Lockheed Martin (*Atlas*)—Russia's Energiya, the ESA community with their *Ariane*, and emerging Chinese and Japanese launch vehicles. While applications and commercial opportunities remain strong, the actual market for launch vehicles is still small. With the specter of failure still plaguing the industry, rising insurance rates make the profitability marginal.

Titan

The *Titan* was developed by Martin Aircraft as a backup ICBM to the Convair *Atlas*, with the first launch occurring in February 1959. Like the *Atlas*, it used liquid oxygen as its oxidizer. The *Titan* was quickly replaced in 1962 by a larger, more powerful version designated the *Titan* II, which used storable propellants to allow more flexible launch options. *Titan* II ICBMs were deployed in 54 silos across the U.S. from 1963 to 1987 and were also used to launch ten two-man Gemini flights.

The *Titan* 23B, 24B, 33B, and 34B were *Titan* IIs with the Agena D upper stage for launching the 7,500 pound KH-8 spy satellites into polar orbits. The 10-foot-diameter first stage was powered by a two-chambered Aerojet General LR-87-AJ-11 using storable hypergolic propellants—Aerozine 50 (hydrazine and UDMH) and nitrogen tetroxide.

The *Titan* III and *Titan* 34D were "heavy lift" versions with various upper stages and two 10.5-foot-diameter solid-fuel rocket motor units (SRMUs) attached to each side of the first stage. These vehicles were assembled in a vertical position in a building remote from the launch pad similar in concept to the Saturn Vehicle Assembly Building. They were then transported to the firing site by a team of diesel locomotives pulling the assembly on a dual set of standard railway track.

The *Titan* IV was a stretched *Titan* III also with SRMUs which could be configured with either the Centaur upper stage, the Inertial Upper Stage (IUS), or without any upper stage. The *Titan* IV was capable of delivering 10,000 pounds to geosynchronous Earth orbit (GEO) or up to 39,000 pounds into a low Earth orbit. It was used for U.S. Military payloads and for commercial and scientific launches such as NASA's Cassini probe to Saturn.

The last *Titan* IV launched was a Defense Meteorological Satellite from Vandenberg AFB on October 18, 2003. It was the largest launch vehicle flying at the time of its retirement in 2005. Its was discontinued because it was extremely expensive to operate (almost $200 million per launch) with preparations requiring up to six months. In all, 27 *Titan* IVs were launched from Cape Canaveral and 12 from Vandenberg.

Atlas

The *Atlas* missile, originally built by Convair Aircraft in the mid-1950s as America's first Intercontinental Ballistic Missile (ICBM), became the backbone of America's space program in the 1960s serving as the launch vehicle for its first manned spacecraft—Mercury. Teamed with the Agena upper stage, it provided for the Ranger lunar program and Mariner Mars/Venus/Mercury and Pioneer planetary missions. Using the high-energy liquid-hydrogen upper stage Centaur, it propelled the Surveyor to the Moon and launched many commercial satellites

First tested in 1957, the liquid-fueled *Atlas* was considered as a "one and one-half-stage" missile as all three of its main engines were ignited on the launch pad. Developing a total of 360,000 pounds of thrust, the two booster engines were jettisoned after 150 seconds of burning, while the single 60,000-pound sustainer continued to propel the payload for another six minutes.

Atlas was phased out as an ICBM beginning in 1962 in favor of the larger and more flexible *Titan* and the quick response solid-fuel Minuteman. Direct descendants of the *Atlas* continued to be used as launch vehicles with the last refurbished *Atlas* vehicle lifting off from Vandenberg AFB in 1995 carrying a Defense Meteorological Satellite.

By the mid-1980s, virtually all of the ex-military *Atlas* E and F boosters that had been placed in storage following their deactivation had been expended, and the production line had to be re-opened. In May 1988, in light of the *Challenger* disaster, the Air Force contracted with Lockheed-Martin (who had absorbed the assets of the General Dynamics *Atlas* program) to develop the *Atlas* II. It was used to launch Defense Satellite Communications System payloads and some commercial satellites.

Atlas II provided higher performance than the earlier *Atlas* I by using engines with greater thrust (490,000 pounds) and longer propellant tanks. The *Atlas* II could lift payloads of 14,500 pounds to LEO and 6,100 pounds into geosynchronous orbit (22,000 miles). This series also used an improved Centaur upper stage to increase its payload capability.

The *Atlas* II had a length of 156 feet, a core Diameter of 10 feet, and a liftoff weight of 414,000 pounds. The first launch occurred in February 1992 and the last on August 31, 2004, with the orbiting of the U.S. National Reconnaissance Office satellite NROL-1. This marked the 73rd consecutive successful flight for the *Atlas* family in any launch configuration, and it was considered the most reliable launch vehicle during its time. It was also the last flight (after 47 years of use) of the Rocketdyne "MA" engine cluster.

The *Atlas* IV was essentially an *Atlas* II with four strap-on solid-fuel rocket boosters that were 37 feet long and 40 inches in diameter. Two of the motors ignited at launch to provide initial thrust. The other two boosters ignited 59 seconds into the flight to sustain the thrust level as the first two solids are jettisoned.

The newest version of the *Atlas* is an *Atlas* in name only as it has a completely new structure and first stage propulsion unit. It no longer uses the pressurized "balloon tanks" nor the 1.5 staging configuration. It has a rigid monocoque framework for its first stage.

Most remarkably, both the *Atlas* III and *Atlas* V use the Russian RD-180 engine. The RD-180 is marketed by RD AMROSS in a partnership between the Pratt & Whitney Company of the United States and NPO Energomash of Russia. It is a two-chamber, gimbaled derivative of the four-chamber RD-170 of the improved R-7. The RD-180 produces 860,200 pounds of thrust at sea level and 933,400 pounds in a vacuum. It has a Specific Impulse 311 seconds to 337 seconds and can be throttled between 47 percent and 100 percent.

The new *Atlas* series can use SRBs built by Aerojet that are 67-feet long and provide 250,000 pounds of thrust each. Various vehicle configurations employ one, two, three, or five SRB boosters. The boosters ignite at liftoff and burn for 85 to 90 seconds. Total liftoff weight varies with the configuration to a maximum of 735,000 pounds with a total length of 195.9 ft depending on the payload fairing. The *Atlas* first stage, called the Common Core Booster, is 106 feet long and 12.5 feet in diameter.

The second stage Centaur is 41.5 feet in length and 10 feet in diameter. It can be configured with the standard two RL-10 LH2 engines, each producing 15,000 pounds of thrust, or a single RL-10 depending on the mission.

Because the *Atlas* V can be configured with a different number of solid-fuel boosters and Centaur LH2 engines, a three-digit vehicle naming designator distinguishes the various versions. The first digit defines the payload fairing diameter in meters (4 or 5) while the second digit specifies the number of solid-fuel boosters. The third digit defines the number of Centaur engines. Thus a 401 has a four-meter payload fairing, no solid-fuel boosters and a single RL-10 engine. It can lift 27,500 pounds into LEO. The 552 is the most capable configuration with the five-meter payload fairing, five solid-fuel boosters and two RL-10 LH2 engines. It can lift 45,200 pounds to LEO—almost equal to the payload of the Space Shuttle.

In a typical *Atlas* V launch, the RD-180 thrust chambers ignite first, and a rapid diagnosis assures the proper functioning of the propulsion systems. The RD-180 engine is then throttled up to 100 percent of rated thrust, and the solid rocket boosters (if attached) are ignited to commit for liftoff.

The flight control system manages the RD-180 throttle settings and throttles down for the transition through Max Q, and then returns to full thrust. At about 100 seconds into the flight, the engine is again throttled down to limit acceleration to 5.5 g's. This is a similar flight profile to that used with the Space Shuttle. The payload fairing is jettisoned during booster operation, approximately 210 seconds into the flight.

The first stage engine cutoff occurs at about T+240 seconds, and the first Centaur burn period lasts about 4.5 or 9.0 minutes (depending on engine configuration) after which the Centaur and its payload coast in a parking orbit. The second Centaur ignition is mission dependant but typically occurs 22 minutes into the flight and burns for about six minutes, at which time spacecraft separation occurs. The 500 series Centaur provides the ability for three burn periods, allowing direct insertion into geosynchronous or geostationary orbits.

Delta

The origin of the *Delta* family of expendable launch vehicles extends back to the *Thor* Intermediate-Range Ballistic Missile (IRBM) developed in the mid-1950s for the U.S. Air Force. The *Thor*, a single-stage liquid-fueled rocket, was modified in 1958 to add the second and third stages of the *Vanguard* and was known as the *Thor Able*. It continued to be upgraded and improved, and the designation progressed to "*Thor Delta*." When solid-fuel boosters were attached to its sides, it was referred to as the "*Thrust Augmented Thor*" (TAT). When the *Thor* first-stage tankage was elongated and its eight-foot diameter continued into the second stage, the reference to its *Thor* origins were dropped, and it was simply referred to as the "*Delta*" rocket.

The *Delta* II can be configured as a two- or three-stage launch vehicle with a

varying number of strap-on solid rocket boosters and two sizes of payload fairings depending on mission requirements. The *Delta* II main engine is the Rocketdyne RS-27A, while the second stage consists of an Aerojet AJ10-118K using storable propellant with restart capability. A third stage is usually required for geosynchronous transfer orbit and planetary missions, and typically uses the Thiokol Star-48B solid-fuel rocket motor.

Delta II rockets have launched all replacement satellites that make up the current constellation for the U.S. Air Force Global Positioning System. Between 1996 and 2003, *Delta* II launched seven missions to Mars, including the Mars Pathfinder that landed on Mars in July 1997, and the Mars Exploration Rovers, "Spirit" and "Opportunity," in January 2004.

The *Delta* III was developed to provide a greater capability launcher and can deliver 13,000 pounds to low-Earth orbit (LEO) and 8,400 pounds into a geosynchronous transfer orbit (effectively doubling the performance of the *Delta* II), with payload fairing sizes up to 13 feet in diameter.

The core first-stage uses the Rocketdyne RS-27A main engine and can be configured with three, four, or nine 46-inch-diameter strap-on graphite epoxy solid-fuel rocket motors (GEM), depending upon mission. When three or four GEMs are used, all are ignited at liftoff. When nine GEMs are used, six ignite at liftoff and three during flight when the first six have been jettisoned. Three of the GEMs are equipped with thrust-vector capability to improve vehicle control. The second-stage uses the Pratt & Whitney LH2 RL10B-2 and carries more propellant than *Delta* II.

The *Delta* IV was developed by The Boeing Company as a part of the Evolved Expendable Launch Vehicle program. It is assembled horizontally and then erected vertically on the launch pad, where the payload is installed. This process reduces on-pad time to less than 10 days and the total time the vehicle is at the launch site to less than 30 days, reducing costs of launch site operations.

All configurations use the Boeing Common Booster Core (CBC) first stage with the Rocketdyne RS-68 LH2 main engine. The *Delta* IV second stage uses the Pratt & Whitney RL10B-2 with two propellant capacities and a carbon-carbon extendible nozzle that enables the RL10B-2 to achieve 465.5 seconds of specific impulse with a thrust of 24,750 pounds.

Delta IV vehicles can launch payloads weighing from 9,300 lb. to 28,100 lb. to geosynchronous transfer orbit and can lift 50,000 lb. to low-earth orbit (LEO), depending on the configuration. The five versions are: medium, medium+(4,2), medium+(5,2), medium+(5,4), and heavy. The first digit in parenthesis relates to the payload shroud diameter (in meters), while the second digit indicates the number of 60-inch strap-on GEM solid-fuel booster units. In the medium configuration, either the 13.1-ft.- or 16.6-ft. diameter composite fairing may be used. The heavy vehicle uses a 16.6-ft.-diameter composite fairing or a 65-ft.-long, 16.6-ft.-diameter aluminum fairing.

All versions use the common booster core powered by a Rocketdyne RS-68 LH2 engine that develops 650,000 lb. of thrust. The RS-68 is the first large, liquid-fuel booster engine developed in the U.S. since the Space Shuttle main engine (SSME) over 30 years ago, with a significantly reduced parts count. The upper stage of the *Delta* IV is powered by the Pratt & Whitney RL-10B2 LH2 engine. The

Delta IV medium+(5,2) has increased upper-stage fuel capacity.

The *Delta* IV heavy uses two additional CBCs, instead of using GEMs, to provide a total of 1,950,000 lb. of thrust. It is capable of launching 50,000 lb. to LEO and 28,950 lb. to geosynchronous transfer orbit (GTO), more than any other currently available launch vehicle.

Russian Angara and Baikal

Following the collapse of the Soviet Union in the early 1990s, the Russians planned a new generation of space boosters that would use launch sites within its borders. The Khrunichev enterprise and RKK Energiya competed for the government subsidies to develop the new booster designated Angara.

Khrunichev proposed a vehicle similar in configuration to the Proton rocket using the four chambered RD-170—the first stage of the Zenit. The second stage was a hydrogen-oxygen engine, from the Energiya heavy-lift rocket. The alternative, RKK Energiya's Angara-2, featured modular architecture by "packaging" several identical rocket boosters (each powered by a two chambered version of the RD-170) to provide a range of payload capabilities.

Khrunichev was awarded the development of the Angara rocket in September 1994 because its design would use the complete engine from the Zenit rocket, while RKK Energiya's proposal would have to split the engine assembly in half to provide the two combustion chambers.

However, the vehicle developed by Khrunichev turned out to be similar to the modular configuration RKK Energiya had originally proposed. This revised version used a two chamber RD-170 engine. It is interesting to note that the engine developer, NPO Energomash, had also built a two-chamber version of the Zenit engine for the US *Atlas V* rocket development which was paid for by the United States

As now planned, Angara's payload can vary from 2,400 pounds to 50,000 pounds, depending on the number of strap-on booster modules mated to a common core stage. The most capable Angara will replace the Proton rocket. The first launch of the core stage is scheduled for December 2006.

Khrunichev is also developing a reusable fly-back booster, designated Baikal, as an adjunct project with the Angara. Initial conceptual designs show the Baikal strap-ons employing a wing that "rotates" from a stowed position for fly-back and air-breathing engines. After depletion of the rocket propellant in the Baikal, the recoverable boosters separate from the Angara core stage at an altitude of about 50 miles. The wing then rotates 90 degrees into the deployed position, and the booster is flown remotely to a powered horizontal landing on a runway, using residual kerosene fuel from the vehicle's rocket engine tank.

The first flight of the Baikal booster was scheduled for 2006, but because its development is being pursued without financial support from the Russian government, this schedule has been delayed and Khrunichev is looking for investors.

Russian Zenit

The development of the Zenit series of launch vehicles started at the beginning of the 1970s in an attempt to create a standardized family of three launchers which would share propulsion and control systems, booster and upper stages, as well as processing and launch facilities. As the military was primarily interested in the medium version as a satellite launcher, it was given priority, and a preliminary design of the vehicle was completed in April 1974. However, this design was abandoned in favor of one that became the strap-on unit for the Energiya super booster that propelled the Buran.

The original plans called for the first test launch of the Zenit in 1982. However, development of the Glushko-designed 1.6-million-pound-thrust RD-170/171 engine of the first stage of Zenit/Energiya was plagued with problems between 1981 and 1983. The four chambered engine uses a single turbopump. The main difference in the versions is the one-plane gimbaling in the RD-170 for the Energiya versus two-plane gimbaling for the Zenit. The high Specific Impulse of 337 seconds (vacuum) was achieved only through exceptionally high chamber pressures and temperatures which caused many of the development problems. The engine was designed for 10 reuses when used as a part of the Energiya vehicle where the strap-on would be recovered by parachute. The LOX/Kerosene engine can be throttled down to 56 percent of full thrust.

In addition to the difficulties with the engine development, the financial problems in the former Soviet Union in the mid-1980s further slowed the program. The first Zenit-2 rocket lifted off from

Baykonur in April 1985, but a problem with propellant consumption on the second stage caused it to shut down prematurely at T+400 seconds. The payload did not reach orbit, and, because of its military nature, the launch was not announced at the time. The second Zenit-2 was launched in June of 1985, but its second stage shut down prematurely. The October 1985 launch of a Zenit-2 was finally successful with Cosmos-1697 being placed in orbit. The initial testing of the Zenit was completed in December 1987, and by 1990 the Zenit-2 was considered operational by the military.

In July of 1998, a single Zenit-2 successfully launched five foreign commercial payloads that included Thailand's TMSAT, Israel's TechSAT-2, the Chilean FaSAT-Bravo, and Germany's SAFIR-2.

Russian R-7: Soyuz ST

The R-7 that was designed by Sergei Korolev in the mid-1950s continues to be Russia's only man-rated booster after almost 50 years of service. It has been upgraded over the years to include a revised propulsion system (the RD-117/118) with new injectors and mixture ratio for improved performance (264/311 seconds of Specific Impulse). Modern guidance and control system electronics allows for in-flight orbital plane (azimuth) changes. Vehicles previously flew a fixed trajectory with the launch pad rotated to the appropriate azimuth (as derived from the German V-2). A new digital telemetry system provides for the launch vehicle monitoring, and a larger 12-foot-diameter payload fairing is available. Finally, a change in other components assured that all critical assemblies are built within Russia, and there is no dependence on any "foreign" suppliers.

The ST-2 second stage engine (RD-0124) produces 66,000 pounds of thrust (Isp 359 seconds) for 300 seconds. The Fregat third stage (used for geosynchronous and planetary probes) uses a single-chamber, 4,400-pound-thrust Lavochkin engine with four clusters of three ten-pound hydrazine thrusters for attitude control. Using UDMH fuel and N2O4 oxidizer, it can be restarted up to 20 times with a total burn time of up to 877 seconds.

Launch weight of the 150-foot-tall vehicle is typically 670,000 pounds with a total thrust of 930,000 pounds. The cost for each launch is about $30 million, exclusive of the payload.

In partnership with the ESA, the R-7, known commercially as "Soyuz ST," can be launched from the European Space Port in French Guiana (CSG) on the coast of South America. This capability virtually assures that this vehicle will be around for many more years to come as its first flight from the European Space Port is scheduled for 2008.

French Ariane

With the creation of the European Space Agency in 1974, one of its first objectives was to build the Spacelab module for the American Space Shuttle to carry into space, as Europe continued its path towards a more united effort in space exploration. But the European nations wanted more than just a ride into space aboard the Shuttle and began a project to compete with it for launching commercial payloads. With France providing the largest share of the funding, the project expanded the technology of the eleven-member organization. The three-stage rocket that resulted, *Ariane* (the mythical daughter of the Goddess Europa), would bring Europe, and France in particular, into a higher state of parity with the United States and the Soviet Union.

The first and second stages used the liquid propellants of nitrogen tetroxide and unsymmetrical dimethel hydrazine in a four-engine cluster. The third stage used the high energy liquid hydrogen (LH2) and liquid oxygen.

The primary launch site chosen for building a major launch complex for *Ariane* and its follow-on vehicles was in French Guiana. The first launch, in December 1979, successfully placed a test satellite in low earth orbit. *Ariane* was used over the next six years to orbit a variety of commercial and scientific payloads. The basic configuration was upgraded with improved engines and lengthened propellant tanks to become the *Ariane 2* and *Ariane 3*.

The *Ariane 4* is a family of six medium-to-heavy-lift vehicles that build on a three-stage liquid-propellant core vehicle designated the *Ariane 40*. This central core uses a single LH2 Vulcan 2 engine of 300,000 pounds of thrust (in the later versions). The *Ariane 5* uses two large Solid Rocket Boosters (SRBs) to augment the first stage core. The original second stage used hypergolic propellants, but the later versions have a re-startable LH2/LOX engine. The configurations employ a naming scheme such that *Ariane 42P* provides two solid strap-on motors while the *Ariane 42L* has two liquid strap-on boosters. The *Ariane 44P* and *Ariane 44L* follow that scheme while the *Ariane 44LP* uses two solid and two liquid strap-on boosters. *The Ariane 44L* is the most powerful with the ability to place 10,000 lb. into a geosynchronous transfer orbit.

Ariane 5, although designed primarily as a commercial satellite and manned spacecraft launch vehicle to compete with the American Space Shuttle, has been used to launch scientific payloads into deep space. The *Ariane's* 39,000 lb. LEO payload capability was to carry the manned HERMES Space Plane which was cancelled for financial reasons. The *Ariane 5* is a large vehicle with a height of 177 feet, which varies depending on the upper stage and payload configuration. The weight at lift-off is 1.7 million pounds lifted by 2,568,000 pounds of thrust.

The first successful test launch took place in October of 1997, while the first operational launch of a commercial satellite occurred in December 1999. With a per launch cost of under $200 million and a good track record of reliability, the *Ariane* has been a cost-effective alternative to both U.S. and Russian expendable launch vehicles.

Chinese Long March 2F: CZ-2E

China began development of the *Changzheng-1* (CZ-1) in 1965 with the objective of launching the first "Asian" satellite in competition with Japan. The CZ-l's first and second stages were derived from the DF-3 Intermediate Range Ballistic Missile (IRBM) with a solid-fuel third stage. On April 24, 1970, the CZ-1 placed the DFH-1 satellite into orbit on its second attempt, and China became the fifth country to launch a satellite—the Japanese had launched their Ohsumi satellite two months earlier on February 11, 1970.

The CZ-2 was the next in the series, begun in 1970, and was to launch the FSW-1 recoverable military reconnaissance satellite using the DF-5 ICBM as a basis. The first stage was composed of four liquid-rocket engines having a total thrust of 600,000 pounds. The second stage had a single engine of 160,000 pounds thrust. The CZ-2C has become the primary launch vehicle for the unmanned Chinese space program since the 1980s. It is from this family of rockets that the three stage, man-rated version of CZ-2E emerged for launch of the *Shenzhou* manned spacecraft. The 200-foot-tall rocket has a first stage thrust of 1.3 million pounds, and it has the ability to place 18,000 pounds into LEO.

With a configuration similar to the Russian R-7, the core first stage is surrounded by four liquid-fuel boosters that produce 183,000 pounds of thrust each (Isp of 291 seconds vacuum) and a burn time of 128 seconds. The core unit has a thrust of 732,000 pounds from four thrust chambers. The third stage is a single 186,000-pound-thrust motor. All the stages use Nitrous Oxide and UDMH as propellants.

Sea Launch: Zenit

The world's only ocean based launch services company, Sea Launch, provides commercial satellite customers the most direct and often the most cost-effective route to geosynchronous transfer orbit. By sailing to equatorial launch sites, Sea Launch rockets can lift a heavier spacecraft for the same energy that it would take from a launch site in more northern latitudes. It can also launch to any desired orbital inclination. Sea Launch is a cooperative venture involving Boeing Commercial Space Company (U.S.A.), Kvaerner ASA (Norway), SDO Yuzhnoye/PO (Ukraine), and RSC-Energiya (Russia), who together began the program in 1993 following the demise of the Soviet Union.

Two unique ships form the Sea Launch system. The first is a custom-built Assembly and Command Ship (ACS), and the second is the Launch Platform (LP), a semi-submersible vessel that is one of the world's largest ocean going launch platforms. The Assembly and Command Ship is specially designed to serve as a floating rocket assembly facility for launches at sea. The ship is 660 feet long, approximately 106 feet wide, with a displacement of more than 34,000 tons.

The Odyssey Launch Platform, a former North Sea oil drilling platform, is a semi-submerged Launch Platform 436 feet long, about 220 wide, with a displacement of 30,000 tons and a submerged draft displacement of 50,600 tons. Special facilities onboard enable the storage of rocket fuels (kerosene and liquid oxygen) sufficient for each mission.

Using the two or three stage *Zenit*-3SL, payload diameters of up to 14 feet can be accommodated. Because of its equatorial launch capability, heavy-lift performance of up to 13,000 pounds and orbital-placement accuracy results in a reduction of on-board fuel consumption for final on-orbit maneuvering.

In 1998, the U.S. State Department suspended work between Boeing and its Russian and Ukrainian partners because of concerns that sensitive aerospace data were being transferred to potentially hostile foreign governments. That suspension was eventually lifted and the first Sea Launch flight test occurred in March 1999. This was followed by the first commercial launch on October 9, 1999, of DIRECTV 1-R geosynchronous satellite.

Air Launch: Pegasus

Several of the early space visionaries conceived the idea of using a winged first stage to climb to altitude, launch the subsequent stages, then return to land for reuse. This was the concept for the first design iteration of the Space Shuttle—a huge *Delta* winged aircraft carrying the Orbiter to 4-6,000 mph and launching it at 100,000 feet. While the Shuttle design retreated to something much less grand (and cheaper), the concept was sound.

In April 1990, a three-stage solid-fuel rocket named *Pegasus* was carried aloft by a B-52 from Edwards AFB and launched over the Pacific at 40,000 feet. Following a short free-fall, the first stage of the Pegasus ignited. The vehicle has short wings to help it through the first portion of its flight in the lower layers of the atmosphere as it accelerates to Mach 8. With a payload of up to 1,000 pounds, *Pegasus* has launched more than 80 satellites in the intervening years. Built by Orbital Sciences Corporation, it is currently launched from the belly of a converted Lockheed L-1011 "Stargazer" and has provided its services to commercial, government and international customers.

As with Sea Launch, *Pegasus* can be transported to any latitude for more economical access to equatorial orbits and represents the first privately funded venture to launch a satellite.

Proposed NASA Heavy Launch Vehicle: HLV

The retirement of the Saturn V at the end of the Apollo program in the early 1970s left NASA (and the world) without a launch vehicle for exceptionally large payloads—those between 50,000 and 200,000 pounds. More than 30 years later, in 2005, NASA Administrator Michael Griffin proposed the development of the largest practical booster that could be created at a reasonable cost using existing technology. Based on the structure of the Shuttle external tank that clusters five Space Shuttle Main Engines (SSME-LH2), augmented by the Solid Rocket Boosters (SRB) and an LH2 second stage, the new HLV, named Ares V, could again provide a weight lifting capability of 200,000. A smaller version, using a single SRB first stage and a single LH2 engine in the second stage would provide for launching the Crew Excursion Vehicle (CEV) that is projected as the follow-on for the Shuttle.

There are several obvious reasons for wanting a large booster (beyond just the ability to launch a bigger payload) such as avoiding multiple launches and the massive complications and delays that would accompany them. Whether the HLV is able to survive the funding cycles and become a real product remains to be seen.

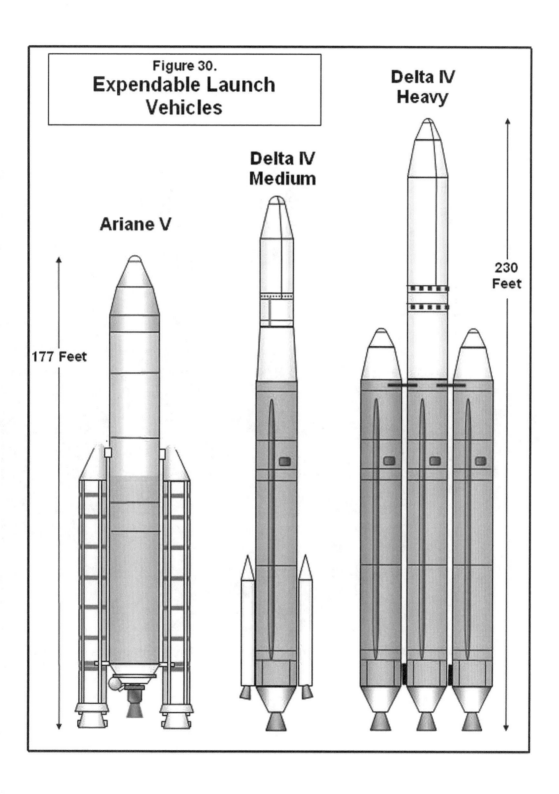

Figure 30.
Expendable Launch Vehicles

Delta IV Heavy

Delta IV Medium

Ariane V

177 Feet

230 Feet

One of the first images sent back from the surface of Mars from America's Viking lander

Chapter 34 — Search for Extraterrestrial Life

Perhaps the most exciting aspect of man's push into space is the prospect that there might be life elsewhere in the universe. The 19th century had abounded with speculation—especially following Percival Lowell's observations of "canals" on the face of Mars—that we are not alone. Mars is the most studied of all the planets because it is the most "Earth-like," leading to the question: Is there life on Mars? The first probes were sent to Mars during the "Space Race" in the 1960s. The Soviet Union sent six large probes to the planet, and all failed. The United States had a string of spectacular successes during that period.

When the Mariner 4 spacecraft flew past Mars in 1965, the images returned by its TV camera were of poor resolution but revealed a cratered surface much like the moon—and no canals. The next two Mariners, 6 and 7, presented a much more detailed surface with volcanoes and a tremendous canyon that stretched thousands of miles across the Martian globe—and definite indications of water-like erosion patterns. Mariner 9 (also called Mariner 71 as it was launched and arrived at Mars in 1971) produced even more detailed images which hinted that perhaps Mars was once a warmer and much wetter planet.

Both Mars and Venus are very important to understanding the Earth and its own ecosystem. If Mars was once more Earth-like, and if no indications of life are discovered, it may cause a rethinking of the uniqueness of life and an examination of other possibilities for the creation of life. In fact, one theory which has gained in popularity is that, in much the same way as wind scatters the seeds of a dandelion, life originated elsewhere in the Galaxy and produced spores that have been scattered throughout space, landing on many heavenly bodies. Only those planets with a unique set of climate attributes, such as the Earth, allowed them to germinate. The hardiness of such a form of life would itself be miraculous—to survive in the hostile temperature extremes and radiation-filled environment of space, without sustenance for millions of years, and then blossom.

Examination of the universe (at least that portion that we can observe and measure) reveals that it is composed mostly of hydrogen (about 99 percent). There are 91 other naturally occurring elements (some in much greater abundance than others) including carbon, nitrogen and oxygen. These can be combined, under the right circumstances, to form methane, ammonia, and water. From these building blocks, laboratory manipulation can produce more complex molecules to include amino acids, sugars, and fatty acids—all components of the carbon based life forms on Earth. Long chains of amino acids can be formed into proteins, and the sugars can form carbohydrates. However, here is where the laboratory process reaches its limits—what actually happens to transform these organic building blocks into a living organism is unknown. Whether science will discover the miraculous process or God will choose to keep it hidden, only time will tell.

The Space Science Board of the National Academy of Sciences published a report in 1965 in which it recommended three primary goals for the national planetary exploration program: the origin and evolution of the solar system, the origin and evolution of life, and the dynamic processes that shape man's terrestrial environment.

As for the possibility of life on Mars, many scientists agree that, from the point of view of Earth type micro-organisms, Mars is self-sterilizing. The solar ultraviolet radiation, lack of moisture and the oxidizing nature of the soil should prevent the formation of living organisms. Of course, it was also possible that the Martian environment might have been considerably different millions of years ago, so there is the possibility of some form of fossil record.

How might life manifest itself on Mars, and for what characteristics should an unmanned probe search? For the purposes of designing the first lander in the 1960s, it was assumed that the cameras carried to the surface of Mars would not observe flowering plants or unusual critters walking around. Life would most likely exist as microscopic organisms. To the scientist, one way of detecting the presence of life is to examine the biological processes associated with it. It was also assumed that these processes would be similar to those of life forms that were already known (not that there couldn't be other unknown life forms).

A 1965 report by NASA, "An analysis of the Extraterrestrial Life Detection Problem," established five characteristics shared by virtually all living things. These included growth, movement, reproduction, metabolism, and irritability (the ability to respond to an external stimulus). Because the first proposed Mars lander would have a short life (90 days), this duration limited the kinds of experiments that could be performed. It was decided to examine the surface soil for metabolism. To accomplish this, the lander would scoop up a handful of Martian dirt and place it in sealed container. A nutrient rich stimulus would be provided, and any detectable waste products (in the form of gases) would be observed.

Excitement ran high, not only for the team that would develop the lander but for the general public as well. Here was the first opportunity for man to examine the surface of a planet that was a likely candidate for life.

The Viking Program

Planning for an unmanned Mars landing began in 1966 as a part of a program named Voyager, which was to scout the Martian environment for a manned landing that would occur by 1980. The project leaned heavily on the Apollo hardware with both an orbiter and lander to be launched by a Saturn V. However, by 1971 as the moon race concluded and the Soviets appeared out of the running, the lack of political emphasis canceled the project. The Voyager name would be resurrected for another deep space program.

A less expensive alternative that could be launched by the Titan III-E using a Centaur upper stage was then proposed and approved. Much of the early work on Voyager found its way into the newly named Viking program. Because of the likelihood of failure on such a complex and long duration mission, two independent spacecraft were built (Viking 1 and Viking 2). Each consisted of an orbiter and lander. The former would remain in orbit around Mars and provide detailed images of the planet as well as acting as a relay for communications between the lander and Earth.

NASA's Langley Research Center had management responsibility from 1968 through to 1978, when the Jet Propulsion Laboratory (JPL) assumed the task. JPL's initial assignment was the development of the orbiters, tracking and data acquisition, and the Mission Control and Computing Center. Martin Marietta Aerospace developed the landers. NASA's Lewis Research Center had responsibility for the Martin Titan-Centaur launch vehicles.

Viking 1 was launched on August 20, 1975, and the second craft, Viking 2, was launched three weeks later on September 9. The fully fueled orbiter-lander pair had a total weight of 7,760 pounds. The orbiter was based on the Mariner 9 spacecraft bus, and had a total weight at launch of 5,121 pounds. The overall height was eleven feet with four solar panels extending outward 32 feet from the orbiter and producing 620 watts of power at Mars.

The main spacecraft propulsion unit was fueled by a gimbaled bipropellant (monomethylhydrazine and nitrogen tetroxide) liquid-fueled rocket engine. The engine was capable of 300 pounds of thrust, which provided a Delta-V capability (change in velocity) of 4,800 feet per second. This allowed for midcourse corrections and Mars orbital insertion. Attitude control used 12 small compressed-nitrogen jets.

An inertial guidance unit with six gyroscopes provided three-axis stabilization using the sun and the star Canopus as a reference. Communications were accomplished through 20-watt S-band (2.3 GHz) transmitters. A two-axis steerable high-gain parabolic dish antenna with a diameter of five feet and a fixed low-gain antenna were configured. A 381-MHz relay radio was also used with data buffered on two tape recorders, each capable of storing 1,280 Mbits. Two identical and independent computers, each with a 4,096 16-bit word memory, allowed for command processing and for storing uplink command sequences and data.

Scientific instrumentation on the orbiter included imaging, measuring atmospheric water vapor, and infrared thermal mapping. These were enclosed in a temperature-controlled, pointable scan platform. The scientific instrumentation had a total weight of 160 pounds.

The lander consisted of a six sided aluminum base supported on three extended legs. Power was provided by two radioisotope thermal generator (RTG) units containing plutonium. Each generator weighed only 30 pounds and provided 30 watts of continuous power. Four nickel-cadmium 28-volt rechargeable batteries were used for peak loads.

Lander propulsion for deorbit consisted of a monopropellant hydrazine ($N2H4$) rocket with 12 nozzles arranged in four clusters of three that provided seven pounds of thrust, giving a Delta-V (change in velocity) capability of 590 feet per second. These nozzles also acted as the attitude control thrusters for the lander.

Terminal descent and landing were achieved by three monopropellant hydrazine engines with 18 nozzles to disperse the exhaust and minimize affects on the ground and were able to be throttled from 60 pounds to 600 pounds of thrust. The hydrazine was purified and the lander sterilized before launch to prevent contamination of Mars with organisms from Earth. An inertial reference unit, with four gyros, an aerodecelerator, a radar altimeter, and a terminal descent and landing radar, provided for control of the thrusters. Total weight of the lander at launch was 1,445 pounds.

The lander computer had a 6,000-word memory for command instructions and a 40-Mbit tape recorder for data storage. It carried instruments to study biology, chemical composition (organic and inorganic), Martian weather, seismology, magnetic properties, and it had two cameras for scanning the appearance and physical properties of the visible environment. The scientific payload of the lander had a total weight of approximately 200 pounds. After separation and landing, the lander had a weight of about 1,320 pounds and the orbiter 1,980 pounds.

The Trip to Mars

The relative positions of Earth and Mars are constantly changing due to the Earth's orbital period of 365 days as compared to the 686 days that Mars takes to complete one revolution of the Sun. This constantly changing relationship means that there is only one opportunity every 25 months to take advantage of these positions to launch a spacecraft that requires the least amount of energy to complete the trip.

There are two types of minimum-energy paths; Type I, the shorter, requires slightly less than 180 degrees of travel around the Sun while Type II requires more than 180 degrees. As such, trips to Mars may take as little as 6 months on a Type I trajectory or as long as 10 months for a Type II. The Vikings were programmed for the longer journey because the Type II allows the greatest payload for the available launch vehicle power. During the Mars launch window there is a period of 40 days when the Type II trajectory can be flown. Thus, it was imperative that the launch occur during this period in mid-1975.

The Viking voyage from the Earth to Mars represented a typical flight profile that would likely be followed by a future manned expedition. Viking used the Titan III-E as its launch vehicle, the most powerful rocket of its time (the Saturn V was retired by then). The basic configuration consisted of a core two-stage liquid-fuel Titan II with large solid rocket boosters (SRB) on either side that generated 1.2 million pounds of thrust each. Unlike most "parallel" launch vehicle configurations, however, the first stage of the Titan did not ignite for lift-off. Only the solid boosters fired, leaving a spectacular billowing smoke trail as they ascended into the heavens.

At T+6.3 seconds after lift-off, the giant rocket rolled slightly to align its flight with the desired azimuth for its initial earth orbit. Just before burnout of the solids, at T+111 seconds, the first stage of the Titan came to life; its two liquid-fuel engines used hypergolic propellants and generated 523,000 pounds of thrust. Eleven seconds later, the SRBs completed their job, and a set of small rockets on their sides thrust the now empty cylinders to the side and away from the core unit. Following a burn of 253 seconds, the Titan first stage shut down and separated from the remaining rocket, and the second stage ignited. It provided 102,000 pounds of thrust and burned for another 208 seconds. Just ten seconds into the second stage burn, the payload shroud that protected the Viking from the destructive air pressure and friction heating of the atmosphere was jettisoned.

The second stage shut down at T+401 seconds into the flight, and it, too, separated as the remaining vehicle was allowed to coast for 76 seconds before the twin LH2 engines of the Centaur came to life and generated 30,000 pounds of thrust. These provided the final push into the initial parking orbit 115 miles above the Earth and shut down after 126 seconds. The Viking was now in earth orbit where it would wait for the exact time to begin its second burn for Trans-Mars Insertion (TMI). This period varied from 6 minutes for Viking 1 to 30 minutes for Viking 2 because of the proximity between the Earth and Mars, which had shifted in the 20 days between the launch of the two spacecraft.

Mars occupies an orbit whose mean distance (average) from the Sun is 141 million miles. A spacecraft starting from the Earth (93 million miles from the Sun) must add to the 66,600 mph orbital velocity of the Earth to cause the spacecraft to extend its solar orbit outward to intercept the orbit of Mars. The Trans-Martian Injection (TMI) burn adds another 7,500 mph to the 17,800 mph earth-orbital speed to achieve escape velocity (25,300 mph). However, the flight is not a direct path to Mars but rather a curved segment of an ellipse (elongated circle) that will intercept the planet 310 days later.

With Reference to Figure 31, this path can be visualized by looking at the face of a clock and placing the Earth at the four o'clock position at the time of launch and Mars at the 6:00 position. Mars will be at the 12:00 o'clock position when the spacecraft arrives, after having traveled 460 million miles to reach it. The Earth meanwhile will have almost completed a full orbit and will be at the 5:00 o'clock position. The Earth will then be 206 million miles from Mars and the spacecraft. This technique is known as the Hohmann Transfer, conceived by Walter Hohmann in 1916, and is applicable to all orbiting bodies, including Earth satellites, lunar probes, and interplanetary flights, which use the basic laws of physics that Hohmann sought to exploit.

A 318-second burn of the Centaur performed the TMI, and the spacecraft was traveling at just over 37,000 feet per second. Following burnout, the spacecraft was separated from the third stage Centaur which then vented residual propellants through the combustion chambers for a small amount of thrust to assure different flight paths between it and the spacecraft.

Viking was then aligned to the proper orientation for radio communication with the Earth, and the Solar panels deployed. The Sun was acquired as an attitude reference as was the Star Canopus. Within three hours of launch, the spacecraft was communicating with the Deep Space Network and ready for its interplanetary cruise period. The speed of Viking at this point dropped off considerably as a result of the gravitational pull of the Earth, but it continued to move outward from the Earth's orbital path towards that of Mars. Two mid-course maneuvers were planned that occurred early in the mission (day 4 and day 30) that adjusted the Viking's speed, based on tracking data, to assure intercepting Mars more than nine months into the future.

During the cruise segment, periodic checks were made of various systems to ensure their operation and the overall condition of the spacecraft. Approximately 30 days before Mars encounter, the final calculations were made regarding the optimum attitude and thrusting needed for orbital insertion around Mars, and the scientific instruments aboard the orbiter were turned on.

The cameras began to image the planet about 200 hours before orbital insertion. Viking 1 arrived at Mars on June 19, 1976, and Viking 2 on August 7, 1976, after a journey of more than 300 days. Both were placed into orbit following a long (40-minute) decelerating burn that slowed them for capture by Mars's gravitational field. The orbit was highly elliptical and ranged from 930 miles to 20,500 miles with an orbital period of 24.6 hours to match the rotation of the planet.

Each of the Vikings orbited Mars for about a month while the TV cameras sent images of the proposed landing sites back to Earth. While these sites had been tentatively selected based on the Mariner 9 flight three years earlier, the cameras of Viking provided much higher resolution. After studying the orbiter photos, the Viking site certification team considered the original landing site for both Vikings to be unsafe. Two alternative sites were chosen; Viking 1 would land on the western slope of Chryse Planitia and Viking 2 at Utopia Planitia.

With the landing sites selected, the lander, encapsulated in its aeroshell thermal heat shield, separated from the orbiter at the high point (Apoapsis), and the six-minute deorbit burn commenced. Because of the high altitude from which the deorbit sequence began, the lander fell for more than three hours before encountering the upper limits of the Martian atmosphere at about 160 miles above the surface. Although the atmosphere is only one percent that of the Earth's, the reentry generated considerable friction, and the aeroshell communications with the orbiter encountered a brief blackout period when the air around it was ionized.

At four miles above the surface, the spacecraft had decelerated to about 600 mph, and the aeroshell was jettisoned and a parachute deployed to further slow the lander to 150 mph. When the radar altimeter determined that the lander was 4,600 feet above the surface, the parachute was released, and the terminal descent propulsion system fired to slow the vertical speed to about 6 mph. On touchdown the engines shut down—Viking 1 lander arrived on Mars on July 20, 1976, and Viking 2 on September 3, 1976.

Viking's Observations of Mars

The two Viking landers conducted four experiments to detect the presence of microbiological life. Soil samples were scooped-up by an extendible arm and placed in a test chamber. The Gas Exchange Experiment (GEX) looked for changes in the makeup of the gases in the test chamber, which would indicate biological activity. There were none.

In the case of the Labeled Release Experiment (LR), if microbes consumed any of a radioactively tagged liquid nutrient, gases emitted by the metabolic processes of these microbes would be revealed. The overall results were inconsistent.

The Pyrolytic Release Experiment (PR) essentially "cooked" a soil sample that had been exposed to radioactively-tagged carbon dioxide to determine if the soil sample had been used by organisms to make organic compounds. Seven of nine runs appeared to show small concentrations, but the results were later discounted. In the Gas Chromatograph—Mass Spectrometer Experiment (GCMS)—heated soil samples also failed to detect any sign of organic chemistry.

While some of the initial results appeared to meet NASA's criteria for the detection of life, further analysis indicated that an unexpected chemical reaction invalidated the data. All the biology experiments encountered puzzling chemical activity in the Martian soil but provided no clear evidence of living microorganisms in the soil near the landing sites.

The NASA report concluded that "Viking not only found no life on Mars, it showed why there is no life there... the extreme dryness, the pervasive short-wavelength ultraviolet radiation..."

Gilbert Levin, a principal investigator for the Labeled Release (LR) experiment, continues to believe that the experimental results appeared to support the possibility of life—although he does admit it is not conclusive. Moreover, as all of the experiments used samples from the top layer of the Martian surface, it is possible that because of the harsh surface conditions life might be found in the deeper subsurface.

Temperatures at the southern landing site (Viking Lander 1) were as high as 7°F at midday and -107°F during the night. The lowest nighttime temperature was -184°F. The mean daily pressure observed by Viking 1 Lander was as low as 6.8 millibars and as high as 9.0 millibars. The pressures at the Viking Lander 2 site were 7.3 and 10.8 millibars. (Standard surface pressure on Earth at sea level is about 1,013 millibars.)

Scientists had expected wind speeds to reach several hundred miles an hour from observing global dust storms, but neither lander recorded gusts over 74 mph, and average velocities were considerably lower. The atmosphere is composed of 95 percent carbon dioxide. The mean gravity on mars is about 40 percent that of Earth—thus a 150-pound person would weight about 60 pounds on Mars.

Viking initially determined that the residual north polar ice cap (that survives the northern summer) is water ice, not frozen carbon dioxide (dry ice) as had been believed. The southern cap probably retains some carbon dioxide ice through the summer. The Martian surface is an iron-rich clay.

More than 4,500 close-up images of the Martian surface were transmitted from the landers. The orbiter cameras provided unparalleled detail of surface features, including color and stereo observations. The Orbiter's 52,000 images mapped 97 percent of the Martian surface. The Orbiters imaged the entire surface of Mars at a resolution of 500 to 1,000 feet, and selected areas to 25 feet.

When Viking set down on Mars and its cameras began to scan the immediate surroundings, almost no one really believed that shrubs and trees would be observed let alone some Martian animal sniffing at the spacecraft looking to relieve himself. However, Astronomer Carl Sagan said that he still had a glimmer of hope for that possibility.

Nevertheless, if life ever is discovered on Mars, would it be indigenous or have we imported it from Earth on the dozen or so spacecraft that have landed or crashed over the past quarter century? Although great care was taken to sterilize the spacecraft, the process is not perfect. Earth may have already contaminated the planet, and life forms observed could be microbes that have grown from that contamination.

Thirty years after the Viking landers touched down, the "life on Mars" debate is still unsettled. While there has been some criticism that the Viking experiments were too "Earth-centric," given the situation, it is apparent that some initial assumptions had to be made. The Viking mission came to an end May 21, 1983, with its final cost in 2004 dollars being $3.9 billion—the most expensive unmanned

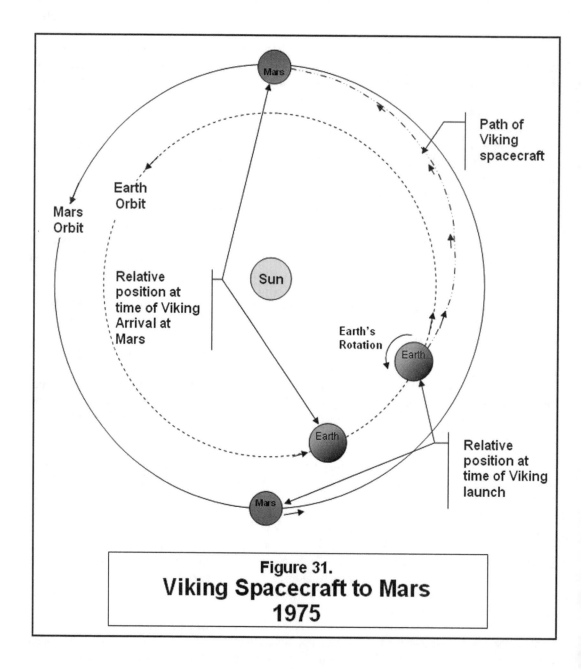

Figure 31.
Viking Spacecraft to Mars
1975

program in the history of NASA. It would be more than two decades before another spacecraft from Earth would continue the quest to demystify Mars.

The Rovers: Sojourner, Spirit, Opportunity

Following the Viking mission, Mars lay undisturbed for the next twenty years, though it was more for lack of money than lack of desire; the Shuttle consumed a growing share of the American space budget. The Soviet Union sent two large probes in 1988. Unfortunately, they continued a 28-year string of failures that have thwarted the Soviet attempts to probe the red planet since their first effort in 1960. The next attempt by the United States was the 2,200-pound Mars Observer launched by a Titan III in September 1992. The objective was to orbit Mars and study the geoscience and climate of Mars using high-resolution images. Unfortunately, contact with spacecraft was lost 3 days before Mars orbital insertion.

When former NASA Administrator Dan Goldin arrived on the scene in April 1992, NASA was faced with significant budget cuts. He addressed the problem of increasingly expensive space projects with a new approach. In traditional project management, the triad of "on-time, within budget, and to specification" is the mantra. Goldin looked at the billion dollar projects of the past (Viking, Voyager, Galileo, Cassini, and Hubble) and asked, "What if we scaled down our expectations so that we fly missions for one-fourth the cost?" NASA could perform four times as many missions—although perhaps not as spectacular as Viking—and, when a failure occurred, not as much would be lost.

The expression he coined was "faster, better, cheaper." However, reflecting back to the project management philosophy, two of the three attributes are mutually exclusive. If cost is reduced, a "better mission" could not be the result. While his expression implied the opposite, Goldin accepted compromises in performance.

Pathfinder, launched in December 1996, was the first mission that epitomized the successful low-budget approach to space that worked—it cost $270 million. Its mission was to deliver a small shoe-box-sized rover to the Martian surface. Recognizing the need to use the less powerful Delta II rocket to save money, the reduction of spacecraft weight caused the design team to look at other alternatives to the soft landing problem.

Arriving at Mars, a spacecraft is typically traveling at velocities of up to 15,000 mph. For a soft landing, virtually 100 percent of this kinetic energy must be absorbed. The Martian atmosphere can be used to dissipate about 98 percent with a high-drag heat shield and a parachute. This still leaves the other 2-percent—several hundred miles per hour, with which to contend. Viking used a sophisticated variable-thrust retro-rocket and shock absorbing legs. However, these added weight—and weight is the enemy of spacecraft.

Pathfinder employed a different technique that combined fixed-thrust solid-rocket motors and air bags. The solid rocket motors ignited two to three seconds prior to impact and slowed the lander from about 300 mph to a virtual stop while still 30 feet above the Martian terrain. The lander, which hung below the retro-package encapsulated in a large inflated airbag, separated and fell the remaining distance.

The air-bag system provided protection of the payload by bouncing over rocks and other surface hazards. Because the lander could right itself from any orientation, the need for attitude control during landing had also been eliminated. Motion sensors assured that the lander had come to rest prior to initiating the remaining sequences of righting itself and rover egress. Inside the airbag was a pyramid-shaped lander with hinged petals that opened to reposition the lander to an upright orientation if it had come to rest upside-down (the Soviet lunar lander had first used this technique in 1966). These petals then flattened out to reveal the enclosed science platform and small mini-rover called Sojourner. Weighing only 23 pounds, Sojourner's task was to examine the nearby rocks. A triple ramp-like system provided the ability of the rover to drive off in any direction if the lander came to rest against the side of a large bolder.

Pathfinder was an electronic base station equipped with cameras, scanners, and communications equipment to relay the rover's findings back to Earth. It landed on July 4, 1997, at Ares Vallis, a washed-out basin believed to have interesting rock—yet providing a relatively flat terrain to reduce the landing risk. The rocks revealed hints of a watery past: small round pebbles similar to a type of rock called "conglomerate" that forms in water. Pathfinder's data seemed to indicate that Mars had been much more watery than was first thought.

Sojourner and Pathfinder (renamed Sagan Memorial Station) functioned for 3 months before succumbing to the harsh Martian winter. The mission was terminated November 4, 1996, when contact with the lander could not be reestablished.

While Pathfinder and Sojourner were concluding their data gathering, the 2,300-pound Mars Global Surveyor was the next attempt by the United States to probe Mars. Launched aboard a Delta II rocket in November 1996, it reached Mars in September 1997. Use of the Delta II in place of a Titan III reduced the mission cost by more than $200 million. However, it resulted in a payload only one-third the size of Viking.

The initial Martian Orbit Insertion (MOI) maneuver used a traditional retro-rocket to establish the spacecraft in a highly elliptical, 186 mile by 35,000-mile, orbit. To reduce the amount of on-board fuel (and its weight) needed to slow the spacecraft for entry into its polar orbit around Mars, the craft used a trajectory that allowed it to dip into the upper layers of the Martian atmosphere. The drag it encountered continually lowered the apoapsis of the orbit over a period of several months. (Apoapsis is the high point in a Martian orbit equivalent to the apogee in Earth orbit.) This aerobraking eventually circularized the orbit to 250 miles after almost a year. Global Surveyor produced a planet-wide survey of high resolution images over the next several years to allow a comprehensive study of the weather and surface of the planet.

The Mars Surveyor '98 Program was composed of two spacecraft launched separately, the Mars Climate Orbiter and the Mars Polar Lander. They were part of NASA's 10-year plan during which launches would occur every 26 months when the Earth and Mars are favorably aligned. These two missions were designed to study the Martian weather, climate, and water and carbon dioxide resources, in the search for evidence of long-term and periodic climate changes.

The Mars Climate Orbiter was launched by a Delta II rocket in December 1998, and was scheduled to arrive at Mars in September 1999. However, an error in the transfer of information between the Mars Climate Orbiter spacecraft team in Colorado and the mission navigation team in California led to the inability of the spacecraft to perform a maneuver correctly to place it in orbit around Mars. The Colorado team used English units (e.g., inches, feet and pounds) while JPL used metric units.

As the failure evaluation team characterized it, "The problem here was not the error, it was the failure of NASA's systems engineering, and the checks and balances in our processes to detect the error. That's why we lost the spacecraft."

After the embarrassing English/Metric conversion mistake, NASA turned its attention to the Polar Lander Mission. The Mars Polar Lander was launched on a Delta II and, after an 11 month cruise, reached Mars on December 3, 1999. The Polar Lander spacecraft saw a return to the heavier and more complex variable thrust retro-rocket all the way to the surface. Initial deceleration used aerobraking on an eight-foot diameter ablation heat shield followed by a parachute at an altitude of about five miles at just under 1,000 mph. The parachute was to jettison at about one mile above the surface and the descent engines would fire. The thrust was varied by the spacecraft control system and Doppler radar. However, the last telemetry from 1,282-pound vehicle was received just prior to atmospheric entry—the cause of failure remains unknown.

The Mars Surveyor '98 program spacecraft development cost was $195 million; with launch costs $92 million and mission operations $43 million. The failure of these two spacecraft in a three month period called into question Administrator Goldin's "Faster, Better, Cheaper" philosophy. Other problems with various NASA programs highlighted the lack of management ability. John Pike, a respected member of the Federation of American Scientists, was quoted as saying, "It's basically Management 101. They're saying there are four things you have to balance: cost, schedule, content, and risk, but they don't say very much about how to balance those."

In October 2001, the Mars Odyssey satellite was launched to Mars carrying three primary instruments. These included THEMIS (Thermal Emission Imaging System) for determining the distribution of minerals, GRS (Gamma Ray Spectrometer) for determining the presence of 20 specific chemical elements on the surface of Mars (including hydrogen in the shallow subsurface), and MAREE (Mars Radiation Environment Experiment) for studying the radiation environment to determine the possible radiation risk to future human explorers. These instruments revealed indications of possible quantities of water ice buried beneath the surface in the Polar Regions.

The Odyssey orbiter also provided a communications relay capability for the next generation of Mars Exploration Rovers by retransmitting about 85 percent of the data from the rovers to Earth. The name "2001 Mars Odyssey" was selected in tribute to the works of science fiction author Arthur C. Clarke.

The Spirit and Opportunity rovers were direct descendants of the Pathfinder/Sojourner system but are much larger golf-cart-sized vehicles. These rovers were self-sufficient dune buggies that (unlike Sojourner) carried their own science, power and communications equipment. The two landed on opposite sides of Mars in January 2004. Their primary objective was to look for geological evidence of flowing or standing water on Mars.

Opportunity detected deposits on rocks similar to concretions that form when iron precipitates out of water. Also found were "lenticular voids," which are often the result of water-borne crystals breaking down from water or chemical changes.

Spirit found tentative evidence of past water on a rock made up of stripes of different minerals. Geologists believe the stripes were cracks filled by minerals carried by water.

Communications and computer problems at times plagued Spirit, but the flexibility of the programming allowed these to be bypassed. Both Spirit and Opportunity explored the planet far longer than the 90 days expected, as each survived more than two years.

In an attempt to provide added data for the "life on Mars" question, the British-built Beagle 2 was sent to Mars in 2003. The spacecraft was successfully ejected from the Mars Express vehicle, but nothing more was heard from Beagle 2, and the mission was presumed lost.

NASA's Mars Reconnaissance Orbiter, launched in August of 2005, arrived on March 10, 2006. Following a retro-fire to establish a highly elliptical orbit, it used aerobraking to lower its orbit. It then examined Mars in even greater detail from low orbit, looking for water distribution (including ice, vapor or liquid) as well as geologic features and selected minerals. The orbiter was designed to provide support for future missions to Mars by scouting potential landing sites and by providing a high-data-rate relay for communications back to Earth.

The Martian Meteorite: To Many Maybes?

On August 7, 1996, NASA scientists announced that they might have found evidence of past life on Mars in a rock. However, the rock they were examining was a meteorite found on Earth; the assumption was that it might have come from Mars. The discovery that was making the headlines suggested possible fossilized organic molecules were enclosed in the meteorite.

NASA scientists believe the rock in question was of possible Martian origin because gases trapped in it closely match the composition of the present day Martian atmosphere (as measured by spacecraft sent to Mars). A total of 31 igneous rocks (a crystallized form of lava) have been placed in a class called SNC meteorites (after the three type samples, Shergotty, Nakhla, and Chassigny). The rock in question was found in 1984 by an American scientist on a government funded search for meteors in Antarctica. As it was the first to be to be picked up that year, the rock was numbered "84001."

NASA estimates that about 16 million years ago, a huge comet or asteroid struck Mars, with enough force to eject these rocks to escape velocity (12,000 mph). Once in space they drifted for millions of years until several encountered Earth's atmosphere 13,000 years ago and one in particular fell in Antarctica as a meteorite.

NASA dated the 4.2-pound, fist-sized "rock 84001" at about 4.5 billion years—the period when it was believed the planet Mars was formed. NASA thinks it may have originated from under the Martian surface where water is thought to have existed. The water could have been saturated with carbon dioxide from the Martian atmosphere, and carbonate minerals could have been deposited within small fractures. NASA went on to speculate that living organisms may have assisted in the formation of the carbonate. NASA believes it is possible that the remains of microscopic organisms may have become fossilized.

In the carbonate deposits, NASA found a number of features that could be construed as suggesting past life. Traces of organic molecules were found called polycyclic aromatic hydrocarbons (PAHs) along with mineral compounds commonly associated with microscopic organisms—and possible microscopic fossil structures that looked similar to bacteria but were 100 times smaller.

NASA's revelation came just before congressional approval of its 1997 budget and at a time when public interest in the space program was ebbing. Some scientists doubt the rock is from Mars, that the PAHs were organically formed, or that the tiny irregularities in the rock are fossils. When pressed, NASA admitted that PAHs are not necessarily the result of organic activity.

While the rather weak "evidence" failed to create a stir in the scientific community once all of the suppositions had been revealed, the story did generate considerable public reaction and a recent poll revealed that most Americans believe that NASA "found life on Mars."

Mars: The Next Ten Years

Over the next decade, more inexpensive "scout" missions to Mars are planned, but NASA will be returning to more sophisticated spacecraft for its mobile science laboratory to launch, possibly by 2009. The definitive mission in the next ten years will be the first return of samples of Mars back to Earth, which will probably be launched no earlier than 2014. Several of these missions may be flown, depending on the results of the first. If no life is found, then the possibility of a manned mission diminishes considerably.

The first 30 years of study have shown that Mars has relatively smooth poles composed mostly of frozen carbon dioxide. The mid-latitudes have dormant volcanoes, deserts, and rocky plateaus with ancient gullies, perhaps carved by running water—in many ways the landscape is similar to the American southwest. However, the liquid water issue is still hotly contested. For example, a region of Mars that some scientists believe was once a shallow lakebed and likely habitable for life may have instead been created by the reaction of sulfur-bearing steam vapors moving up through volcanic ash deposits.

The European Space Agency has begun a program that will include an instrument for testing deep soil samples on Mars in a European mission called ExoMars. The ExoMars rover will drill for soil samples up to six feet into the Martian surface in search of past or present life. The craft is named for the study of the origin and distribution of life in the universe known as exobiology.

A Molecular Organic Molecule Analyzer (MOMA) developed by ESA and funded by NASA will be included as part of the ExoMars mission scheduled for 2011 and be launched from Kourou, French Guiana, aboard an Ariane 5.

A Closer Look at Venus: Magellan

Although Mars has always been the traditional favorite in the search for extraterrestrial life, Venus still intrigues scientists. Its close proximity to the sun has resulted in a hot surface that probably pre-cluded any possible emergence of life. But the history and evolution of the planet could still provide many clues to the origin of the solar system. The Russians had better luck with their early Venus probes in the 1960s than they did with Mars and had actually landed a spacecraft which sent images back from the surface. The United States also had good experiences with their early spacecraft to Venus. But the cloud shrouded planet demanded a different approach for imaging the surface that was not readily available until the 1980s.

Magellan was the first interplanetary spacecraft to be launched by a space shuttle when it was car-ried to low earth orbit by Atlantis (STS-30) in May 1989. Released from the shuttle's cargo bay, a solid-fuel Inertial Upper Stage (IUS) then fired, sending Magellan on an extended 15-month journey that orbited the Sun 1-1/2 times before it arrived at Venus in August, 1990. A solid-fuel motor then fired to slow the spacecraft into a highly elliptical polar orbit, ranging from 182 miles to 5,296 mi., that completed one revolution every 3 hours and 15 minutes.

The Magellan spacecraft was actually built with many of the "spare" parts from other spacecraft. To create the 15-foot-long, 7,612-pound structure, the high-gain antenna was inherited from the Voyager program along with the 10-sided main structure and a set of thrusters. The computer system, attitude control, and power distribution came from the Galileo program, and its medium-gain anten-na was left over from Mariner 9. Power was provided by two 8-foot solar panels that supplied 1,200 watts of power.

During its low passes around the planet, Magellan's radar mapped a 15-mile wide surface segment several hundred miles long—peering through the cloud shrouded planet's covering with its radar. As it ascended back to its high point in the orbit, it transmitted the images back to Earth. As the planet rotated under the spacecraft's orbit, Magellan was able to reveal more of the planet's surface features. By the end of its first eight-month orbital cycle (September 1990 to May 1991), Magellan's radar imaged 84 percent of Venusian surface. The spacecraft then conducted a second mapping on another eight-month cycle and a final third session that concluded in September 1992. It mapped 98 percent of the planet, and, because the "look angle" of the radar was slightly different from one cycle to the next, a three-dimensional view of Venus's surface revealed the elevations of various surface features.

Magellan's high-gain antenna directed the microwave radar energy toward the planet and received the returned echoes off Venus's surface. These pulses were directed at a slight angle using a technique similar to "side-looking radar." Special processing of the returns yielded higher resolution with a technique called "synthetic aperture radar."

A study of the Venusian surface as revealed by the radar mapping showed many volcanic surface features including large lava flows—some over 4,000 miles long—lava domes, and the volcanoes themselves. Like the Earth, there were few impact craters on Venus, indicating that it is probably less than one billion years old. In contrast with Mars which has a much thinner atmosphere but shows significant wind erosion patterns, Venus had little to indicate significant wind activity.

During Magellan's fourth 8-month cycle beginning in May 1993, flight controllers experimented with the aerobraking technique, using atmospheric drag on the spacecraft to lower its orbit. In September 1994, Magellan's orbit was again lowered for a test called the "windmill experiment." The spacecraft's solar panels were rotated slightly to give them a "pitch" like a propellor or windmill. As the spacecraft descended into the denser layers of the atmosphere, flight controllers were able to measure the torque from the thruster control that was required to maintain Magellan's orientation and to keep it from spinning. This experiment provided data on the upper atmosphere.

On October 11, 1994, Magellan's orbit was lowered a final time, and the spacecraft entered the atmosphere and was intentionally destroyed.

The Hubble Space Telescope

Since the days of Galileo, the optical telescope has been the primary means of exploring the universe. With each new increase in its technology, the telescope has expanded mankind's look into the cosmos and revealed new wonders. It has also been a motivating tool for the young, as it allows the backyard novice the opportunity to experience the thrill of seeming closer to the moon and the planets.

One of the many astronomers who gained fame by his observations was Edwin Hubble. In the 1920s, Hubble used the 100-inch Hooker Telescope at the Mt. Wilson Observatory in Pasadena, California, to measure the distances and velocities of other galaxies. Until Hubble's pioneering work, it had been assumed that the Milky Way was the only galaxy in the universe and that it was perhaps a few thousand light years across. While he worked with others at the Hooker facility, thousands of galaxies were discovered. His work ultimately led to the current theory of an expanding universe.

At the same time that Hubble was doing his pioneering work, rocket scientist Hermann Oberth proposed placing telescopes in orbit. The principle problem with earth based telescopes is the atmosphere. The 50-mile thick band of gases that surround and give life to our planet also present an impediment to receiving the faint light from far off objects in space. It is like trying to see through water to the bottom of a pond. The undulations in the flow of the atmosphere are what make stars appear to twinkle. To avoid the atmospheric disturbances, large observatories are built on mountains away from cities to get as clear and dark a viewing of the heavens as possible.

In 1946, astronomer Lyman Spitzer, motivated by the dramatic advances in rocketry during WWII, wrote a paper entitled *Astronomical Advantages of an Extra-terrestrial Observatory*. He highlighted the primary advantages of a space-based observatory over ground based telescopes. With a space based telescope, "angular resolution" (the smallest separation with which objects can be distinguished) would be limited only by natural diffraction of the light, rather than by the diffusion and turbulence of the atmosphere and the effects of gravitation on the instrument. This would provide resolution about ten times greater for the same size telescope. Another advantage would be the ability to observe infrared and ultraviolet light, which are almost entirely absorbed by the atmosphere before reaching the Earth's surface.

With the advent of the first satellites, astronomers began lobbying for a space based viewing platform, and by 1969 the National Academy of Sciences became an advocate of such a project. However, it was not until 1977 that Congress approved funding for what would become the Hubble Space Telescope (HST)—named in honor of the astronomer. As a part of the program, the Space Telescope Science Institute was formed as a center for astronomical research and planning.

Much of the technology for the HST was derived from that used in military reconnaissance satellites. The reflecting mirrors and optical systems of the telescope were designed to very precise specifications—being polished to an accuracy that is expressed as 1/20th the wavelength of visible light, about one millionth of an inch. The Perkin-Elmer company, who had been contracted to build the mirrors, used extremely sophisticated computer-controlled polishing machines to shape the mirror. They began in 1979, using ultra-low expansion glass. But the polishing part of the project experienced a series of schedule slippages and significant budget over-runs and was not completed until the end of 1981. Despite its reputation in optics, Perkin-Elmer's competence to not only complete the project but perform the work with precision, supported by appropriate quality control checks, was often brought into question.

Development of the spacecraft involved many engineering challenges, not the least of which was that it would have to endure the frequent passage (every 46 minutes) from intense heat of the sunlight into the cold shadow of the Earth while remaining stable enough for extremely accurate pointing of the telescope.

In 1983 the Space Telescope Science Institute (STSI) became involved in a power struggle with NASA, who wanted to keep control of the HST within its domain. STSI ultimately was granted the responsibility for the scientific operation of the telescope and delivery of data to the scientific community.

More delays continued to postpone the scheduled launch until September 1986, but before that could occur, the Shuttle Challenger met its fate the preceding January. The disaster brought much of the U.S. space program to a standstill, and the launch of Hubble would have to wait for almost four more years.

Finally, on April 24, 1990, the HST was placed in orbit by the Shuttle Atlantis (STS-31). Among the five scientific instruments on board was the Wide Field and Planetary Camera (WF/PC) which provided high-resolution imaging for optical observations. Also a part of the instrumentation was the Goddard High Resolution Spectrograph (GHRS), the High Speed Photometer (HSP), Faint Object Camera (FOC) and the Faint Object Spectrograph (FOS). From an initial cost estimate of $435 million (in FY77 funds), the product had ballooned to $2.5 billion to get to orbit. But worse news was yet to come.

Within weeks of the launch, the images being returned showed a serious problem with the optical system. Although these images were about as sharp as ground-based viewing, the telescope failed to achieve a final sharp focus, and the best image quality obtained was drastically lower than expected—not the anticipated ten fold increase.

Analysis of the flawed images showed the cause of the problem to be the primary mirror—it had been ground to the wrong shape. It was determined that the conic constant of the mirror (its curvature) was -1.0139, not the intended -1.00229. It was 78 millionths of an inch out of specification!

A critical instrument used by Perkin-Elmer to shape the mirror had been incorrectly calibrated. Two other instruments that were used to cross-check the precision of the polishing work had detected the problem during the fabrication process, but their findings were ignored since the company believed that these instruments were less accurate than the primary one. Criticism was also leveled at NASA for poor oversight of the quality control process. In its flawed state, the Hubble could not obtain data any better than ground based telescopes.

It was not possible to replace the mirror in orbit, or return the telescope to Earth for a repair. However, the HST design incorporated provisions for being serviced in orbit, and a plan was devised to install a "corrective lens"—much like fitting a human with a pair of glasses. The design to correct the problem was called the "Corrective Optics Space Telescope Axial Replacement" (COSTAR). However this required that one of the other instruments had to be removed—the High Speed Photometer.

The first servicing mission to Hubble required extensive training of the astronauts and the use of many specialized tools. The mission (STS-61) took place in December 1993 and involved installation of several instruments and other equipment with five spacewalks over a 10-day period. The longest of these EVAs lasted almost 8 hours.

Also replaced were the solar arrays and their drive electronics, four of the gyroscopes used in the telescope pointing system, and the onboard computers. After the repairs and upgrades were completed (and tested), the HST was boosted into a higher orbit (368 miles) by the Shuttle. Service Mission 1, with its extensive EVAs, was one of the most complex ever attempted. While NASA had taken considerable criticism for originating the problem, it bathed in the accolades that followed when the HST began sending the long awaited images from space—and they were impressive.

Data and images from Hubble has addressed many debates in the astrophysics community and uncovered information that has required changes in existing theories and the formulation of new theories to explain. Distances to specific stars and their relative speeds allowed the value of the "Hubble Constant" (a component of "Hubble's Law"), which had varied by up to 50 percent, to be determined to within 10 percent . Hubble's law relates to the "red shift" in light frequencies coming from distant galaxies that is proportional to their distance. While helping to define the age of the universe, the HST also uncovered evidence that instead of decelerating under the influence of gravity, the universe may actually be accelerating.

The collision of the Comet Shoemaker-Levy 9 with Jupiter in 1994 occurred just months after Servicing Mission 1 had made Hubble fully operational. Its images were pivotal in studying the dynamics of that collision. Overall, the importance of the HST to astronomy cannot be overstated.

It has been pointed out, however, that the HST cost about 100 times more to build and maintain than a typical world class observatory. Moreover, advances in recent years in Earth based telescopes, that are coupled to computers, can rapidly compensate for atmospheric distortion—a technique called adaptive optics. Using deformable mirrors or material with variable refractive properties, earth based viewing has approached some of the capabilities of the HST for far less cost.

Viewing time on the HST is not restricted to any nationality or academic affiliation, and competition is intense. However, the actual viewing periods are significantly limited by the location of the HST in its low-Earth orbit which occults (limits the view of) most astronomical targets. Less than half of each orbit (about 40 minutes) is available for active viewing, and this is further restricted when the satellite is over the South Atlantic due to the high radiation levels of the lower Van Allen belt. There is a continuous viewing zone (CVZ) that is 90 degrees to the plane of Hubble's orbit.

Scheduling viewing time is also affected by minute variations in its orbital period. Thus observation schedules cannot be established until a few days in advance. Hubble images are initially stored on the spacecraft and transferred to the ground via the Tracking and Data Relay Satellite System. The data is eventually made available via a public archive at http://archive.stsci.edu/hst.

Continued maintenance was performed every few years with Servicing Mission 2 (STS-82) in February 1997 (again boosted Hubble's orbit), Servicing Mission 3A (STS-103) in December 1999 (replaced all six gyroscopes), and Servicing Mission 3B (STS-109) in March 2002 (replaced the solar arrays for a third time). The new arrays were much more efficient and only two-thirds the size of the old arrays, resulting in less drag from the fringes of the upper atmosphere. The new solar panels also provided 30 percent more power that allowed all instruments on HST to be run simultaneously, and they reduced a vibration problem that the old arrays had induced when the satellite entered and left direct sunlight.

Few spacecraft have captured the public's imagination as has the Hubble. NASA understands this and has been careful to get as much public relations mileage from it as possible. But with the Shuttle Columbia's demise, the maintainability of Hubble has come into question. Future Service Missions to the HST by the Shuttle have been put on hold. Some in NASA feel the risk of flying the Shuttle is so severe that it should only be used to visit the ISS. There, its exterior can be inspected, and if damage by debris during launch was critical, the repairs could be made or a rescue mission launched.

Because a given Shuttle mission cannot reach both the HST and the ISS, future manned service missions to the HST have been canceled. While the decision has come under attack by many in the scientific community, others believe that allocating the money to the new James Webb Space Telescope (JWST) would be a better priority. The JWST is in the early development stages and is not intended to be maintained. This smaller spacecraft will have an expected lifetime of 5 to 10 years and will be launched no earlier than June 2013. Its mirror will be five times larger than that of the HST, and it will be placed in an orbit further from the Earth, called L2, to permit a greater viewing area.

An official panel from the National Academy of Sciences recommended that the Hubble telescope be preserved regardless of the perceived risks. The current NASA administrator, Griffon, has also taken issue with the decision to abandon the HST and has authorized the Goddard Space Flight Center to prepare for one more manned Hubble maintenance flight—but has not committed to the actual flight. Any mission to the HST will cost $1 billion, and could not take place before 2007.

In 2005 it was decided to switch the HST to using only two-gyroscopes for regular telescope operations as a means of extending the lifetime of the four gyroscope platform. Estimates of the failure rate of the gyros indicate that Hubble may be down to one gyro by 2008, after which the telescope would be rendered unusable.

If it is not re-boosted by a shuttle or other means, it will re-enter the Earth's atmosphere sometime between 2010 and 2012, depending on solar activity and the on-board attitude control consumables. However, an uncontrolled reentry could be hazardous as some large components would not be completely consumed by the heat. NASA has considered attaching an external propulsion module to deorbit the HST when it is no longer economically useful.

Over its lifetime, Hubble has taken more than 400,000 separate observations of more than 25,000 astronomical targets. Its cumulative costs are approximately $14 billion (adjusted to FY2005).

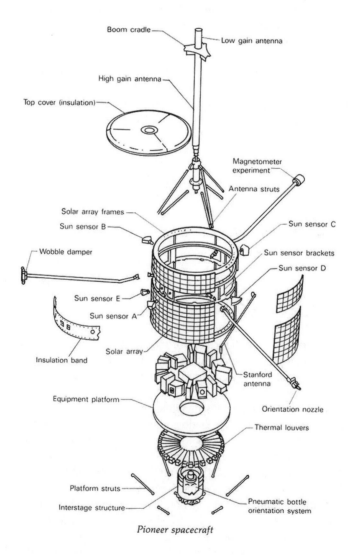

Pioneer spacecraft

Chapter 35 — Deep Space Missions

One of the goals put forth by the National Academy of Sciences in 1965 was planetary exploration *"aimed at understanding the origin and evolution of the solar system."* With this objective in mind, NASA extended its reach to the outer planets—those beyond Mars. However, travel to these remote regions of the solar system required new design considerations for the spacecraft. The orbits of Venus and Mars, being less than 50 million miles from Earth, are typically within four to eight months of flight time. On the other hand, the gas giants ranged from 430 million miles to Jupiter and over 1.7 billion miles for Uranus. The ability of the spacecraft to survive over these distances required more attention to the types of systems employed.

Pioneer 10 and 11: Jupiter and Saturn

The first truly deep-space probe was the 570-pound Pioneer 10 spacecraft which departed Earth in March 1972 on an Atlas-Centaur. To reach Jupiter, it was given an initial velocity of almost 32,000 mph—about 6,000 mph more than a Mars probe—and quickly passed our own Moon's orbit in just eleven hours. It was one of two probes launched a year apart to investigate the planet.

While Mars held the attraction of possible life, Jupiter has the allure of spectacular colored bands that can be discerned with powerful earth based telescopes. Its atmosphere is composed of methane, ammonia, and hydrogen, all believed to be components of the Earth's environment when it formed an estimated four billion years ago. A giant "red spot" in its southern hemisphere is also visible from Earth. Jupiter has its own internal heat source—it radiates about four times as much energy as it receives from the sun. As the largest planet in the solar system, it is more than eleven times the diameter and 1,000 times the volume of the Earth. Because of its distance from the sun, Jupiter takes almost 12 earth years to complete one orbit.

Despite its massive size, the planet rotates on its axis in only 10 hours, which creates a turbulent atmosphere. However, Jupiter is also an enigma because its density is only about one-fourth that of the Earth—slightly more than water. At the time of launching the probe, Jupiter was known to have 15 satellites—this number would grow to 63 by 2004. One, Ganymede, is larger than the planet Mercury.

Solar power for spacecraft electronics to the outer planets is not feasible. The energy provided by the Sun at the distance of Jupiter is only four percent of that received at the Earth. The weight and size of a large solar array and the degradation of solar cells over time from exposure to micrometeorites ruled them out. Nuclear electric was the obvious choice. Four Space Nuclear Auxiliary Power (SNAP-19) radioisotope thermoelectric generators (RTGs) were used. The heat generated by the plutonium-238 dioxide provided 155 watts of power at launch and 140 watts by the time of the Jupiter encounter. The RTGs were designed to provide usable power for 30 years

Pioneer 10 was the first spacecraft to travel through the Asteroid belt—cosmic debris composed of asteroids and dust—that lies between the orbit of Mars and Jupiter. At the time, this was an unknown hazard that turned out to be not as great a threat as once imagined.

As Pioneer 10 passed within 81,000 miles of Jupiter in December 1973, it obtained close-up images of the planet and its moons. Traveling at the speed of light, its 8-watt radio signals required more than 45 minutes to reach the Earth. When these pictures were linked as a series of motion picture frames, the circulation of the Jovian atmosphere could be seen. A series of image filters captured the yellow-orange and blue-gray bands of the atmosphere along with the circulation patterns and movement of the giant 30,000-mile-long "red spot" generated by the 22,000 mph equatorial rotational speed. Pioneer 10 took measurements of Jupiter's radiation belts, magnetic field, and atmosphere. The measurement of the intense radiation environment near Jupiter was important in the design of subsequent spacecraft that would visit Jupiter later—Voyager and Galileo.

Because it would be the first man-made object to leave the solar system, it was decided to include a plaque aboard Pioneer 10 that would graphically portray the location of its launch—the third planet from the Sun. The plaque, designed by astronomer Carl Sagan, also had nude figures of a man and a woman to illustrate the basic physiology of the Earth's inhabitants. Its objective was to convey information as to its origin to potential extraterrestrials who might encounter it in the distant future.

Pioneer 10 officially ended its scientific investigations of the outer regions of the solar system in March 1997—twenty-five years after its launch. However, it continues to be tracked and it sent its last communication in January 2003. NASA engineers then reported that its radioisotope power source had decayed to the point where NASA's Deep Space Network (DSN) could no longer detect a signal.

Pioneer 10 held the record as the most remote object ever made by man (over 8 billion miles away) until February 1998. That was when Voyager 1's distance from the Sun equaled that of Pioneer 10 at 69.4 AUs (an AU is the distance from the Earth to the Sun—93,000,000 miles). It then outdistanced Pioneer 10 at the rate of 1.02 AUs per year. Pioneer 10 will continue to travel through deep space into interstellar space, heading in the general direction of the red star Aldebaran (the eye in the constellation Taurus, The Bull) 68 light years away. It will take Pioneer over 2 million Earth years to reach Aldebaran.

Pioneer 11 was launched in April 1973, and followed Pioneer 10 to Jupiter. Confirming the data acquired by Pioneer 10, it was targeted to pass within 13,000 miles of the planet to acquire additional information. However, the closer pass also meant that it encountered more intense radiation and was at a higher risk of failing.

The close Jupiter encounter accelerated the spacecraft and redirected it by gravity assist for the second part of its planned journey to Saturn. Pioneer 11 made the first encounter with Saturn in 1979 as it passed under the rings and within 13,000 miles of that planet.

The Pioneer 11 Mission ended in September 1995 when the last transmission from the spacecraft was received. It is traveling toward the constellation Aquila (The Eagle), where it will pass close to one of the nearest stars in that constellation in about 4 million years.

Voyager 1 and 2: The Grand Tour

The spectacular Viking missions to Mars in 1975 were followed closely by two more expensive ($1 billion) probes to the outer planets. Voyager 2 was launched in August with Voyager 1 following in September of 1977. Because Voyager 1 followed a slightly different trajectory it would be the first to arrive at Jupiter even though it was launched 16 days later than Voyager 2—thus the rationale for their naming. The alignment of the three gas giants of the outer planetary system dictated the launch timing as that opportunity would not occur again until 2157. The position of Jupiter, Saturn, Uranus and Neptune made it possible for a single space vehicle to visit all four, using gravity assist to accelerate and redirect the spacecraft to its next intended target.

Driven into the heavens by the Titan III E-Centaur, the two spacecraft were almost identical—weighing 1,590-pounds and powered by Radioisotope Thermal Generators (RTGs) that provided 420 Watts of electrical power. Voyager 1 arrived at Jupiter in March of 1979 where it found three previously undiscovered moons. Voyager 2 arrived four months later in July 1979.

Jupiter, the largest planet in the solar system, has a complex structure consisting of an atmosphere about 300 miles thick composed of about 82 percent hydrogen and 17 percent helium. Beneath the gaseous layer is a transition zone about 600 miles deep where the hydrogen forms into a liquid ocean, perhaps 15,000 miles in depth, and then there is a liquid metallic hydrogen region that generates pressures on the interior of up to 3 million earth atmospheres. The center of the planet is believed to be a rather small rocky core where the temperatures may be as high as 54,000 degrees F.

Voyager images showed Jupiter's moon Io to have active volcanism, the only solar system body besides the Earth to be thus confirmed. It verified that Jupiter did indeed have faint rings of dust but not nearly as spectacular as those of Saturn.

Following their encounter with Jupiter, both spacecraft were directed on to Saturn with arrival times in November 1980 (Voyager 1) and August 1981 (Voyager 2). Voyager 1 passed 77,000 miles from the ringed planet and 2,500 miles from its largest moon, Titan. It discovered that Saturn's rings, which have seven major divisions labeled A through G, contained spoke-like structures in the B-ring and a braided structure in the F-ring.

The atmosphere of the planets can be examined when the spacecraft travels behind them and the radio signals pass briefly through their atmosphere enroute to the Earth. This period, called Earth occultation, allows the attenuation of the radio energy from the spacecraft to provide measurements of the density and height of the atmosphere.

Saturn requires almost 30 years to travel around the sun in its orbit, but, like Jupiter, it rotates rapidly on its axis in only 10 hours and 39 minutes. This high rotational rate results in a turbulent atmosphere much like that of Jupiter. Because of its distance from the Sun, Saturn receives less than one percent of the sunlight as does the Earth. Again, like Jupiter, it is a gaseous/liquid planet with perhaps a small iron core slightly larger than the Earth.

After its encounter with Saturn, where it discovered three additional moons, Voyager 2 was then redirected by Saturn's gravitation pull for a fly-by of Uranus (January 1986) and Neptune (August 1989). During this first ever flyby of Uranus, the third largest planet in our solar system, Voyager 2 passed at a distance of 50,000 miles and discovered 10 previously unknown moons and two rings.

It imaged the planet that is the only one whose rotational axis lies on its side and which completes an orbit of the sun once each 84 years. Because of the relative motion of the spacecraft and the planet, and the fact that there is little light at such a great distance from the sun, long exposures were needed to capture each image. A technique called "image-motion compensation" slowly rolled the spacecraft in an effort to keep a constant view of the subject while the aperture of the camera was open. Earth based computers then processed the received image and used enhancement methods to further adjust the clarity.

Voyager 2 passed 3,000 miles from the surface of Neptune in August 1989. This planet orbits the sun every 165 years and has eight known moons of which Voyager discovered six. The cloud covered planet rotates every 16 hours yet has the highest recorded winds of any planet—over 1,200 mph, but these are opposite the planetary rotation. Images of its moon Triton showed geyser-like eruptions of nitrogen gas several miles into its thin atmosphere. With a surface temperature of -391 degrees F, it is the coldest known body in the solar system. Neptune's rings, thought to be only ring segments, were shown to be complete.

Data collected by Voyager 1 and 2 were not limited to the encounters with the outer gas giants, as particle experiments and an ultraviolet spectrometer collected data continuously during the interplanetary cruise phase. Data transmissions continued as the Voyager Interstellar Mission passed through the edge of the solar wind's influence (the heliopause) and exited the solar system. Voyager 2 continues its travels into interstellar space at a speed of 49,000 mph away from the sun.

Ulysses: Out of the Ecliptic

The Ulysses spacecraft was launched from the Shuttle Discovery on October 6, 1990. Its mission was to reach into higher solar orbital inclinations—out of the planetary ecliptic. To accomplish this with minimal energy, the spacecraft was initially targeted to Jupiter so that it could use "gravity assist" to accelerate Ulysses out of the ecliptic plane to higher solar "latitudes." Ulysses encountered Jupiter in February 1992, and the transition was achieved to a maximum Southern latitude (inclination) of 80 degrees, which it reached in September 1994.

With the new orbit established, Ulysses provided the ability to study conditions of the Sun's environment—the heliosphere. The spacecraft is radiation-resistant and spin stabilized and weighed 814 pounds at launch. Like all deep space probes to date, it uses an RTG for electrical power. Dornier Systems of Germany built the spacecraft for ESA, which is responsible for cruise and encounter operations. NASA provided the Space Shuttle Discovery, the IUS and PAM-S upper stages to send it on its way, and the RTG.

Galileo: Jupiter and its Satellites

The Pioneer and Voyager missions to Jupiter revealed a dynamic planet, a virtual mini-solar system, which had much to teach scientists about the characteristics of early planetary formation. In cooperation with Great Britain, Germany, France, Canada and Sweden, NASA launched the Galileo spacecraft to Jupiter in October 1989, from the Shuttle Atlantis (STS-34), for an in-depth investigation of the planet and its moons.

Following release in earth orbit by the Shuttle, a two-stage solid-fuel rocket accelerated the spacecraft to escape velocity towards the first of three gravity assist planetary fly-bys. The complex journey was the direct result of the Shuttle Challenger disaster three years earlier. The Galileo mission had planned to use the Centaur stage carried aboard a Shuttle for a more direct route to Jupiter. However, following the Challenger explosion, it was deemed too risky to carry the LH2 fueled Centaur in the payload bay of the Shuttle. Because of the weight of Galileo, there was no vehicle capable of powering it directly to Jupiter—thus the need for the gravity assist technique.

The 17-foot spacecraft weighed 4,900 pounds at launch—almost one-half of it (2,000 pounds) was propellant for the onboard thrusters. A solid-state imaging camera, a variety of spectrometers and particle detectors, a magnetometer, plasma investigation apparatus, dust detectors, and a heavy ion counter were a part of the 260 pounds of scientific instrumentation. The RTGs provided 570 watts of electrical power at launch.

The route Galileo took to Jupiter first brought it into the orbit of Venus for a flyby, in February 1990, when it came within 10,000 miles of that planet. Using gravity assist for added velocity, it then swung back outward past the Earth (where it flew by within 600 miles) and then completed yet a second pass by the Earth two years later in December 1992 at an altitude of only 188 miles for the final acceleration to Jupiter.

The spacecraft had been oriented to point its high-gain antenna (which remained closed like an umbrella at the time) towards the sun to help shield the spacecraft from the direct rays which could cause overheating of the instrumentation while it was close to the sun on its passage near Venus. As it began the final leg of its journey, it was reoriented to point the big antenna back towards the Earth. However, the umbrella-like structure jammed as it was being unfolded and the antenna failed to deploy fully. An extensive effort was made for almost three years as the craft moved further from the Earth and on towards Jupiter. During this time, when it became apparent that the use of this antenna would not be possible, new communication software was developed and sent by way of the low-gain communications system. This reprogramming allowed transmission of most of the critical data, using the low speed and less capable antennas.

Galileo was unique in its attitude control in that it used a "dual-speed" structure that allowed a part of the spacecraft to spin for stability (at three rpm). The other segment remained in a fixed position (using gyroscopic control) for instrument orientation and imaging. The onboard propulsion system, built by the German firm of Messerschmidt-Bolkow-Blohm, consisted of twelve 22-pound thrusters and one 90-pound engine. All used monomethyl-hydrazine as a fuel and nitrogen tetroxide as the oxidizer.

During its journey to Jupiter, Galileo came within 1,000 miles of the asteroid Gaspra in October 1991, and passed within 1,400 miles of asteroid Ida in August 1993. In imaging Ida, a small moon was observed—the first ever discovered orbiting an asteroid. While en route, Galileo also imaged the impact on Jupiter of fragments of the Comet Shoemaker-Levy 9 in July 1994. The spacecraft arrived at Jupiter and achieved orbital insertion in December 1995.

While in orbit, the spacecraft made multiple flybys of Jupiter's moons Io, Callisto, Ganymede, Europa, and Amalthea. Many of these encounters were one-thousand times closer than those made by the Voyager spacecraft. The imaging of each provided spectacular and enigmatic views of these very different structures. Galileo probed the intensity of volcanic activity on Io and found indications of subsurface saltwater on Europa, Ganymede, and Callisto.

Galileo also carried a 750-pound 50-inch diameter, 36-inch high atmospheric probe, released in July of 1995, to enter the Jovian atmosphere six months later in December. Using an aeroshell to protect it during its plunge into the atmosphere, it decelerated to a speed that allowed a small 8-foot parachute to be deployed. The probe descended until the rising temperatures and pressures finally caused it to cease transmitting. It recorded many large thunderstorms with electrical discharges (lightning) up to 1,000 times more powerful than those that occur on the Earth.

The primary mission of Galileo extended from its launch in 1989 through December 1997. Because of its exceptional performance, the Galileo mission was given two extensions. The first, "Galileo Europa," was for the period of 1997 to 2000, when it conducted an extensive study of the Jovian moon Europa with eight close encounters. The second was called "Galileo Millennium" during the period of 2000 and 2001which allowed its observations to be teamed with the Cassini probe that arrived at Jupiter during that period. It was then deorbited to descend to its destruction into Jupiter's crushing atmosphere at a speed of 108,000 mph in September 2003 to avoid impact with (and possible contamination of) the moon Europa.

The Galileo spacecraft had traveled 2.8 billion miles in its journey and had made 34 orbits of the gas giant. The total cost of the mission was $1.39 billion with international partnerships contributing an additional $110 million. Galileo had a profound effect on our understanding of our solar system.

Cassini-Huygens: Exploring Saturn's Moon Titan

Composed mostly of hydrogen and helium, Saturn, with dozens of moons, is perhaps the most fascinating planet in our solar system. This second largest gas giant has a vast magnetosphere and a turbulent atmosphere with winds of up to 1,000 mph near the equator. These high-speed winds coupled with the convective currents from within the planet's hot interior create a series of yellow and gold bands around the planet similar to Jupiter.

However, what sets Saturn apart from any other planet is the most extensive and complex ring system in our solar system. Orbiting tens of thousands of miles out from the planet are billions of particles of ice and rock, ranging in size from grains of sand to huge boulders. These are arranged as thin discs around the planet, almost like an angelic halo, prompting the moniker "Jewel of the Solar System."

First observed in 1610 by the astronomer Galileo, who could not clearly see the rings themselves, he was unsure why the planet appeared different over the course of several years. The answer was that, depending on where the planet was in its orbit, the rings presented a different silhouette. The Dutch astronomer Christian Huygens in 1655 was the first to observe that the "distortions" viewed by Galileo were actually rings around the planet. Italian astronomer Giovanni Cassini was the first to observe four of Saturn's moons: Iapetus, Rhea, Tethys, and Dione (between 1671 and 1684). As earth based telescopes improved in their power and resolution, in 1675 he discovered the separations in the ring system of Saturn now known as the Cassini division. He was the first to suggest that the rings were composed of large numbers of small moons.

The first spacecraft to visit Saturn, Pioneer 10 and 11 in 1979, and Voyager 1 and 2 in 1981, sent back startling images of the complex ring structure that actually numbered in the hundreds. This debris may be pieces of a moon perhaps destroyed by a passing comet or asteroid. Although one of the most studied planetary structures, many questions remain about their formation.

Of Saturn's 34 known moons, Titan is larger than the planet Mercury, and one of the few moons in our solar system with an atmosphere. Scientists believe Titan, enveloped in a dense haze, may hold clues to the past environment of the Earth as it existed several billion years ago. Saturn has many smaller ice covered satellites such as Enceladus. Although only about 300 miles in diameter it is one of the brightest objects in the solar system because of the way its ice reflects the sunlight. Some moons, like Pan, Atlas, Prometheus, and Pandora, are called "shepherd moons" because they appear to herd Saturn's orbiting particles into its distinct rings. The gravitational interplay of these moons may be the cause of the twisting wave patterns in some of the rings. The moon Iapetus has one very black side while the other side is quite light in color.

To help answer some of the perplexing questions posed by previous spacecraft flyby encounters, the Cassini mission (named in honor of the Italian astronomer) was designed specifically to explore the rings and moons from orbit. It also carried a lander that was built by the European Space Agency and the Italian Space Agency and called Huygens (to honor the Dutch astronomer).

Launched by a Titan IVB with a Centaur upper stage on October 6, 1997, the 12,593-pound spacecraft carried more than 4,000 pounds of fuel for attitude control and for its entry into orbit around Saturn.

Cassini began its long trek by way of Venus—using two encounters with that "inner" planet (April 1988 at 190 miles and June 1999 at 950 miles) to provide gravity assist. It then returned to the vicinity of the Earth (passing by just 500 miles away) in August 1999 before performing a distant (6 million mile) Jupiter flyby in December 2000.

Cassini arrived at Saturn and entered a highly elliptical orbit on July 1, 2004, following a 94-minute firing of its two 100-pound-thrust engines. After traveling 2.2 billion miles, it began making scientific measurements and imaging the planet, its rings, and its moons.

The European Space Agency's 705-pound Huygens Probe was released from Cassini in November 2004. It began its descent towards Titan where it entered into the thick atmosphere in January 2005. Protected by a heat shield, Huygens survived the 12,400 mph fiery plunge and then descended under a ten-foot-diameter parachute to the surface where it arrived on January 14, 2005. It is the farthest from Earth that a man-made spacecraft has successfully landed.

Scientists were unsure whether they would find a solid or liquid surface. The probe was built to determine that aspect. As it descended under the parachute, Huygens emerged from the cloud cover at about five miles above the surface and sent back images (with a resolution of about 65 feet) of the terrain, which showed definite signs of river channels possibly carved by liquid methane. The instruments revealed the presence of complex organic-like chemistry, which reinforces the theory that Titan could harbor molecules that may have been the precursors of the building blocks of life on Earth.

At touchdown, Huygens's sensors discovered what scientists called a "muddy" surface. A single image taken on the surface revealed smooth stone like objects about a foot in diameter that scientists believe to be frozen water, as the temperature recorded was -180 degrees C. While the actual composition of the surface was yet to be determined as of January 2006, Huygens has been a resounding success with the completion of an extraordinarily complex mission.

The "New Millennium Program"

Deep Space 1 was the first mission in NASA's New Millennium Program—whose objectives were to test new technologies using a series of deep space and earth-orbiting vehicles. The spacecraft, launched on Oct. 24, 1998, aboard a Delta rocket from Cape Canaveral, was the first deep space NASA mission to focus on technology, rather than science.

Among the technologies tested were an autonomous navigation system, intelligent self-repairing software, a miniaturized radio system, a compact light-weight camera, and a new set of smaller scientific instrumentation. Deep Space 1 also had a set of secondary goals that included a flyby of the asteroid Braille and the Comet Borrelly before being turned off in December, 2001. Also aboard the spacecraft was the NSTAR electrostatic ion thruster which provided the primary method of "cruise" propulsion for the 8-1/2-foot, 1,000-pound spacecraft.

Another mission in the New Millennium program was the $212 million Stardust spacecraft launched in 1999 towards a comet named "Wild 2"—some 500 million miles away. It is believed that comets are frozen bodies of ice and dust that formed from "leftovers" at the same time as the planets about 4.6 billion years ago. The dust grains streaming off the comet may contain many of the organic molecules necessary for life.

As it flew through the comet's coma (the halo of gas and dust that extend outward from the nucleus) in January 2004 at a distance of 150 miles, it used a tennis racket-sized collector to scoop up some of the dust. (A science fiction movie from the 1950's performed a similar mission). The cosmic particles were then sealed inside a reentry capsule. The spacecraft also provided 72 black-and-white images of Wild 2 that showed craters and canyons.

As it swung to within 69,000 miles of the Earth two years after the encounter, the Stardust "mothership" released the sample return capsule. The shuttlecock-shaped capsule's fiery passage into the Earth's atmosphere in January 2006, at about 29,000 mph, was the fastest Earth re-entry of any human-made probe. Parachutes eased the final phase of the descent to a landing in the Utah desert—completing the first successful return of fragments from a deep space object. The Stardust mothership remained in orbit around the sun, and its camera and scientific instruments continued to function.

NASA sent yet another probe into the path of a comet when it launched Deep Impact aboard a Delta II rocket in November 1999. Its payload's objective was a rendezvous with a comet whose orbit lay outside that of the Earth's. On July 4, 2005, the Deep Impact spacecraft arrived at "Comet Tempel 1" and released an 820-pound "impactor" that was directed into the path of the Comet while cameras and spectrometers on the mother craft observed the resulting collision. The impactor was a battery-powered spacecraft that operated independently of the mother craft for the one day of its life. After its release, the "impactor" (which also contained a camera) maneuvered into the path of the comet. The impact produced a crater to expose the comet's primordial interior and debris that scientists examined with their remote instrumentation in hopes of better understanding both the solar system's formation and implications of comets colliding with Earth.

Efforts to bring back samples from other heavenly bodies continued with the Genesis probe launched in August of 2001 aboard a Delta II rocket. Its mission was to capture samples of solar wind particles in an effort to provide scientists with information on the composition of the Sun in order to explore the origins of our solar system.

After a short stay in low Earth orbit, its upper stage restarted to move the spacecraft towards a "halo" orbit that is called the L1 Lagrangian Sun-Earth libration point—about 239,000 miles from Earth (a point that places it in a special orbit with relation to the Earth and the Sun). Genesis reached the L1 point after a leisurely cruise of three months and fired its hydrazine thrusters for 268 seconds to insert itself into the halo orbit. Like the moon, its orbital speed was about 2,800 mph. It completed five halo orbits over the next 30 months to collect the required samples.

In May 2005, the 500-pound return capsule was released and deorbited to begin its descent towards the Earth to be recovered under parachute descent by a helicopter. The faulty installation of gravity switches designed to trigger the parachute release caused the returning capsule to crash at the Dugway Proving Ground in Utah. Scientists salvaged the remains which, it was hoped, contained atoms from the solar wind.

In January 2006 yet another probe, New Horizons, (launched by an Atlas V/STAR) was sent into deep space to what at that time was considered the ninth planet—Pluto. The probe achieved a speed greater than any previous spacecraft when it left Earth at 35,000 mph. It passed the moon's orbit just 9 hours after launch—in contrast to the Apollo manned missions which took almost three days to reach our nearest neighbor.

Pluto and a yet unnamed "tenth planet" are now the only major bodies in the solar system not visited by a spacecraft. Pluto lies in an orbit that is more inclined to the planetary ecliptic than the others. It is 6 billion miles from the orbit of the Earth where it receives only 1/1600th the solar energy as does the Earth. The spacecraft, powered by an RTG providing 250 watts of electrical power, is expected to last more than 20 years. After the Pluto encounter in July of 2015, the spacecraft will explore Kuiper Belt Objects. The Kuiper Belt is a disk-shaped cloud of debris beyond the orbit of Neptune. It is believed that short-period comets are formed there, and objects called ice dwarfs and minor planets have been detected.

Artist's impression of the Pluto - New Horizons deep space probe

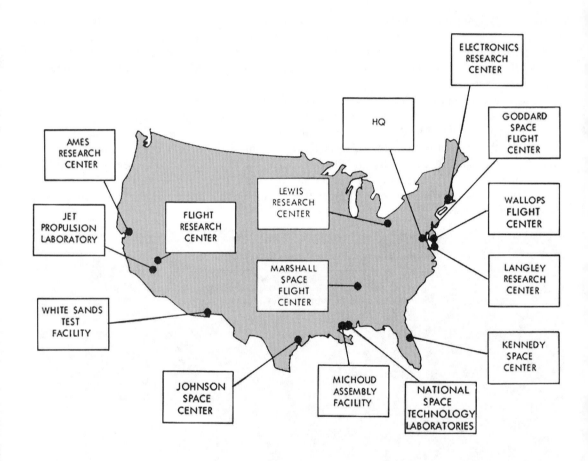

NASA centers around the United States

Chapter 36 — Competitive Partners in Space

While America has taken center stage in the exploration and exploitation of space, several other countries and partnerships have been established over the years to effectively compete and periodically collaborate—including Russia. While governments are the dominant purveyor of space projects because of their deep pockets, several private enterprises have become established. Moreover, although the financial resources of a country are far more viable than any corporation, the inability of such bureaucratic monoliths as NASA to effectively create and manage a large space project has become legend. Coupled with a Congress (in the case of the United States) that cannot seem to provide coherent and stable support for a project, the future of space based operations is uncertain.

United States: NASA

For the United States, the National Aeronautics and Space Administration (NASA) has taken the lead role in space exploration since its inception in 1958. Given a broad mandate, its operations currently include four principle organizations or mission directorates: Aeronautics provides for new technologies that involve aircraft. Exploration Systems examines both robotic and human capabilities in the aerospace environment. Science provides for investigation of the solar system, including the Earth. Space Operations involves those activities in support of on-going programs such as the International Space Station and its supply infrastructure.

NASA Headquarters is located in Washington, D.C., where it provides management oversight of a vast array of space flight centers, research centers, and other installations.

NASA's $16 billion annual budget represents about 7/10 of one percent of the national budget. Its staffing has stabilized at about 21,000 direct employees and more than 200,000 contractor employees. This compares to the high point of 34,000 civil servants during the peak years of the Apollo project in the mid-1960's and more than 400,000 contractors. (The staffing levels and budgets noted here are approximate values that vary with the programs being supported)

To accomplish its goals, NASA has evolved dedicated centers that have developed a high level of specialization. They continually examine and push the state-of-the-art in aerospace sciences and technology.

The Ames Research Center in Sunnyvale, California, a part of the predecessor NACA since 1939, has a series of large wind tunnels for simulating atmospheric flight environments. Being in the "silicon Valley," it has also become a prime developer of NASA's information technology with expertise in supercomputing and networking. Employing 4,000 people, it has an annual budget of more than $750 million.

Dryden Flight Research Center is the flight research hub located in Edwards, California. An outgrowth of the Muroc High Speed Research Station that was established by NACA in 1946, it is most noted for its collaborative research with the military in the X-series aircraft, as well as being the primary recovery point for Shuttle flights.

Glenn Research Center was created in 1941 as the Lewis Research Center in Cleveland, Ohio. It was renamed in 1999 to honor John H. Glenn, the first American to orbit the Earth. Specializing in Exploration Systems, to include micro-gravity and the biosciences, the center employs 3,300 people. Its budget is more than $1.3 billion.

Goddard Space Flight Center was established in 1959 as the first NASA center. Located in Greenbelt, Maryland, it is responsible for operating the space flight tracking facilities among its many responsibilities. With a budget of almost $3 billion, Goddard employs 10,000 people.

Jet Propulsion Laboratory in Pasadena, California is one of the best known of the NASA centers. Formed in the late 1930s as an outgrowth of the rocket experiments being performed by the California Institute of Technology, JPL has played a pivotal role in the American satellite and deep space programs since 1958. It is NASA's lead center for robotic exploration of the Solar System.

Johnson Space Center has been the primary location for Mission Control of manned spacecraft operations since 1965. Named for President Lynden B. Johnson, who was an early supporter of the space program, JSC employs 16,000 persons and has an annual budget of over $4 billion.

Kennedy Space Center provides the primary launch operations for the vast majority of America's space operations. Its annual budget is more that $1 billion and it employs 14,000 civil servants and contract personnel.

Langley Research Center is the oldest of the predecessor NACA facilities, having been founded in 1917 as the nation's first civilian aeronautics laboratory. More than half of its operations support aeronautics and atmospheric sciences as opposed to space activities. The facility, with an annual budget of $400 million, employs 4,000 people.

Marshall Space Flight Center in Huntsville, Alabama, is primarily responsible for space transportation and propulsion systems. Named for Army General George C. Marshall, it was originally the Army Ballistic Missile Agency and the home of the former Peenemünde rocket team under Wernher von Braun.

The Stennis Space Center was originally constructed in the early 1960s to manufacture the first stage of the Saturn V Booster. Today it continues its specialization with the fabrication of the Space Shuttle External Tank (ET). It was named for Senator John C. Stennis an early congressional supporter of the space program. Total workforce is about 4500 while the annual budget is $650 million.

NASA's biggest problems are not related to science or technology. As a vast bureaucracy, it suffers from the inability to manage its own administrative and political structure. This, in turn, affects its ability to make technical decisions. The chaos at NASA is the result of almost 50 years of political pressure, technology development challenges, and budgetary considerations. These aspects of NASA were its strong points in the early days under its second administrator, James Webb, but, following the Apollo project, most authorities agree that there has been a steady downward spiral.

The Government Accounting Office (GAO) has consistently been critical of NASA's decision-making process—especially with respect to its overall funding estimates and schedules. One GAO report pointed out that NASA's own financial reporting system, into which it put over $180 million, never proved usable.

NASA's handling of the safety issues that allowed the Challenger and Columbia disasters are now a matter of record. The inability to avoid simple but costly and embarrassing mistakes, such as the Mars Climate Orbiter and Mars Polar Lander, again highlights its lack of awareness.

The GAO admonished NASA for its failure to recognize the appropriate technology maturity levels before attempting to implement them. They also noted that the overall project management flow failed to stabilize a design before attempts to fabricate and assemble a project. Finally, the traditional project management triad of cost, schedule, and quality deliverables has consistently been jeopardized by the failure to take advantage of appropriate design reviews.

Representative Mark Udall (D-Col), the ranking member of the U.S. House Subcommittee on Space, and Aeronautics (Committee on Science) recently commented, *"The GAO report cautions that several of NASA's major acquisitions have been marked by cost, schedule, and performance problems, yet, the challenges NASA faces in the future are likely to far exceed those it has faced in the past. I want NASA to succeed on all of its important missions in science, aeronautics, and human space flight."* Many of NASA's critics have called for the agency to return to its former role in research and leave the development projects to the private sector.

Former NASA Administrator Dan Goldin once said, "We want failures. If you don't, you don't have lofty enough goals." However, NASA critics point out that it is one thing to have failures attributable to pushing the state-of-the-art; it is quite another to fail because of "simple mistakes" involving rubber and foam materials and metric conversion. With the Shuttle, NASA turned aside valid safety issues, yet during the recovery process, became risk-averse—spending literally billions of dollars on issues that could and should have been resolved years earlier for far less money.

There is no doubt that NASA and its associated contractors have accomplished significant, almost unbelievable, feats of exploration in space. The ingenuity and resourcefulness of mankind in the space sciences has had a profound affect on our daily lives. However, the problems of managing a large organization must be addressed.

As alluded to in President Eisenhower's farewell address in 1961, "big science" projects have loomed large over the landscape. Over the intervening years, the scientific community has gained what many consider "undo influence" because of the spectacular technological events of the 20th century.

"Science" has often been put forward as a cure-all for many of the world's problems. However, the failure of science to produce alternative energy sources or to create the vaunted "Strategic Defense Initiative" (Star Wars), after decades and billions of dollars, has impaired public confidence. Some large scientific projects, such as the defunct "superconducting supercollider," were seen by the populace and the politicians as a diversion for the intellectual curiosity of the elite scientific community without a legitimate return for the tax monies invested. Some have termed these types of projects as "white-collar welfare." As a result, curiosity driven science for the primary benefit of academics has continued a slow and steady decline in importance to the average person and the elected representatives in Washington.

NASA has likewise fallen prey to the limitations of science as well as its inability to manage large bureaucracies. Its incapacity to "think outside the box" has caused many a proposed program to suffer an early death. While most critics agree there is a role for NASA in the future of space travel, NASA must continually demonstrate that it can establish and control effective research and development programs.

European Space Agency: ESA

While the Soviet Union had presented America with a period of hostile competition in space beginning in 1957, other nations (France and the United Kingdom) looked to that arena as an object of scientific interest as well as commercial opportunities and national pride. The Second World War had drained much of the European scientific talent to the United States and to the Soviet Union, and the European economic recovery extended well into the 1950s. Despite the early efforts of the French and British, most Europeans recognized that they would not be able to compete on an equal basis with the two superpowers. However, following the impetus of the first Sputniks, they agreed in 1958 to establish two agencies: one for developing launch vehicles (European Launch Development Organisation — ELDO) and another to address scientific research (European Space Research Organisation—ESRO). The organizational structure was completed in the early 1960s, and several satellites were successfully launched—but all used American launch vehicles.

ESRO was merged with ELDO to form the European Space Agency (ESA) in 1974 with eleven member nations and launched its first significant scientific mission in space in 1975. Many of the ESA projects were performed in cooperation with NASA—primarily for the launch vehicle. In order to avoid its dependence on NASA, which was moving to use the Shuttle exclusively, ESA constructed a series of rockets, called Ariane, for unmanned scientific and commercial payloads. Ariane 1 and its successors, including the Ariane 5, established ESA as a serious competitor in the commercial space-launch market with 25 successful launches by 2005. ESA's spaceport, Kourou, in French Guiana on the South American coast, was chosen for its ability to efficiently launch satellites into equatorial orbits—the basis for geosynchronous applications.

In the 1980s France pressed for an independent European manned launch vehicle. In 1985 ESA decided to pursue a reusable spacecraft model and in 1987 began a project to create a mini-shuttle named Hermes. The spacecraft was a small, reusable vehicle that was to carry 3 to 5 astronauts and 3 to 4 metric tons of payload into LEO for scientific experiments. With a total maximum weight of 21 metric tons it would have been launched with the Ariane 5 rocket which was being developed in parallel. The design phase concluded in 1991, but production was never implemented, in part because of the fall of the Soviet Union. With the opening of the Iron Curtain, ESA looked to a cooperative effort with Russia to built the next-generation human space vehicle. The Hermes program was cancelled in 1995 after about 3 billion dollars had been spent.

In recent years ESA has partnered with the Russians to launch the R-7/Soyuz from Kourou in French Guiana. While awaiting the development of a reusable Russian spacecraft, now called Kliper (the Russian equivalent of the American Crew Excursion Vehicle), ESA has paid for its astronauts to fly on the Shuttle and Soyuz spacecraft. Europe will probably finance much of the development costs of the Kliper (an estimated 3 billion dollars) and participate in its construction. It will be launched from French Guiana and Baykonur.

The ESA budget, coupled with the independent space science allocations of the member countries, has risen steadily and reached almost $5 billion dollars for 2005 (as compared to NASA's $16 Billion).

Twenty-two percent of the budget is currently used for launch vehicles, with 20 percent for human space flight. The three largest ESA contributors are France (29 percent), Germany (23 percent) and Italy (14 percent).

Russian Space Agency: Roskosmos

The Russian Space Agency (RKA-Roskosmos) was formed, following the economic and political fall of the former Soviet Union, to provide centralized control of Russia's civilian space program. Although historically plagued by lack of funding, the outlook for the future of the Russian space program has improved considerably with the Russian government approving a $15 billion dollar budget for the 2006-2015 period starting with almost one billion for 2006—up from $200 million in 2001. However, this is considerably less than the $104 billion, 13-year long-range plan for NASA.

Although RKA employs only about 300 people, much of the work is performed by contractors such as Energiya Rocket and Space Complex, which owns and operates the Mission Control Center in Korolev. The military equivalent of the RKA is the Military Space Forces (VKS). The VKS operates the Plesetsk Cosmodrome launch facility, while RKA and VKS share control of the Baikonur Cosmodrome.

With NASA struggling to return the Shuttle to flight, the Russians now enjoy an international reputation of providing relatively safe manned spaceflight and inexpensive access to the ISS. While it had been almost strangled out of space by lack of funding, the next decade may see a renewed oil-based Russian economy that could propel it into a position of leadership in space technology.

China National Space Administration

With the orbiting of Shenzhou 5, on Oct. 15, 2003, with "taikonaut" Yang Liwei aboard, China became the third country to send man into space.   The Chinese astronauts are known as "yuhangyuan"—travelers of the universe. Their English nickname "Taikonaut" is derived form the Chinese word for space—Taikong. The launch followed four unmanned test flights of the Shenzhou spaceship between 1999 and 2003. Yang was followed two years later with two taikonauts, Fei Junlong and Nie Haisheng on October 12, 2005, aboard Shenzhou-6 spacecraft for a five-day mission.

Recognizing the importance of "public relations," the Shenzhou 6 mission was broadcast live on national Chinese television. It was reported that the loudspeakers set up for visiting dignitaries viewing the launch serenaded them to the theme music from Battlestar Galactica. Yang's flight however, was launched without any publicity.

Although much of its starting technology was imported from Russia and the United States, China is no stranger to innovative engineering. The configuration of the Shenzhou follows the basic design of the Soyuz. It consists of an orbital module that contains scientific equipment, the crew-carrying ascent/decent module, and a service module with attached solar panels to provide electrical power, attitude control and orbital maneuvering. Because Shenzhou is a close replica of the Soyuz, it has the ability to carry a crew of three. It is launched by the Long March 2F booster from China's Jiuquan Space Launch Center in northwestern Gansu Province.

One difference between Soyuz and Shenzhou is the use of the "orbital module." As it is equipped with its own set of solar panels, it becomes an independent unmanned satellite after separation from the descent module and continues in orbit, providing scientific data for several months.

As with the former Soviet Union, China's initial efforts were hindered by its poor economic conditions. However, this is in the process of changing as America has become an important trade partner— boosting the economy significantly over the past decade. While China has been actively pursuing the exploration of space with unmanned satellites since 1970, the manned project did not receive state approval until 1992.

Chinese officials have implied that their space program may include space station construction, and lunar exploration by 2020. It is possible that China's entry into space will have far-reaching effects. It may be one of the reasons that President Bush initiated a new vision for space exploration that returns the U.S. to the Moon and promises manned expeditions to Mars. It will be interesting to see if China's move into space represents another iteration of the space race.

The essence of the Chinese space program was created in the late 1950s by a U.S. educated visionary—Tsien Hsue-Shen. Tsien was a co-founder of the Jet Propulsion Laboratory in Pasadena, California, in the 1940s. At the end of World War II he was considered one of the few "rocket scientists" capable of understanding the significance of the technology of the V-2 created in Nazi Germany.

In 1950 Tsien's security clearance was revoked because of his close ties with the Communist Party. When it became clear that he would no longer be allowed to work on U.S. defense programs, he returned to China in 1955. Tsien led China's space program until his retirement in 1991.

Launching an unmanned satellite is no longer considered a significant technological achievement and can now be accomplished by relatively impoverished countries along with the creation of an ICBM. Manned spaceflight, however, is an order of magnitude greater in effort both technologically and financially. China's commitment to a manned program says much about its political leadership as well as its financial strength.

As China has the 40 years of observable experience from the Soviet and American programs, they will avoid some of the pitfalls that cost lives and money. However, whether they have the will to endure a long and costly program remains the question. If China moves towards democracy, the desires of its people could undermine an extensive and costly space exploration program.

Judging from the wall-to-wall "Made in China" products marketed in the United States by megachains like Wal-Mart, China may become a formidable economic giant to rival the United States. If that potential is achieved, it is conceivable that the huge sums necessary to create a vast space program might propel China into technological leadership. Like the former Soviet Union and the United States, China recognizes that to be acknowledged as a world power it must participate in space exploration and its commercial exploitation.

Japan Aerospace eXploration Agency

The Japan Aerospace eXploration Agency (JAXA) is Japan's aerospace organization with annual funding of about $2 billion. This compares with the Chinese and Russian budget of about 1 billion per year.

In 1965, Japan began development of a satellite launcher derived from the solid-fuel Lambda-3 sounding rocket (similar to the U.S. Scout rocket). The four-stage Lambda-3H was capable of orbiting up to 220 pounds. Four unsuccessful attempts were made between 1966 and 1969 from Uchinoura near Kagoshima on the island of Kyushu until success was achieved in 1970. Following this launch, use of the Lambda was discontinued.

Since that time, Japan developed a very advanced technology in both satellites and launch vehicles. Two series of rockets were created, designated the N- and H-series, with more than 30 launches over the past 30 years. The rockets are launched from the small island of Tanegashina off the southern tip of the main island of Kyushu.

The high-energy liquid-hydrogen-powered (LH2) two stage H2-A, which can lift two tons to GEO, provides the current heavy lift capability. As with large rockets of most of the other nations, the H2-A augments the liquid fueled core with solid-fuel rockets attached to the periphery to provide added power during the early burn of the first stage.

The 174-foot-tall H-2A rocket, built by Mitsubishi Industries is an outgrowth of its predecessor, the H-2, but is more capable and less costly. Japan had attempted to use the H-2 to sell launch capabilities to other nations, but the rocket was too costly and unreliable to compete effectively in the international market. Japan is hopeful that the H-2A will address those problems.

Recently launched satellites include applications for basic science, natural resources, weather, and military reconnaissance. The later is a result of the provocative developments made by Japan's neighbor to the west. North Korea has been testing long-range ballistic missiles and is known to have a nuclear warhead development program. These spy satellites represent a significant portion of Japan's space budget. In justification of its multi-billion dollar spy program, Japan's Prime Minister Junichiro Koizumisaid, said, "It's very unfortunate, as our country needs to boost intelligence capability to increase readiness for… national security."

While enduring the criticism of the international community, Japan is attempting to position itself as the technological leader in the Far East. With the world's second largest economy, that position would not have been questioned until the advent of the Chinese manned space program. In some respects, the competition between the two Asian countries is reminiscent of the space race of the 1960s between America and the Soviet Union.

Japan received a significant setback in its scientific program when its Hayabusa spacecraft, on a mission to collect samples from an asteroid, failed.

India Space Research Organization: ISRO

India, surrounded by the often-militant nations of Pakistan and China, has worked diligently to maintain a technology base for defense. To this end, its development of missiles has led it to a capability to launch unmanned satellites. The Indian Geostationary Launch Vehicle (GSLV) is capable of sending 5,500 pounds to LEO. The first stage is a large solid rocket augmented by four liquid-fueled strap-on boosters that have a longer burn time than the solid center stage. The second stage also uses storable liquid propellants. The third stage is the Russian KVD-1M, a LOX engine originally developed for the Proton launch vehicle. It will be replaced by an Indian built engine in the near future.

India's first astronaut, Rakesh Sharma, was launched aboard the Soviet Soyuz to the Salyut 7 in April 1984. Indian-American Kalpana Chawla was killed on her third Space Shuttle mission along with six others in the Columbia Shuttle disaster in February, 2003.

Despite its achievements, India is undecided as to the future of its space program. Although pressured by the presence of a Chinese manned program, India is hesitant to invest more of its small discretionary budget in a venture whose return for the Indian people is nebulous. However, India is eager to use its space technology to benefit its national development goals. As its economy continues to grow, it may find more reasons to expand its role in the international space community.

Private Ventures

SpaceX is one of several private corporations developing relatively large expendable rockets for both commercial and military satellite launch capabilities. The economic failure of the Shuttle has opened the door to many more companies seeking to use the vast technology base that has been established. Unlike Boeing and Lockheed Martin, SpaceX is developing a reusable family of launch vehicles (the Falcon series) designed to reduce significantly the cost of access to space. The capabilities range from the light (Falcon 1), to medium (Falcon 5) and heavy lift (Falcon 9) for low earth orbit, geosynchronous orbit, and planetary missions. The numeric designation defines the number of engines in the recoverable first stage.

On a smaller scale are the efforts of several companies to develop access to the space environment for tourism. To this end, the Ansari X-Prize has provided a financial motivating factor. A $10,000,000 award was offered in 1996 by a consortium of benefactors to encourage the space tourism industry. In much the same way as the Ortiz Prize offered in 1919 sparked competition for the flight by Charles Lindbergh from New York to Paris in 1927, the Ansari X-Prize was established to reward the first privately financed team who could build and launch a spacecraft to an altitude of 62 miles. The altitude was chosen by the Fédération Aéronautique Internationale (FAI) as the point at which space officially begins for the purposes of defining activities that represent records. Initially called the "X Prize," it was renamed in 2004 in recognition of the contributions from Iranian-born entrepreneurs Anousheh Ansari and her brother-in-law Amir Ansari, who now reside in the United States. The prize required that a manned spacecraft, capable of carrying three people, had to be launched twice within a two week period—obviously returning safely each time.

Scaled Composites, a company headed by the renown aircraft designer Burt Rutan, announced publicly in 1996 that they would compete for the prize. Among the more than 20 companies who registered, the Scaled Composites' effort was financed by Microsoft co-founder Paul Allen and build in secret by a bevy of volunteers. The completed vehicle was rolled out in April 2003. The spacecraft was one of two parts—the first being a twin jet powered mother ship (the White Knight) that carried the craft for an air launch. The 8,000 pound winged spacecraft was powered by a 17,000-pound-thrust hybrid motor that used a solid-fuel core and a liquid oxidizer that produced Specific Impulse 250 of seconds.

Launched from the White Knight carrier at about 40,000 feet, the ship climbed into the vertical attaining a velocity of just over Mach 3 during the 87 seconds of powered flight. It then coasted to its maximum altitude of 65 miles where it executed a metamorphosis into what is called the "shuttlecock" mode. The twin tails of the ship rotated to present a high drag configuration to assure that the reentry was aerodynamically controlled and the speed build-up was minimized.

When the ship reached the denser layers of the atmosphere, the tail was transitioned back to a normal attitude, and it performed as a glider with the pilot executing an unpowered landing in the conventional manner. Sixteen unpowered and powered test flights were made to test the various systems before first the X-Prize attempt on September 29, 2004 with pilot Mile Melvill at the controls. The second flight, which secured the prize, was made one week later on October 4, 2004 by Brian Binnie.

Launch Abort System --
emergency escape during launch

Crew Module –
crew and cargo transport

Service Module –
propulsion, electrical power, fluids storage

Spacecraft Adapter –
structural transition to launch vehicle

Early artist's impression of the Project Orion Crew Exploration Vehicle

Chapter 37 — The Next Generation of Manned Spacecraft

Shuttle Replacement Technology: An Elusive Goal

When commercial air travel is considered, the first 50 years of flight (from the Wright Brothers to the Boeing 707), saw aircraft speed increase from 40 mph to 600 mph. However, the second 50 years have not seen any increase! In fact, airlines have slowed their cruise speeds slightly to conserve fuel. While the British-French Concord is the exception, that aircraft (now retired) was not widely accepted, and only twelve were built for a very special clientele. Simple economics, the value of time, and a concern for the environment have put progress on hold in this portion of the aerospace arena.

In the early 1980s, then President Ronald Reagan proposed that the United States embark on building not just a successor to the Concord but a revolutionary craft. He saw it as one that could, "by the end of the next decade, take off from Dulles Airport and accelerate up to twenty-five times the speed of sound, attaining low-earth orbit or flying to Tokyo within two hours....." Coined "The New Orient Express," the craft was intended to fill two roles—a super high-speed transport for long distance flight, and a single stage to orbit spacecraft. The implications of such a craft are significant.

From a military point of view, the roles of reconnaissance and weapons delivery stand out—as a successor to the SR-71 and a follow on to the B-2. This aspect appeared quite prominent when the initial funding showed the Defense Research Projects Agency (DARPA) picking up 80 percent as a part of a project called "Copper Canyon" that ran through 1985. NASA, presumed to acquire a successor to the Space Shuttle, did not want to devote any significant part of its budget to the program. A single-stage-to-orbit vehicle that could operate in much the same way as an airliner was also a key part of the Strategic Defense Initiative (SDI) for a space based ballistic missile defense system. However, just as the Space Shuttle was not able to achieve the "airliner turnaround" promise, the NASP seemed destined not to achieve it either.

The primary enabling technologies involved—the Scramjet and the thermal protection system—when coupled with the logistics of maintaining and launching a hydrogen fueled vehicle appeared to doom the project. While the concept of a single-stage-to-orbit craft is still appealing, like the Shuttle, its complexity, cost (projected at $20 to $30 billion), and volatility seem beyond reason.

Further analysis reveals a gap between the velocity provided by the air-breathing power plant, and orbital velocity. A small second stage rocket would be required. In addition, the promised technology of thermal protection and the efficiencies of the Scramjet never materialized—at least publicly. The myth persists that the military have continued the push and that the breakthrough has been made. The presence of the super secret Area 51 in the Nevada desert continues to reinforce credibility to the presence of a hypersonic aircraft called "Aurora."

The NASP never advanced to the building of hardware despite the designation X-30 given to the craft. The Hypersonic Systems Technology Program (HySTP), was initiated in 1994 to consolidate the research in hypersonic technologies generated by the National Aerospace Plane (NASP) program and others into a long-term development effort.

Another of NASA's efforts to develop a follow-on to the Shuttle grew out of its Space Launch Initiative program of the mid-1990s that continued the quest for a Reusable Launch Vehicle (RLV). The X-33 was a cooperative project between NASA and the Lockheed Martin Aeronautics Company to build a technology demonstrator for a single-stage-to-orbit vehicle. Announced in 1996 by then U.S. Vice President Al Gore, the unmanned X-33 was scheduled for a series of suborbital test flights by 1999 but ran into several technical difficulties that continued to escalate the costs and extend the schedule.

The X-33 design was based on a lifting-body shape with two innovative "linear Aerospike" rocket engines and a metallic thermal protection system. The lifting body configuration was to be supplemented by abbreviated horizontal control surfaces coupled with short vertical fins. The vertical take-off, single-stage-to-orbit, fully reusable craft struggled against several technologies as it fought to achieve a significant lowering of the cost to low earth orbit (LEO). Its proposed lightweight structure and components emphasized the importance of weight control to achieve performance that had not been previously attainable. If successful, Lockheed Martin was poised to grow the X-33 into a commercial product called the VentureStar—a true single-stage-to-orbit airliner to space that would be flying by 2003.

Like its predecessor, the X-33 ran into numerous technical difficulties, not the least of which was the Aerospike engine. The ability to achieve a stable gliding platform of the wedge shaped craft at the wide range of speeds was also a major challenge as its weight continued to grow.

The X-34 unmanned vehicle was yet another attempt to demonstrate reusable technology. More conventional in appearance than the X-33, having short stubby wings that spanned 27 feet, the 58-foot-long ship was to reach speeds of Mach 8 and altitudes of 50 miles. A joint project between NASA and Orbital Sciences Corporation, the X-34 was a test bed for a full sized vehicle that could be supported by a much smaller ground crew for servicing and that could provide a two-week turnaround time between flights.

A review of the X-33 and X-34 programs in 2000 determined that there was no end in sight for several of the technology intensive problems, and the costs were beyond the ability of NASA to absorb. In 2001 NASA discontinued both programs after a combined investment of $1.5 billion. These and other projects seemed to point out that a single-stage-to-orbit vehicle was not yet possible as a follow-on to the Space Shuttle. There would be no "spaceliner" to replace the Shuttle—the VentureStar project would not fly—at least not in the foreseeable future.

Crew Exploration Vehicle—CEV

The situation with a Shuttle replacement began to take on critical proportions with the failure of the X-30, X-33, and X-34 projects. In the first few years of the new millennium, the aging Shuttle continued to be pressed to shoulder the burden in support of the ISS. But the Columbia disaster in February 2003 brought the problem to a head.

The Crew Exploration Vehicle (CEV), announced by President George W. Bush in January, 2004, was a response to the Columbia disaster. He stated that the United States would "develop and test a new spacecraft, the Crew Exploration Vehicle, by 2008, and… conduct the first manned mission no later than 2014." Its primary purpose would be to ferry crews to the Space Station after the shuttle is retired. But it would also be a possible component of the proposed lunar and planetary exploration program.

With respect to capability, the Shuttle is a hard act to follow with its ability to carry the crew, payload, and the primary propulsion system—and return it like an airplane. But the cost of that capability and the lack of a crew abort system caused NASA to revert to a ballistic recovery design for America's next (and fifth) generation spacecraft. Currently being referred to as the Crew Exploration Vehicle (CEV), the spacecraft, named Orion, will probably be cone-shaped in design, similar to, but larger than, the Apollo and will be reusable for up to 10 flights. To keep the development costs to a minimum, Orion will use many of the on-board Shuttle systems and an expendable launch system to achieve LEO. Orion will be capable of returning to either land or water. The CEV contractor, Lockheed Martin, was selected by NASA in 2006 after intense competition with the Northrop Grumman and Boeing team.

Orion will have various configurations that could carry from 3 to 6 persons to Earth orbit—or beyond. To accommodate the various configurations, a Service Module attached to the CEV will be cylindrical in shape (like Apollo) and provide for the various life support, communications, orbital maneuvering, and attitude control functions. However, in place of the Apollo-like fuel cells, the CEV will use deployable solar panels for electrical power generation. An abort system (an escape tower as in Mercury and Apollo) would allow the crew to survive a launch vehicle malfunction during ascent.

By returning to a more modular approach to the overall configuration, Orion can grow with future missions planned for lunar or Martian exploration. Using a combination of earth orbit rendezvous (EOR) and Lunar Orbit Rendezvous (LOR), it could rendezvous and dock with a Mission Module such as a Surface Access Module (SAM) for lunar or Martian exploration and an Earth Departure Stage (EDS) to achieve escape velocities.

The SAM would be docked to the forward end of the CEV for extended periods in space such as a lunar mission, and would provide for the extra consumables and living space. The EDS would be attached to the aft end. The configuration would be strikingly similar to the Apollo CSM/LM. A lunar mission would therefore be composed of modules launched independently by at least three expendable vehicles.

By stacking all the stages in tandem rather than in parallel (as is now the case with the Shuttle), the CEV concept would avoid debris from one stage impinging on another. It would also allow for more abort scenarios.

The CEV development has been proposed as an "interactive spiral" in which five distinct phases would move the CEV through progressive capabilities. Exploration Spiral One (CEV Earth Orbit Capability) would be completed by 2014, and would validate the basic systems in LEO while serving as the primary crew transportation system to the ISS. Exploration Spiral Two (Extended Lunar Exploration) would be ready by 2018 and provide for returning a four-person crew to the moon for short stays of up to two weeks. Exploration Spiral Three (Long Duration Lunar Exploration), sometime after 2020, would provide routine manned missions of up to several months on the lunar surface. It would also be used to validate technologies and operational techniques for a manned Mars mission. Exploration Spiral Four (Crew Transportation System Mars Flyby) would perform a Mars flyby mission using improved elements of the CEV by 2030. Exploration Spiral Five (Human Mars Surface Campaign), sometime after 2034, would send a manned mission to Mars.

President Bush's budget request for Fiscal Year 2005 included $428 million for Project Constellation (the name of the iterative development program). The plan provides for $6.6 billion over the next five years to develop the CEV. The development cost of the CEV through 2015 is estimated at $15 billion while the follow-on Mars missions could be as high as $230 billion.

The launch vehicle for the CEV, currently named ARES I, is projected to be derived from Shuttle components. A single SRB as a first stage and an LH2 upper stage will provide the propulsion to orbit. The larger ARES V, with a payload equivalent to the Saturn V, would be used to lift the Mission Modules into LEO for rendezvous with the CEV.

The Russian Kliper

In early 2004 the Russian Energiya Rocket and Space Corporation revealed the Kliper manned spacecraft project which it said had been under development since 2000. Energiya is but one of three Russian companies vying to build their third generation spacecraft which would be reusable; the Khrunichev Space Center and the Molniya Research and Production Association are the other competitors.

The proposed Kliper (the English translation of the Russian word for Clipper), is designed to replace the now 40-year-old Soyuz—and reduce the cost of manned space flight. As with the CEV, it is a partly reusable craft that would have a disposable service module to house some of the support systems. The current estimates are that each Kliper would be capable of 25 flights before retirement.

The Kliper would employ aerodynamic lifting body characteristics that would allow it to perform reentries that expose the occupants to less G forces than the Soyuz. The short-winged Delta planform would land like the Shuttle. The Kliper provides for a launch escape system that would allow it to abort during the powered ascent.

The Kliper would be primarily a manned spaceship, carrying up to six cosmonauts, with payloads limited to 1,600-pounds of equipment, to LEO. This capability highlights the change in space transportation philosophy that has also been accepted in the American planning for the CEV—the cargo would be separated from the human crew and would use different launch vehicles. The Kliper, when operational, will provide up to 15 days of independent operation and will have an on-orbit life of up to one year when docked to the ISS. This is a clear advantage over the Soyuz which has an on-orbit life of only 90 days. As with the CEV, there have been some thoughts circulated that the Kliper may be able to expand to a manned lunar or planetary role.

During reentry, Kliper's lifting body design would allow some maneuvering, and, with the advances in computer guided de-orbit techniques, it would permit targeting to any pre-selected major airport (i.e. 10,000 foot runway) in the world. Its relatively small size (compared to the Shuttle) would allow it to be easily transported back to its launch base in any of the large cargo carrying aircraft such as the Antonov 124.

The ESA has shown interest in collaborating with the Russians to produce the craft, which would be similar in size and configuration to the cancelled European mini-shuttle Hermes, but there has been no commitment. The biggest impediment to ESA participation has been that the Russians want the ESA to simply be an industrial contributor—and not involved with the design.

In its initial design, the Kliper would weigh about 30,000 pounds (beyond the lifting ability of the R-7 which has a current limitation of about 16,000 pounds). This would have required the use of the new Angara-A3 or the Zenit. But in an effort to become less reliant on the new designs, the RKA reworked the Kliper so that it could use the R-7 derivatives.

In the redesign, the Kliper would be a part of a two launch assembly—dividing the weight between the two components. This approach would allow the Russians to retain the ability to use the venerable R-7, which is now referred to as the "Soyuz vehicle," for launch. The second part would be a true "space tug," being called the Parom. The Parom would be placed in orbit to service several successive Klipers. The Kliper would rendezvous and dock and then use the Parom to provide orbital maneuvering and to boost the Kliper to higher orbits. The Parom would remain in orbit until it exhausted its propellant supply, and then it would be deorbited to destruction—being replaced by another Parom tug.

The Parom would essentially replace the Progress spacecraft, but like a true "tug," it would not carry any supplies itself. Any Russian or foreign launch vehicle could rendezvous with a Parom. The Parom would be able to maneuver cargo modules weighing up to 60,000 pounds—twice the mass of the largest station sections carried into orbit by the Shuttles.

Under the current schedule, the Kliper would launch by 2010—perhaps not an unplanned timing with the retirement of the Space Shuttle. However, it may be another two years after that date before the Kliper can be operational.

Both the Kliper and CEV represent the next generation of manned spacecraft, and the United States and Russia are competing to see who can garner the most support from among the European communities and Japan. The competition could be intense as there is a lot at stake in the aerospace business community as well as the desire by both the Russians and the U.S. to share the financial burden with others. Russia has indicated that it will develop the Kliper even if it cannot find a partner. Many believe there is room for both programs as a safeguard against another "Achilles heel" syndrome as evidenced by the inadequacy of the American Shuttle Program.

It is important to consider the historical aspect of both the CEV and Kliper—that the Shuttle was the only manned access to space for the United States for a period of 30 years, and Soyuz will have served the Russians for more than 40 years. With the development costs being so high, these new spaceships will probably serve well into the mid-century. They will not only support the ISS but its successor as well as any projects for a return to the Moon or for the first excursions to Mars.

Chapter 38 — Advanced Propulsion Systems

As mankind looks toward the future of spaceflight, there are a variety of new and innovative propulsion methods currently within his technological grasp. These may be categorized by one of three regions of space in which they are most suited to operate. The first region is from the surface of the Earth into Low Earth Orbit (LEO). These sources of power must have the ability to move large payloads and need to generate literally millions of pounds of thrust over relatively short periods (minutes) to overcome the force of gravity and transit through the atmosphere. This is the traditional zone of operation that has characterized contemporary rocketry.

Once in LEO, the atmosphere no longer presents frictional drag and a second form of power can accelerate at less than 1-G over longer periods of time (months or even years). This second region can be divided into two sub-areas: the inner solar system (out to the planet Mars) and the outer solar system (beyond Mars). Operating within the inner solar system allows access to the Sun as a source of power; beyond Mars, the effects of the Sun diminish considerably.

Spacecraft require much higher escape velocities when exploring the outer planets due to their great distances from the Sun. Previous missions to this region of the solar system have typically used gravity assist from the inner planets. This technique was needed to accelerate larger spacecraft to the higher velocities with minimal launch vehicle power and has added literally years to the mission, as was the case with Voyager and Galileo. This added time also means larger quantities of consumables, such as attitude control fuel, and it presents a higher risk of systems failure over the extended flight times. These spacecraft are also limited to the use of radioisotope thermoelectric generators for electrical power generation because they travel too far from the sun to use solar power. If man is to personally explore the solar system and beyond, new methods of propulsion must be developed to reduce the transit time and increase the payload.

New Life for Chemical Propulsion?

While it is recognized that conventional chemical propulsion rockets are limited to a Specific Impulse values that are dwarfed by other more exotic forms of spacecraft propulsion, their engineering is closer to the current state-of-the art (present less technical risk), and they can provide very large thrust profiles. One promising improvement in chemical rocket construction is the Aerospike engine. Aerospikes have been around for almost 50 years but have never been able to achieve the high thrust levels beyond the 200,000 pounds needed for first stage applications to LEO—the very application for which they are best suited.

Designing an engine nozzle is a compromise of several factors—most notably the altitude at which the rocket will operate. Engines that power the first stage of a rocket represent the most difficult compromise. They must provide for the initial liftoff at essentially sea level and take the rocket into areas of rapidly lowering atmospheric pressure. Because the expansion ratio of the nozzle is calculated to provide the optimum performance based on the surrounding atmospheric pressure, the nozzle of traditional rocket engines can operate at peak efficiency only in a relatively narrow altitude range.

The expanding nature of the exhaust can be observed as the rocket travels into the upper atmosphere, when the exhaust plume widens dramatically over a short period of time. Second stage engines and those built to function entirely in the vacuum of space have a much greater exhaust nozzle flare and length and produce a noticeable higher Specific Impulse for the same fuel consumption.

The biggest difference between an Aerospike and conventional rocket engines is the shape of the nozzle. With the traditional bell shape, exhaust gas expansion is controlled within the nozzle. The Aerospike engine is shaped like a "V" with combustion taking place in a series of small chambers around the top of the "V" (see Figure 33.) The converging sides of the "V" are called the ramp. The Aerospike's exhaust plume is open to the atmosphere on one side and with the ramp on the other. The exhaust is expelled down the outside of the ramp using ambient atmospheric pressure on the open side to constrain the exhaust and act as one side of the engine nozzle. This allows the engine to operate efficiently at different altitudes because the atmospheric pressure provides for a varying constraint on the

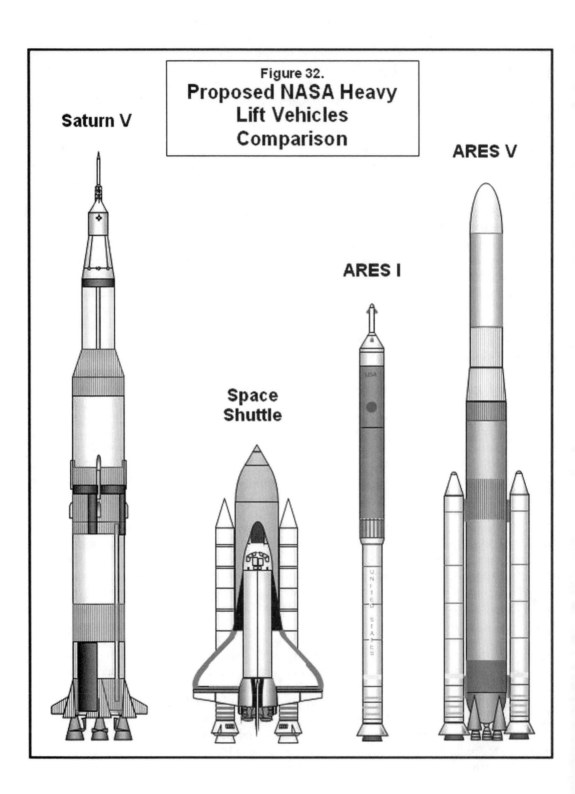

Figure 32.
Proposed NASA Heavy Lift Vehicles Comparison

Saturn V

ARES V

ARES I

Space Shuttle

exhaust gases. The engine may have either a conical configuration or (flat sided) configuration.

The Aerospike also has the ability to vector the thrust without the need to move or "gimbal" the engine. This may be achieved by employing a pair of box-shaped (linear) Aerospikes as a set of four thrust chambers and varying the thrust of each. The box-like arrangement of the Aerospike engine also lends itself to the aerodynamic wedge shape of some of the proposed aerospace craft such as the X-33.

Chemical propulsion systems, concentrated on LH2 and storable fuels, will continue to be the primary means to LEO for the next 25 years. However, the great distances involved in traveling to the outer limits of the solar system (and beyond) demand new sources of energy far greater than can be obtained from chemical systems for upper stages and for the "cruise" period.

Ion—Electric Propulsion

Of all the advanced systems that may provide for faster travel in space, the ion engine, electronic propulsion (EP), has always been considered one of the most likely to succeed. Goddard mentioned electric propulsion as early as 1906, while Tsiolkovsky wrote in 1911: *"It is possible that in time we may use electricity to produce a large velocity for the particles ejected from a rocket device."* EP is also very popular with the science fiction writers of the present. An ion is simply a charged particle— an atom (nucleus) of a specific element that has either lost or gained a net charge of electrons. An ion thruster uses beams of accelerated charged particles for propulsion based on Newton's action/reaction principle.

While the method for accelerating the ions may differ, all designs use the high charge-to-mass ratio of ions and accelerate them to very high velocities (up to 300,000 fps as opposed to chemical exhaust velocities of 16,000 fps), and in so doing, they achieve a high Specific Impulse. EP uses a much smaller amount of reaction mass for the power generated as compared to chemical rockets and can deliver greater efficiencies by a factor of ten. But, because the engine is expelling charged particles at a much lower volume than chemical engines, the thrust derived is very low—typically measured in ounces rather than hundreds of thousands of pounds. However, with high-propulsion efficiency it is possible to reduce the propellant mass by up to 90 percent—significantly reducing the spacecraft weight and the corresponding costs. With current energy sources of a few tens of kilowatts, and given a typical Specific Impulse of 3,000 seconds, current ion thruster technology can provide only extremely modest forces resulting in accelerations in the milli-G range.

With that factor in mind, ion engines are currently not practical for initial take-off from the Earth. Their primary use would be on-orbit attitude control, station keeping, or for long duration missions to the outer planets where the engine would provide thrusting for months or years at a time.

The first truly practical ion engines were developed in the early 1960s by NASA researcher Harold R. Kaufman, a physicist who made use of the Duoplasmatron invented by Manfred von Ardenne. The duoplasmatron is an electronic process similar in many ways to the simple electronic vacuum tube of the 1940s. A cathode filament emits electrons into a vacuum chamber which ionizes a gas (such as argon) through interactions with free electrons from the cathode to form a plasma. A plasma is essentially a fourth state of matter (solid, liquid, and gas being the more common states). This plasma is then accelerated through a series of charged grids, and becomes a beam of ions (electrically charged particles), moving at very high speeds, which exit the engine through a nozzle. The thrust is generated by the reactive force (Newton's third law) of electrically repelling the ions. The ion engine requires a significant source of electrical power and a quantity of a fuel such as one of the inert "noble" gases of argon or xenon.

Electric rocket engines can produce about 6-pounds of thrust for each million watts of power. Thus, ion engines are often referred to more by their electrical power consumption than by the thrust produced. Because it takes significant electrical energy to generate an ion beam, nuclear generated electricity is often considered as a primary source of power, although early modest applications made use of batteries recharged from solar cells.

There are varieties of ion thrusters, besides those based on the duoplasmatron, which have been developed and successfully used in spacecraft applications. In 1995, the Solar Electric Propulsion Technology Application Readiness (NSTAR) system was developed. This was a variant of the duoplasmatron called an "electrostatic ion thruster." As the ions approach the negative grid, they are electrostatically focused through the apertures of the negative grid and out into space at a high speed to improve the efficiency and reduce grid erosion.

NSTAR was used with the 1,000-pound Deep Space 1 spacecraft launched to initial escape velocity by a Delta II rocket in 1998. This was the first time an ion engine performed as a method of "cruise propulsion" in a deep-space mission. Deep Space 1 carried about 200 pounds of xenon propellant, which provided for 20 months of continuous thrusting, accelerating the spacecraft by about 10,000 miles per hour over that period.

NASA has used the Xenon Ion Propulsion System to perform station keeping on geosynchronous satellites. To prevent the craft from gaining a net negative charge, electrons are expelled from the cathode (called the neutralizer) towards the ions behind the ship to equalize the amounts of positive and negative charges. The Japanese Space Agency's Hayabusa spacecraft was powered by two xenon ion engines. It successfully rendezvoused with the asteroid "Itokawa" and performed station keeping with it for several months.

The power of the electrical source of an ion thruster determines the exhaust velocity of the beam, and the "particle accelerator" form of ion engine can achieve an exhaust velocity approaching the speed of light. This can result in a theoretical Specific Impulse of 30,000,000 seconds. The Hall Effect thruster (also know as a plasma thruster) is yet another type of ion engine that has been used for decades for station keeping by the Soviet Union.

NASA is currently ground-testing an Electron Cyclotron Resonance (ECR) ion thruster called the High Power Electric Propulsion, or HiPEP. The xenon ions are produced using both microwave and magnetic fields. The thruster consumes 20 to 50 KW of power and produces a Specific Impulse of 6,000 to 9,000 seconds. The goal of the project is to achieve a practical application unit by 2008.

Ion engines appear to offer significant utility for the future. However, unless there is a dramatic break-through in electrical power generation in space, they remain too limited in their thrust generation for propelling very large spacecraft.

Nuclear Thermal Power—NTR

Nuclear thermal rocket (NTR) engines use a working fluid (such as hydrogen) heated to very high temperatures by a relatively conventional nuclear fission reactor (the equivalent of the chemical rocket combustion chamber). The expanding gas exits through a rocket nozzle to create thrust. Due to the higher energy of the nuclear reactions, the resulting efficiency of the engine is typically twice that of chemical engines. Thus, the Space Shuttle's chemical SSME LH2 engine produces about 400 seconds of Specific Impulse while a nuclear engine can produce 800 to 1,200 seconds—and in some theoretical designs—twice that much.

The exhaust of chemical rockets is constrained by the chemical reaction, but in an NTR engine, the heat source is not based on the propellant, so an NTR engine can use a low molecular weight propellant, such as hydrogen, to improve performance. The high Specific Impulse levels of an NTR engine offer opportunities for missions with shorter trip times and greater payloads than those that can be accomplished using chemical propulsion.

A nuclear thermal rocket can be categorized by the construction of its reactor—from a relatively simple solid-core reactor (described above) to a more complex but efficient gas core reactor. The weight of a flight-ready nuclear reactor is so great that solid-core designs would probably not achieve a thrust-to-weight ratio of 1:1—thus solid-core engines are considered only for upper-stages where the required thrust is lower.

There are several design techniques that may be used to increase the operating temperature of the reactor and thus the Specific Impulse (to greater than 1,000 seconds), as with particle-bed, rotating-bed, or pebble-bed reactors. With these types of reactors, the fuel is placed in several containers which "float" inside the working fluid (hydrogen or water). Spinning the engine forces the fuel elements out to walls that are then cooled by the hydrogen. The increased complexity of such concepts makes their practicality problematical.

Even greater efficiencies can be achieved by mixing the nuclear fuel with the working fluid—allowing the nuclear reaction to take place in the liquid mixture itself. These "liquid-core" engines can also be operated at temperatures beyond the melting point of the fuel and can provide Isp in the range of 1,300 to 1,500 seconds.

Another "liquid-core" design, the nuclear salt-water reactor, uses water as the working fluid. The nuclear fuel is not retained but is discharged—resulting in large quantities of radioactive waste and would be practical only well outside the Earth's atmosphere.

The "gas-core engine" is similar to the liquid-core design and uses rapid circulation of a fluid to create a pocket of gaseous uranium fuel in the middle of the reactor, surrounded by hydrogen. In this case the fuel does not touch the reactor wall at all, so much higher temperatures could be achieved, providing Specific Impulses of 3,000 to 5,000 seconds.

Only the "solid-core" NTR engine called Kiwi, has actually been built and tested as a part of the Atomic Energy Commission's Project Rover, begun in 1956. Kiwi was the first tested in July 1959, but it was not a flight rated reactor (thus its name). Its core was an assembly of uncoated uranium oxide plates onto which liquid hydrogen (LH2) was pumped. The Kiwi B was a more complete system that used uranium fuel, in the form of uranium dioxide, through which LH2 flowed for cooling. A third iteration used uranium carbide as a fuel before the Kiwi program was concluded in 1964. The next series of tests used a larger reactor and was called Phoebus. It was first fired in June 1965 for over 10 minutes.

Developed in parallel with Kiwi was NERVA (Nuclear Engine for Rocket Vehicle Applications). It was intended to produce a usable 75,000-pound-thrust engine for the upper stages of the Saturn V. The design, known as NRX (for Nuclear Rocket Experimental), underwent successful testing in September 1964. However, the NERVA project was cancelled in 1972 with the decision by NASA to abandon the manned mission to Mars. With the space race assuming lower priorities in the early 1970s, the significant safety issues of nuclear power became paramount. These involved the dispersal of radioactive material in the form of atmospheric fallout should a launch fail catastrophically. While some theoretical work had been performed over the intervening years, it was not until the advent of the Project Prometheus in 2003 that NASA's thoughts again turned to nuclear.

Originally called the "Nuclear Systems Initiative," the Prometheus propulsion systems currently under consideration include conventional NTR, Bimodal, and Tri-modal. "Bimodal" uses the reactor to create electrical power when not being used to produce thrust. "Tri-modal" provides for an afterburner-like operation where liquid oxygen (LOX) is injected into the exhaust nozzle for increased thrust. The project's objectives include both forms of spacecraft power systems: the NTR (similar to NERVA), and the nuclear reactor to provide electrical power for ion engines and spacecraft electrical power.

The initial plan for Prometheus was to use small nuclear reactors as the primary electrical power source. This would allow increased power generation, compared to RTGs, and would permit more flexibility in mission design. Higher communications transmission power and data transfer speeds would extend the distance over which the spacecraft could communicate and would increase the life of those systems.

The first mission to be considered for Prometheus was the exploration of the Jovian moons Europa, Ganymede, and Callisto. But the cost was prohibitive, and it was cut from the FY 2006 budget. NASA considered a scaled down system that would be tested on a short range mission, probably within the vicinity of the Earth, to verify reactor systems in the space environment before committing to an expensive long duration mission, but the prospects for Prometheus powering a mission in the near future are not good. While nuclear energy is still meeting with opposition, it currently represents the most powerful capability in mankind's possession.

While fission is the nuclear process that uses Uranium or Plutonium for obtaining the release of heat in current reactors, the fusion reactor holds the promise of considerably more energy release. Using heat and pressure to initiate a fusion reaction of Deuterium and/or Tritium, this process provides the highest energy-to-mass ratio of any known substance. There are many technological problems with the fusion reactor and, after more than 30 years of intense research, science has failed to harness effectively the reaction. The scientific community is divided as to whether this type of energy can ever be made practical.

The Solar Thermal Rocket (STR) uses the same essential principles as NTR except that the heating of the propulsive fluid is achieved by solar energy rather than nuclear fission. This effectively limits the use of this type of power to the inner solar system.

Supersonic Combustion Ramjet—Scramjet

While much of the attention for advanced propulsion systems has been focused on achieving high-speed flight outside the Earth's atmosphere, the requirement still exists to improve the ability to accelerate heavy payloads from the surface to orbital speed. As noted, some of the advanced propulsion units are relatively low-thrust and cannot provide orbital boost. The air-launched satellite concept, as demonstrated by Pegasus, still provides visionaries with the dream of using the oxygen in the Earth's atmosphere to reduce the weight of propellants that must be carried during the early part of powered flight. Rockets are only about 2 percent as efficient as jet aircraft because the oxidizer represents about 75 percent of the propellant weight.

Pegasus uses a conventional Lockheed L-1011 (it originally used a B-52) powered by jet engines. The speed (600 mph) and altitude (40,000 feet) are modest compared to the numbers that were envisioned for the original design for the Space Shuttle carrier. The jet engine is limited to speeds of up to Mach 3—the top speed of the SR-71. However, there are several concepts in the research stage that can greatly increase the "first stage" boost that the air launch gives to the remaining portions of the vehicle.

The Ramjet has been around since before World War II but has not found a niche for its unique capabilities. Operating as simply a tube with no moving parts, the Ramjet uses air "rammed" into its intake by the speed of the vehicle and sprayed with fuel. On ignition, the expanding combustion "pushes" (Newton's third law) against the "wall" of incoming air to provide thrust. However, the dynamics involved belie the simplicity of its construction. The ramjet requires an initial movement through the air to get the needed volume of intake to begin productive combustion—typically a speed approaching Mach 1. Its second conceptual shortcoming is that it begins to lose efficiency above Mach 6, as the combustion process is defeated by the shockwave of the incoming air. The Ramjet performs best at a specific altitude for which its intake has been configured unless it employs a variable geometry inlet.

Immediately following WWII, there was much investigation of the ramjet that resulted in two notable projects. The Air Force Bomarc F-99 unmanned area defense interceptor of the 1950s was launched vertically using a liquid-fuel Aerojet General LR-59 booster. As supersonic speed was achieved, two Marquardt RJ-43-MA-3 ramjets ignited, and the missile cruised to its target at Mach 2.8 and 65,000 ft. for distances of up to 250 miles. The project was headed by Boeing with the ramjets provided by Michigan Aerospace Research Center (MARC)—thus the acronym Bomarc. The first sites became operational in 1959, and the designation of the missile was changed to IM-99.

The U.S. Navy used the ramjet in its Talos Fleet Defense Missile (RIM-8) that became operational on ships in the early 1960s and remained until 1979.

Despite its drawbacks, the ramjet has continued to find favor with visionaries, and several recent research projects have evolved. NASA has funded the X-43, a modified ramjet that uses supersonic combustion of hydrogen fuel and has acquired the acronym of Scramjet (Supersonic Combustion Ramjet). With this new technique, the unmanned X-43 achieved a world-record Mach 10 flight for an air-breathing vehicle in November 2004.

A parallel development project by the Defense Advanced Research Projects Agency and the Office of Naval Research (ONR) successfully flew a hypersonic scramjet-powered vehicle from the Wallops Flight Facility in Virginia in December 2005. This was the first free flight of a scramjet-powered vehicle using conventional JP-10 liquid hydrocarbon jet fuel. The launch and flight-test were part of the "Freeflight Atmospheric Scramjet Test Technique" (FASTT) program designed to explore speeds up to Mach 10.

The FASTT vehicle integrates the Scramjet into a vertically launched missile configuration that is approximately nine feet long and one foot in diameter. After separating from its booster rocket at 60,000 feet, the Scramjet propelled the vehicle to Mach 5.5 for about 15 seconds. Engineering data was transmitted to ground stations before the vehicle was intentionally crashed into the Atlantic Ocean.

While the scramjet seeks to use the "free" oxygen in the atmosphere, at Mach 10 the aerodynamic drag is extreme, as is the resulting heat generated. Thus, any future use of these engines will probably be for short durations to launch upper stages into space. Long-range cruise scenarios such as the National AeroSpace Plane (NASP) were unable to mitigate some of the Scramjet drawbacks. Hybrid Scramjets called "air augmented rockets" have been conceived as a part of a single-stage-to-orbit vehicle. If this technology can prove workable, the reality of "inexpensive access to space" may be a step closer.

Mass Drivers

Science fiction has long portrayed the "mass driver" as an exotic means of escaping the Earth's gravity. In reality the basic concept is one of the simplest of electronic mechanisms—a linear electric motor. In the conventional electric motor, the magnetic lines of flux are used to generate a rotating motion. In a mass driver the "'stator" component of the motor is an "in-line" arrangement rather than circular, producing an electromagnetic catapult.

Two basic design categories provide for low-acceleration and high acceleration applications. Low-acceleration applications include magnetic levitation transportation systems that have seen limited implementation in futuristic rail lines such as the 18 mile segment built in Shanghai that has achieved speeds of up to 270 mph. But it is the high-acceleration applications which have initially been directed towards weapons that hold promise for spacecraft. In this latter scenario, a short segment of a linear accelerator propels a mass to a relatively high velocity. In the theoretical world this would be to orbital or escape velocity.

Of course, the electrical energy expended in such a device is significant, and typical electrical generators are not sufficient. The capacitor banks needed to store the quantities of energy (typically in the hundreds of Giga-joule range) would be extremely large and not practical. The most effective way of generating the electrical power may be with "homopolar" generators (invented by Michael Faraday in 1831). This technique uses a large conducting flywheel that is spun up to a high speed over a long period of time (hours or even days) in a magnetic field. When the electrical current is needed, the commutator is energized to convert the kinetic energy of the spinning mass into electricity generated by the magnetic lines of flux. The efficiencies of the mass driver are relatively poor, but the use of superconducting coils could increase the efficiencies to perhaps 50 percent. While there are no theoretical limits to the size of a mass driver, the efficiencies drop off at higher velocities.

In the current and foreseeable future (baring a technological breakthrough, which is not anticipated), mass drivers are effective for relatively small payloads of up to one ton. When the Earth's atmosphere and gravitation force are considered, their application is very limited. A hybrid application could involve the mass driver for an initial acceleration, and then the vehicle would have a "second stage" consisting of another power source such as a nuclear powered rocket.

However, if the mass driver were located on the Moon or in space, it would be much more practical. In such an environment, electrical power could be generated from large solar arrays or reflectors, if operated in the inner solar system or nuclear power for deep space missions. When used aboard a spacecraft, the mass driver would be a part of the craft and expel some form of mass in the opposite direction of travel to satisfy Newton's third law. However, the craft has to have a significant quantity of some material to react against. In the theoretical realm, this mass could be collected as "space debris" during the journey. This type of engine could easily produce a Specific Impulse of 3,000 seconds.

While military applications dominate the current thread of research and development, practical mass driver devices for space propulsion will almost inevitably result.

Beam Power: You Can Leave Home Without It

Virtually all of the propulsion systems that have been considered require that some form of mass (fuel) be carried along (or picked up along the way) during the journey to provide for Newton's third law. For high-velocity flight, this imposes a significant weight penalty for some forms of propulsion. However, in the case of a beam powered spacecraft, the source of the propulsive mass remains behind, and only a high intensity beam of energy is directed to the spacecraft. This energy is converted into a reactive force by the spacecraft.

There are several forms of energy that are emitted from our own star (the Sun) that can be used. The first is the steady stream of protons ejected from the Sun by the fusion process. In the simplest application, a solar sail is a large expanse of a very light weight material, unfurled and erected in space, that acts much like a traditional "wind" sail of a boat. The mass of the protons impacting on the sail provides a reactive force, moving the sail (and the attached spacecraft) away from the Sun. By positioning the sails relative to the radially outward moving beam of protons, the spacecraft could be "vectored" to a desired course much the same way as a sailing ship "tacks" with the wind. However, because of the relatively small force of each proton, a very large sail would be needed— an area of many square miles.

A second method is to concentrate the radiant-heat of the sun to create a high working temperature for some fluid. This could be used to drive a turbine for generating electrical power that might then be used in an ion drive—with the fluid being recovered for reuse.

However, both the solar sail and the radiant heat collector become less efficient as the spacecraft moves into the outer solar system; the energy of the Sun dissipates to a level that would be unusable. Of course, depending on the effectiveness of the technique used, the spacecraft could have worked up to a considerable speed by that time.

A third use of beam power is to use a directed beam of energy such as laser light or microwave radiation. This would provide a much higher concentration of energy. The power generating station would have to be in space (or perhaps on the Moon) as the atmosphere would attenuate the beam. Again, the cost of generating such an intense beam of power dictates that the system would be a part of a high-use corridor—perhaps commuting between two points in the solar system.

As can be deduced from the descriptions of the possible advanced power sources, there are many technology hurdles that remain before achieving truly effective propulsion systems. And none of the sources described have the ability to produce velocities approaching one percent of the speed of light. The fastest spacecraft has achieved a speed of 8 miles per second—the speed of light is 186,000 miles per second. Travel beyond the solar system will require new technologies yet undiscovered.

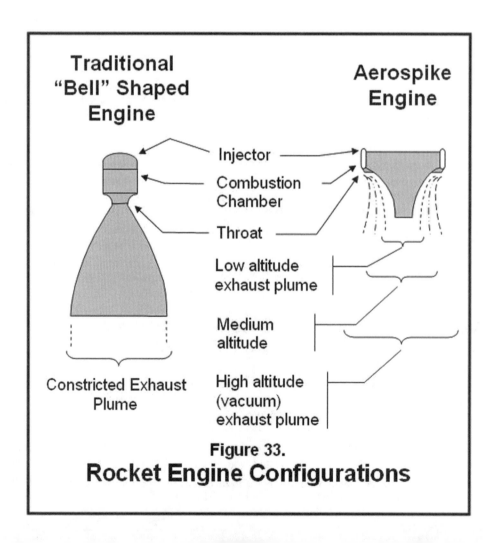

Figure 33.
Rocket Engine Configurations

Chapter 39 — What Will the Future Bring?

The technological advances of the past century have often eclipsed even the dreams of the visionaries in some respects. The simple conveniences that we now take for granted, such cell phones, were beyond the imagination of the science fiction writer of a few decades past. Who could have conceived a wireless phone which allowed for the receipt of email and real-time video while transmitting digital pictures and providing GPS navigation? In just the past 50 years, the electronic storage of a single binary digit (bit) of information has progressed from the several cubic inch vacuum tube, to a chain of molecules a trillion times smaller. Would any reputable scientist have foretold of such advances in 1945?

However, there have also been plateaus as in the case of commercial aviation which allowed travel at 600 mph 50 years ago and has not increased since (the Concord was a singular exception that no longer exists). Likewise the speed and altitude capability of fighter aircraft have not advanced during that time—the F-106 currently holds the speed record for a single engine turbojet that was set in 1959. We went to the Moon in 1969 and have not traveled any further in the almost 40 years that has elapsed.

Thus while some enabling technologies, such as integrated circuits, have outdistanced the visionary, aerospace has not seen any dramatic advances.

A Closed Biosphere: Life Support

In considering long duration space travel or planetary colonization, most visionaries agree on the need to replicate the Earth's biosphere—that part of the Earth's outershell that includes the air, water, and soil on which the sustenance of life is based. The term came into widespread usage with the creation of Biosphere 2, a privately funded manmade ecological system constructed in Tucson, Arizona in 1987. The purpose of the project was to investigate the problems of duplicating the Earthly living environment in a 'Closed Life Support System' (CLSS); a totally self-contained enclosure without any interaction with the Earth's ecosystem.

A CLSS is an environment in which there is no exchange of matter with any part outside the system. While the Earth is often considered as a closed system, in reality it receives energy from the sun and in a like manner, the first artificial Biospheres were not energy self-sufficient. One of the critical aspects of a closed system is the need to recycle all waste back into the life cycle. If space colonization or interstellar travel is to be achieved, this factor is considered essential—everything from the exhaled carbon-dioxide to human excrement must be processed and reused.

For a 24-month period in 1991-1993 and a second six-month period in 1994, eight persons lived and worked within the enclosed Biosphere 2 structure. However, the oxygen levels steadily diminished and external oxygen had to be pumped in to continue the experiment. It is believed that some form of micro-organism depleted the oxygen.

It was also later revealed that other 'supplies' had been furnished periodically which led to diminished credibility of the $150 million experiment. But one important lesson was revealed—that many aspects of the complex interaction within the natural Earth's biosphere are still unknown.

Another closed ecosystem (now called Biosphere-3) was actually the first such facility and was constructed in secret by the Soviet Union between 1965 and 1972. The 315-cubic meter facility housed up to three persons for periods of up to 180 days. These experiments were concluded in 1984.

It has been estimated that a population of at least 150 persons (male and female) would be the minimum to permit normal reproduction for 60-80 generations to cover a 2000 year time span (for a voyage to another solar system) and avoid inbreeding, although some estimates place the initial number at 500 for long term reproduction.

CLSS is a critical technology to the habitation of space, yet a completely closed system has yet to be achieved. The Biosphere experiments have reinforced the notion that closed ecologies are difficult and expensive to achieve and may be dangerous. No long-term plans can be made unless, and until, a functioning and robust system can be adequately demonstrated.

There are several options to the CLSS problem. The first is the traditional approach in which the human is completely isolated from the inhospitable environment through the use of an earth-like biosphere. The second is to change the environment to become less hostile to the biological functions as would be the case for influencing the atmosphere of a planet such as Venus or Mars—terraforming. The third is changing the human body to adapt to the environment of the planet. This latter process could be through genetic engineering, transhumanization, or the cybernetic process of using synthetic parts to augment the human structure—or a combination of all three.

Although it is important to understand the affect of micro-gravity on the human body and on the properties of liquids, the visionaries of the past seemed to anticipate the problems of weightlessness and the creation of artificial gravity (by a slow rotation of the spaceship) has always been a relatively easy answer to that problem. It does impose a more robust structure but keeps the human element in an orientation that appears to offer both a physiological and psychological advantage. It also offers a more appropriate foundation for the other ecological systems (such as the growing of plants).

No doubt, one of the objectives adjunct to the CLSS will be to engineer plant genetics to provide shorter growing periods as well as enhanced nutritional value. However, the ultimate in efficient human nutrition would be food processed directly from waste products.

Interstellar Travel: An Unreality?

Because of the 'Genesis Factors' (the conditions of time, energy, distance, and mass) the ability to travel between planetary systems in other galaxies is beyond credibility. The fastest object propelled by man to date is about eight miles per second—the speed of light is 186,000 miles per second, more than 23,000 times faster. While science fiction may assert that travel faster than light (FTL) will be achieved, for the present, the laws of physics make that goal an impossibility. One light-year equals 5,874,601,673,407 miles (almost six trillion). Although many of the advanced propulsion systems may increase attainable speeds by several orders of magnitude (factors of ten), they will still leave us far short of even one percent of the magic FTL value. The interaction of energy and mass coupled with the element of time, present a theoretical barrier that cannot even be approached.

Assuming that a power source could accelerate a huge million ton-plus spaceship to a speed 10 times faster than we travel today (say 370,000 mph), it would still take more than 6,000 years to make the trip to Proxima Centauri (the closest star) which is 4.28 light-years and 25 trillion miles away. A trip to the closest planetary system of Epsilon Andromeda would take about 14,000 years. Increase that speed ten fold and it still will take 1,400 years. To fly across our Galaxy from one side to the other at the speed of light would take 100,000 years. In the final analysis, intergalactic or even interstellar travel may never be a reality—but moving beyond the bounds of reality is what makes dreams so fascinating.

Assuming that a planet is found in another solar system on which the conditions are favorable to life, the spaceship would also have to decelerate to the orbital speed of the selected planet, which would take a considerable amount of energy (converted from mass). Clearly, a different paradigm will be required for this type of travel. Long-term solutions will require continued scientific study to discover if there are other dimensions of time and space in which mankind can operate.

When interstellar flight is contemplated, it will probably be a one-way trip unless the paradigm of travel in dimensions yet discovered is achieved. Within our current realm then, such a flight would be in many ways more daunting than Noah's task of provisioning the Ark. How would such a journey be equipped, knowing that all of the knowledge and critical tools to build a new civilization must be taken into account? All the replenishable life support necessary to sustain the initial inhabitants during the voyage (as well as the expanding generations to come) would have to be provided. Also included must be the essence of life, as we know it—and as we might want it continued on some far-distant planet. While selected animals and plant life might be included, the ability to carry the DNA structure of thousands of species would accompany the expedition and might allow the regeneration of an earth-like environment—if an earth-like planet were encountered.

While the dreamer might equate travel to another galaxy with the simplicity of a Robinson Caruso story, the planning and provisioning of such a trip would be the most complex task ever attempted by mankind. The size of such an expedition would doubtless tax (literally) the economic resources of the Earth and its citizens. While the physiology of the travelers is a primary consideration, the psychological impact of the journey would need considerable thought. Who would select the colony and by what criteria? Moreover, if the impetus for such a voyage were the imminent destruction of the Earth by some catechistic force, how would the populace to be left behind respond? While we can understand that mankind's ability to travel through space is limited by the physical dimensions in which we are currently constrained, will new scientific revelations present added dimensions beyond time and space?

Realistic Expectations

The past 50 years have seen phenomenal exploration and exploitation of the space environment. No doubt the primary unmanned applications (communications, weather, Earth resources, and navigation) will continue. They are now inseparable from our economy and culture. These near-earth satellites have proven their usefulness. Likewise, the unmanned probes to the far reaches of our solar system have been highly successful—but what about man in space? NASA plans to return to the Moon by 2020 and has overtures towards Mars. However, many people do not believe that the Moon or Mars has any scientific or commercial value to warrant the expense of manned voyages. While it is impossible to predict the future, the possibilities are made more apparent with an understanding of the three primary enabling forces composed of the economic, technological, and motivational factors.

The current space initiative put forward by President Bush seeks to return to the Moon by 2020 and then use the Moon as a stepping stone to Mars. There are proponents and opponents of the plan. Several proposals have been made over the past 40 years for a permanent base on the Moon and excursions to Mars. All have been abandoned—virtually before any funding was expended. Why have we failed to build on the legacy of the Apollo program?

The Apollo program was unique for several reasons. It was born of the space race that saw the apparent technological superiority of the Russians posing a threat to the existence of the free world. It was projected over short time frame (8 ½ years—"before this decade is out") that allowed intervening milestones to be played out on a periodic basic—virtually every few months. Thus, a new "first" in space was achieved at relatively close intervals. The first space walk, rendezvous, and then docking, as well as the first space emergency all took place within a year. This kept the American people (who were paying the tab) interested and involved. Apollo was also able to weather the politics of three administrations, the financial squeeze of the Vietnam War, and the turmoil of the inner cities—primarily because of the efforts of a very capable Administrator—James Webb.

The public appears to be somewhat indifferent (some would say fickle) with respect to space. While the Smithsonian Air & Space Museum is the most visited museum in the world, interest in the Moon landings that followed Apollo 11 (the first landing) gradually drew less and less enthusiasm. A recent poll revealed that few Americans have any idea that the ISS has been inhabited continuously for over 6 years and almost no one has any idea of its goals.

A return to the Moon will find little support from the general populace. Its closeness makes it more desirable than Mars but the "been there, done that" syndrome works against popular support. A Mars mission planned for 30 years into the future will not only be yawned at, but will have a tough battle keeping any funding through four or five administrations. In addition, a price tag that stretches into the hundreds of billions will not find favor with those who feel we have more pressing earthly agendas.

However, a mission to Mars has far more popular support than the return to the Moon. Mars represents a truly challenging opportunity to explore a planet that may have harbored life. When NASA was given the task of assessing a proposed Mars Project back in the early 1990s, it came up with a price tag of about a half-trillion dollars over a thirty year period. Not only did this present "sticker shock" to Congress, but the time frame assures that the changing whims of Congress and the vagaries of NASA would doom the project before it ever left the launch pad.

More innovative and timely proposals were put forth by such engineering fundamentalists as Robert Zubrin. He looks at the Mars project in the way that NASA's John Houbolt viewed the lunar landing in the early 1960s. "Take only what you need as far as you need it" was the mantra that resulted in the Lunar Orbit Rendezvous technique that allowed the Kennedy goal to be achieved within the decade of the 1960s. To this Zubrin adds, "And use the Martian environment to supply critical resources." His plan provides for the generation of methane to power the return flight as well as the Martian rover vehicles. Zubrin's book, "The Case for Mars," provides a compelling analysis of a means to achieve, not just a mission to Mars, but an ongoing scientific base. His critical assessment of the NASA approach to mission planning is a revelation. A basic premise of Zubrin's Mars project is that it can be completed in ten years and for perhaps one-tenth the cost estimated by NASA.

The justification for the pursuit of an aggressive space program has many themes. These range from spreading life throughout the universe to ensuring the survival of our species in the event of a global calamity. Some estimates are that, over the next 100 years, an asteroid-like object has a one-in-a-hundred chance of impacting in the ocean and creating a tsunami that could destroy one or more major cities along the coast. There is a one-in-a-thousand chance of an impact ruining our civilization and killing a large percentage of the population—and a once-in-a-million likelihood of a mass extinction impact.

A remote space colony could 'reverse-colonize' the Earth and restore human civilization following such a catastrophic event. More 'down to earth' motives include the desire to preserve the terrestrial environment as well as to create new business opportunities for economic gain. However, it may be that only a survival based motivation coupled with a set "reasonably priced" short term goals has a chance for succeeding.

The controversy is not simply within the United States; Russia, China, Japan, and the European community may figure prominently in any new space initiative. The costs involved in these projects may be more than one nation can afford—even the United States.

Where to build a Space Colony? Mars has the greatest following because of the possibilities that it might have harbored life and has resources that might allow a colony *to create a Closed Biosphere (CLSS).* CLSS is a critical technology to the habitation of space—yet a completely closed system has yet to be achieved. No long-term plans can be made unless, and until, a functioning and robust system can be adequately demonstrated.

However, Dr. Gerald K. O'Neill, a space enthusiast of some standing, asked "Is the surface of a planet really the right place for an expanding technological civilization?" He proposed the Lagrangian L-5 point. If a large space city were constructed at L-5, it could serve as a relatively isolated outpost to prove and improve the closed life-support system—yet be within the range of help from the Earth until it could achieve total independence. A manned Mars missions is essentially a multi-year task that would require a CLSS—unless transformation of Mars' abundant carbon dioxide can not only provide methane, but oxygen and water. If the decision were made to embark on a manned Mars mission, it would accelerate development of those technologies needed for permanent space habitation.

While many of these initial proposals can be achieved with existing technology, any further steps will require advanced propulsion and power systems. Chemical propulsion will continue to be the primary means to LEO for the next 50 years. However, virtually all plans to explore, colonize, or mine space require a more economical access to low-earth orbit.

At the advent of the new millennium, NASA re-evaluated the prospects for developing advanced propulsion systems in the near-term and at one point defined four phases to reduce the costs. It had placed the Shuttle as being the 'first generation' and positioned each of the successive generations as reducing the cost by a factor of ten while increasing the reliability by a like order of magnitude (the Shuttle ultimately was defined as one loss in each 100 launches). The fourth generation is seen as an operational spaceship in the year 2040, capable of taking-off and landing horizontally and making 10,000 flights per year. Clearly, NASA still projects space flight to become as commonplace as airliner flights of today. *But, is "inexpensive access to space" an oxymoron? NASA's track record is not encouraging.* The nation who achieves this milestone will most likely become the dominant space and terrestrial power.

Ion, Nuclear and Solar Thermal, Mass Drivers, and Beam Power all hold some potential but have technological hurdles. Moreover, none has the ability to produce velocities approaching one percent of the speed of light. As a result, when interstellar flight is contemplated, only a paradigm of travel in dimensions and by means yet to be discovered will make this possible. Even if the known laws of physics have a loophole that can allow for velocities at the speed of light—the prospects for future manned space flight face significant technological, economic, and political barriers.

Despite the boundaries of the Genesis Factors, we must not allow the imagination to be constrained by our current understanding of the laws of physics.

Epilogue

The past 100 years have been an exciting and challenging period for mankind as we have raced to explore the heavens above and the sea below. Technology has enabled a dramatic expansion of man's reach. Do the known laws of physics have a loophole (in the manner of anti-matter or perhaps time-warps) that can allow for manned spacecraft velocities at the speed of light (or greater)? Can the dangers of radiation and space debris be overcome, and the psychological behaviors of the spacefarer controlled? What are the prospects for the future of manned space flight?

Some visionaries contend that the exploration of space and the subsequent colonization of a distant planet might provide a renewed opportunity for mankind to shed the 'baggage' of the past and start anew—to leave behind the worst and bring forward only the best. While this goal may sound appealing, it begs the question of what is considered 'old baggage'. Mankind's own behavioral attributes have not changed since creation and it is the human race that will establish itself in this new frontier. We cannot run away from our past as it forms the foundation of our future. Unless we are re-programmed in some manner, perhaps by one of the bio-technologies that we seek to master, to become trans-human—to overcome our natural instincts, the 'seven deadly sins' (Pride, Envy, Anger, Sloth, Greed, Gluttony, and Lust) will persist in any new civilization.

While we ponder the fanciful projections of a new civilization that might be built on Mars or some even more distant point in space, it is important to understand that most likely none of the earthly environment that we have known and feel at home with may exist. There will probably be no soft warm summer breezes, no organic soil, and no abundance of water. Temperature and pressure extremes will most likely demand an artificial living environment that is the antithesis of the human desire for openness and freedom. If then the exploration of the cosmos reveals that mankind is restricted to this isolated blue outpost on the rim of the Milky Way galaxy for eternity, perhaps we must learn to use our technologies more diligently to preserve our privileged planet and ensure the security of our freedoms. Possibly the line from T.S. Eliot's "Little Gidding" says it best:

We shall not cease from exploration
And the end of all our exploring
Will be to arrive where we started
And know the place for the first time.

List of Acronyms

All acronyms are defined in the text with their first usage.

AAS	Alternative Access to Station
ABMA	Army Ballistic Missile Agency
ACS	Attitude Control System
AEC	Atomic Energy Commission
AFB	Air Force Base
ALSEP	Apollo Lunar Surface Experiment Package
AMU	Astronaut Maneuvering Unit (AMU)
APS	Ascent Propulsion System (Apollo)
ARPA	Advanced Research Projects Agency
ASTP	Apollo Soyuz Test Program
ATDA	Augmented Target Docking Adapter
ATV	Automated Transfer Vehicle
AU	Astronomical Unit
CAPCOM	Capsule Communicator (USA)
CBC	Common Booster Core
CEV	Crew Excursion Vehicle
CIA	Central Intelligence Agency
CLSS	Closed Life Support System
CSI	Constellation Services International
CSM	Command and Service Module
DARPA	Defense Research Projects Agency
DoD	Department of Defense
DOI	Descent Orbit Initiation
DPS	Descent Propulsion System (Apollo)
ECR	Electron Cyclotron Resonance
ECS	Environmental Control System
EELV	Evolved Expendable Launch Vehicle
ELDO	European Launch Development Org
EOR	Earth Orbit Rendezvous
EP	Electronic Propulsion
ESA	European Space Agency
ESRO	European Space Research Organization
ET	External Tank (Shuttle)
EVA	Extra-Vehicular Activity
FASTT	Freeflight Atmospheric Scramjet Test Tech
FTL	Faster Than Light
GAO	General Accounting Office
GATV	Gemini Agena Target Vehicle
GCS	Guidance and Control System
GEM	Graphite Epoxy Motors (solid fuel)
GEO	Geosynchronous Earth Orbit
GEX	Gas Exchange Experiment
GSE	Ground Support Equipment
GTO	Geosynchronous Transfer Orbit
GT-x	Gemini Titan Launches (number 1-12)
HiPEP	High Power Electric Propulsion
HRSI	High temp Reusable Surface Insulation
HST	Hubble Space Telescope
HySTP	Hypersonic Systems Technology Program
IBM	International Business Machines Corp.
ICBM	Intercontinental Ballistic Missile
IRBM	Intermediate Range Ballistic Missile
Isp	Specific Impulse
ISS	International Space Station
IUS	Inertial Unit Stage
JPL	Jet Propulsion Laboratory
JWST	James Webb Space Telescope
KGB	Soviet Secret Police
KORD	Engine Operation Control System (N1)
KSC	Kennedy Space Center
LEM	Lunar Excursion Module (Apollo)
LM	Lunar Module (Apollo)
LEO	Low Earth Orbit
LEV	Lunar Excursion Vehicle (Apollo)

LH2	Liquid Hydrogen
LO2	Liquid Oxygen
LOI	Lunar Orbit Insertion
LOR	Lunar Orbit Rendezvous
LOS	Loss of Signal
LOX	Liquid Oxygen
MA-x	Mercury Atlas flights (number 1-7)
MDS	Malfunction Detection System
MIT	Massachusetts Institute of Technology
MLP	Mobile Launch Pads
MOI	Mars Orbit Insertion
MOL	Manned Orbiting Laboratory
MR-x	Mercury Redstone Flight numbers (1-4)
MSC	Manned Spacecraft Center
MSFC	Marshall Space Flight Center
NAA	North American Aviation
NASA	National Aeronautics & Space Administration
NASDA	Japanese Space Agency
NASP	National Aerospace Plane
NERVA	Nuclear Engine for Rocket Vehicle Applications
NSTAR	Solar Electric Propulsion Technology Appl Readiness
NTR	Nuclear Thermal Rocket
OAMS	Orbital Attitude and Maneuvering System
OMB	Office of Management and Budget
OMS	Orbital Maneuvering System (Shuttle)
ONR	Office of Naval Research
OPF	Orbiter Processing Facility (Shuttle)
OSP	Orbital Space Plane
P&W	Pratt & Whitney
PAH	Polycyclic Aromatic Hydrocarbons
PDI	Powered Descend Initiate
PERT	Program Evaluation Review Technique
PGNCS	Primary Guidance and Control System
PLSS	Portable Life Support System
RCC	Reinforced Carbon Carbon
RCS	Reaction Control System
REP	Radar Evaluation Pod (Gemini)
RKA	Russian Space Agency (Roskosmos)
RLV	Reusable Launch Vehicle
RMS	Remote Manipulator System
RTG	Radioisotope Thermal Generator
SAM	Surface Access Module
SCRAMJET	Supersonic Combustion Ramjet
SDI	Strategic Defense Initiative
SPS	Service Propulsion System (Apollo)
SRB	Slid Rocket Booster (Shuttle)
SRMU	Solid-fuel Rocket Motor Units
SSME	Space Shuttle Main Engine
STA	Shuttle Training Aircraft
STG	Space Technology Group
STS	Space Transportation System
STSI	Space Telescope Science Institute
STS-x	Space Transportation System flight numbers
TDA	Target Docking Adapter (Gemini)
TDRS	Tracking and Data Relay Satellites
TEI	Trans Earth Injection
TLI	Trans Lunar Injection
TMI	Trans Mars Injection
UDMH	Unsymmetrical Dimethyl Hydrazine
USSR	Union of Soviet Socialist Republics
VAB	Vehicle Assembly Building

The following is a list of references used to create and validate Astronautics.

Aldrin, Edwin E., McConnell, Malcolm, Men From Earth, Bantam Books, New York, 1989

Aldrin, Edwin E., Warga, Wayne, Return To Earth, Random House, New York, 1973

Alway, Peter, Rockets of the World, Saturn Press, 1999

Arnold, H.J.P., Man in Space, Smithsonian Publications, New York, 1993

Ashford, David, Collins, Patrick, Your Spaceflight Manual, Crescent Books, London, 1990

Baker, David, The History of Manned Spaceflight, Crown Publishers, 1981

BardwellSteven J., Et al, Beam Defense, Aero Publishers Inc., 1983

Bergaust, Erik, Wernher von Braun, National Space Institute, 1976

Bilstein, Roger E., Stages to Saturn, Univ. Press of Florida, 1996

Bizony, Piers, The Man Who Ran the Moon, Thunder Mouth Press, New York, 2006

Bloomberg, Linlow P., Outer Space-Prospects for Man and Society, Prentice Hall, Inc., New York, 1962

Borman, Frank, Serling, Robert J., Countdown, Silver Arrow Books, New York, 1988

Bower, Tom, The Paperclip Conspiracy, Little Brown & Co, 1987

Boyd, R.L.F., Space Research by Rocket and Satellite, The MacMillan Company, 1960

Brodie, Bernard, Strategy in the Missile Age, Princeton University Press, Princeton, 1959

Bucheim, Robert W., Et al, Space Handbook, Random House, New York, 1959

Burrows, William E., This New Ocean: The Story of the First Space Age, 2001

Burrows, William E., Exploring Space, Random House, New York, 1990

Caiden, Martin, The Astronauts, E.P. Dutton & Company, New York, 1959

Caiden, Martin, War For The Moon, E.P. Dutton & Company, New York, 1959

Chapman, John L., Atlas The Story of a Missile, Harper & Brothers, 1960

Clarke, Arthur C., The Exploration of the Moon, Harper & Brothers, 1954

Clarke, Arthur C., The Making of a Moon, Harper & Brothers, New York, 1958

Clary, David A., Rocket Man - Robert H. Goddard, Hyperion Books, New York, 2003

Collins, Michael, Carrying the Fire, Farrar, Strauss, and Giroux, New York, 1974

Day, Dwayne A., Spaceflight Vol. 36, 1994

Dickson, Paul, Sputnik - Shock of the Century, Walker & Company, New York, 2001

Dornberger, Walter, V-2, Ballentine, New York, 1954

Durant, F.C., Robert H. Goddard - The Roswell Years, National Air & Space Museum, 1973

Dyson, Gerald, Project Orion, Henry Holt & Company, New York, 2002

Editorial Staff, The Viking Mission to Mars, Martin Marietta Company, Denver, 1975

Gatland, Kenneth, Robot Explorers, The MacMillan Company, 1972

Gatland, Kenneth, Manned Space Flight, The MacMillan Company, 1967

Gavaghan, Helen, Something New Under the Sun: Satellites and the Beginning of the Space Age, 1998

Gibson, James M., The Navaho Missile Project, Schiffer Publishing Ltd, Atlanta, 1996

Glenn, John H., Taylor, Nick, John Glenn - A Memoir, Bantam Books, New York, 1999

Godwin, Robert, Whitfield, Steven, Deep Space - The NASA Mission Reports, Apogee Publications, Toronto, 2005

Hall, Al, Et al, Man In Space, Peterson Publishing Co., Los Angeles, 1974

Harford, James, Korolev, John Wiley & Sons, 1997

Hart, Douglas, The Pictorial History of World Spacecraft, Bison Books, New York, 1988

Hart, Douglas, The Pictorial History of NASA, Gallery Books, New York, 1989

Harwood , William D., Raise Heaven and Earth, Simon & Shuster, New York, 1993

Heppenheimer, T.R., The Space Shuttle Decision 1965-1972, Washington DC, 2002

Jenkins, Dennis R., The Space Shuttle The History of Developing the National Space Transportation System, Motorbooks Intl. 1996

Jenkins, Dennis R., Launius, Roger D., To Reach the High Frontier, University Press of Kentucky, 2002

Joels, Kerry Marc, The Space Shuttle Operators Manual, Ballantine Books, New York, 1982

Johnson, David, V1 V2 Hitler's Vengeance Weapons, Stein & Day, New York, 1981

Killian, James R., Sputnik Scientists and Eisenhower, MIT Press, 1977

Kraft , Chris, Flight - My Life in Mission Control, Penguin Putnam, New York, 2001

Krantz, Gene, Failure Is Not An Option, Berkeley Books, New York, 2000

Levine, Alan J., The Missile and Space Race, 1994

Ley, Willy, Rockets Missiles and Space Travel, Vail-Ballou Press, 1959

McDougall, Walter A., The Heavens and the Earth, Basic Books Inc. 1985

McNamara, Bernard, Into the Final Frontier, Harcourt College, 2001

Medaris, John B., Countdown For Decision, G.P. Putnam, 1960

Miller, Walter James, The Annotated Jules Verne - From Earth to Moon, Thomas Y. Crowell Associates, New York, 1970

Muirhead, Brian, Et al, Going to Mars, Simon & Schuster, 2004

Newkirk, Dennis, Almanac of Soviet Manned Space Flight, Gulf Publishing, 1990

Oberg, James E., Red Star in Orbit, Random House, 1981

Ordway, Frederick I., Sharpe, Mitchell R., The Rocket Team, Crowell, 1979

Parkin, Charles M. Jr, The Rocket Handbook for Amateurs, The John Day Company, New York, 1959

Pisano, Dominick A., van der Linden, F. Robert, Winter Frank H., Chuck Yeager and the Bell X-1, Smithsonian Air and Space Museum, 2006

Riabchikov, Evgeny, Russians in Space, Novosti Press, 1971

Russell, John L. Jr., Science Year, Popular Science, 1959

Shepard, Alan, Slayton, Deke, Moonshot, Turner Publishing, Atlanta, 1994

Siddiqi, Asif A., Sputnik and the Soviet Challenge, University Press of Florida, 2003

Siddiqi, Asif A., The Soviet Space Race with Apollo, University Press of Florida, 2003

Stafford, Thomas, We Have Capture, Smithsonian Press, Washington, 2002

Thompson, Neal, Light This Candle, Crown Publishing, New York, 2004

Varfolomeyez, Timothy, Soviet Rocketry that Conquered Space, Spaceflight Vol. 37, 1995

Wagener, Leon, One Giant Leap, Tom Doherty Associates, New York, 2004

Wendt, Guenter, The Unbroken Chain, Apogee Publications, Toronto, 2005

Wilson, Andrew, The Eagle Has Wings, Unwin Brothers Ltd, London, 1982

Yates, Raymond F., Russell, M.E., Space Rockets and Missiles, Harper & Brothers Publishers, New York, 1960

Yeager, Chuck, Janos, Leo, Yeager, Bantam Books, New York, 1988

Yenne, Bill, Secret Weapons of the Cold War, Berkley Books, New York, 2005

Zaehringer, Alfred J., Rocket Science, Apogee Books, 2004

Zaehringer, Alfred J., Solid Propellant Rockets, American Rocket Company, 1955

Zimmerman, Robert, Leaving Earth, Joseph Henry Press, Washington DC, 2003

Zimmerman, Robert, Genesis - The Story of Apollo 8, Dell Books, New York, 1998

In addition, the following URLs have been referenced:

http://www.unitedstart.com/
http://www.worldspaceflight.com/
http://www.astronautix.com/
http://www.hq.nasa.gov/
http://www.aerospace.ru/
http://www.spacedaily.com/
http://science.nasa.gov/
http://www.unitedspacealliance.com/
http://www.colonyfund.com/Viewpoints/Archives/
http://www.xprizefoundation.com/
http://science.ksc.nasa.gov/shuttle
http://www.centennialofflight.gov/essay/SPACEFLIGHT/
http://web.wt.net/~markgoll/
http://www.spacetoday.org/History/
http://liftoff.msfc.nasa.gov/
http://www.shuttlepresskit.com/
http://www.nasm.si.edu/
http://www.designation-systems.net/dusrm/
http://www.biography.ms/
http://www.fas.org/spp/military
http://www.spaceandtech.com/

Index

Apogee Books THE Space Book Company.
Available Titles List • October 2007

Use this sheet to check your inventory and to place your order

Title	ISBN 13 (new)	US$		Order
Friendship 7	9781896522609	18.95		
Sigma 7	9781894959018	19.95		
Gemini 6	9781896522616	18.95		
Gemini 7	9781896522801	19.95		
Gemini 12	9781894959049	19.95		
Apollo 8	9781896522661	18.95		
Apollo 10 1st Ed	9781896522524	14.95		
Apollo 11 Vol 1	9781896522531	18.95		
Apollo 11 Vol 2	9781896522494	15.95		
Apollo 12 Vol 1	9781896522548	18.95		
Apollo 12 Vol 2	9781894959162	24.95		
Apollo 13	9781896522555	20.95		
Apollo 14	9781896522562	18.95		
Apollo 16 Vol 1	9781896522586	19.95		
Arrows To Moon	9781896522838	21.95		
Astronautics Book 1	9781894959636	23.95		
Astronautics Book 2	9781894959667	27.95		
Beyond Earth	9781894959414	27.95		
Columbia A. I. R.	9781894959063	25.95		
Conquest ofSpace	9781896522920	23.95		
Creating Space	9781896522869	30.95		
Deep Space NMR	9781894959155	34.95		
Getting Off Planet	9781894959209	18.95		
Go For Launch	9781894959438	29.95		
High Frontier	9781896522678	21.95		
How NASA Learned	9781894959070	25.95		
Interstellar Space	9781896522990	24.95		
ISScapades	9781894959599	23.95		
Lost Spacecraft	9781896522883	30.95		
Lunar Expl Scrapbk	9781894959698	28.95	Nov 07	
Mars NMR Vol 1	9781896522623	23.95		
Mars NMR Vol 2	9781894959056	28.95		
Moonrush	9781894959100	24.95		
New Moon Rising	9781894959124	33.95		
On To Mars 2	9781894959308	21.95		
Real Space Cowboys	9781894959216	29.95		
Ref Guide to ISS	9781894959346	21.95		
Reflections	9781894959223	23.95		
Return To Moon	9781894959322	22.95		
Rocket Team	9781894959001	34.95		
Saucer Fleet	9781894959704	TBA	Feb 08	
Simple Universe	9781894959117	21.95		
Space Shuttle NMR	9781896522692	23.95		
Space Tourism	9781894959087	20.95		

Title	ISBN 13 (new)	US$		Order
Space Trivia	9781896522982	19.95		
Spaceships	9781894959506	19.95		
To End of Solar System	9781894959681	TBA	Jan 08	
Unbroken Chain	9781896522845	29.95		
Virtual Apollo	9781896522944	24.95		
Visions of Future	9781896522937	27.95		
Voice of von Braun	9781894959643	22.95		
Women Astronauts	9781896522876	23.95		
Women of Space	9781894959032	22.95		
X20 Dyna Soar	9781896522951	32.95		
	6" X 9" Books			
Project Mars vonBraun	9780973820331	12.95		
Sex in Space	9781894959445	17.95		
	8½" X 11" Books			
Kids To Space	9781895949421	29.95		
Pocket Space	**Guides - 4" X 7"**			
Apollo 11 First Men	9781894959278	9.95		
Deep Space PSG	9781894959292	9.95		
Hubble Space Tele	9781894959384	9.95		
Launch Vehicles	9781894959285	9.95		
Mars	9781894959261	9.95		
Project Apollo Expl	9781894959377	9.95		
Project Apollo Test	9781894959360	9.95		
Project Constellation	9781894959490	9.95		
Project Gemini	9781894959544	9.95		
Project Mercury	9781894959537	9.95		
Russian Spacecraft	9781894959391	9.95		
Space Shuttle	9781894959520	9.95		
DVDs w Bonus Book				
Apollo 11 (2 Discs)	9781897421000	24.95		
Apollo 12 (2 Discs)	9781897421017	24.95		
Project Apollo (1 D)	9781897421024	17.95		
Einstein Theory (1 D)	9781894959513	17.95		
DVDs				
Dr.Tellers Bomb (1 D)	9781897421031	24.95		
Rocket Science (3 D)	AP DV 42900	24.95		
Posters				
Apollo/Saturn Launch	Vehicles (19" X 24")	9.95		

Toll Free Order Line: 1-888-55-SPACE (888-557-7223)
Fax: 905 637 2631   email marketing@cgpublishing.com